LITERARY FUTURISM

LITERARY FUTURISM

Aspects of the first avant garde

JOHN J. WHITE

Clarendon Press · Oxford
1990

Oxford University Press, Walton Street, Oxford OX2 6DP
Oxford New York Toronto
Delhi Bombay Calcutta Madras Karachi
Petaling Jaya Singapore Hong Kong Tokyo
Nairobi Dar es Salaam Cape Town
Melbourne Auckland
and associated companies in
Berlin Ibadan

Oxford is a trade mark of Oxford University Press

Published in the United States
by Oxford University Press, New York

British Library Cataloguing in Publication Data
White, John (John James), 1940–
Literary futurism: aspects of the first avant garde.
1. Literature. Avant-garde writers, 1850–1950
I. Title
809.034
ISBN 0–19–815813–0

Library of Congress Cataloging in Publication Data
White, John J., 1940–
Literary futurism: aspects of the first avant-garde / John J. White.
p. cm.
Includes index.
1. Literature, Modern—20th century—History and criticism.
2. Futurism (Literary movement) I. Title.
PN771.W45 1989 809'.91—dc19 88—36447
ISBN 0–19–815813–0

Set by Joshua Associates Ltd., Oxford
Printed and bound in
Great Britain by Biddles Ltd,
Guildford and King's Lynn

Preface

No one who has worked on such a comparative and interdisciplinary subject as the one addressed by the present study can fail to have accumulated numerous intellectual debts to the greater expertise of others in various fields. I hope to have satisfactorily documented my main areas of scholarly indebtedness in the footnotes to the chapters that follow. But I should like to take this opportunity to express my particular gratitude for the assistance of various kinds which I have received at different stages in the writing of this work from the following: Jaroslaw Balan, Mara Cambiaghi, Roger Ellis, Derek Glass, Peter Mayer, Katie Padilla, Dietmar Rösler, Christopher Wagstaff, Mieczyzław Wallis, and Ann White. Jeremy Adler read most of the material in this book and commented extensively on it at a time when his own research on figured poetry was putting him under great pressure. For his invaluable advice and support and that of Siegbert Prawer, who read some of these chapters and enthusiastically encouraged the project from the outset, I am extremely grateful. Hilary Walford has been a scrupulous and thoughtful copy-editor, and I am indebted to her for all the rigour she has demonstrated in attempting to bring order to the Futurist disorder of my typescript. Both she and Virginia Llewellyn Smith, of Oxford University Press, have given invaluable assistance with the production of this study. During the period while *Literary Futurism* was being prepared and written I have had the pleasure and privilege of being a member of the German Department at King's College London. No one could ask for a more congenial atmosphere within which to teach and research. To my colleagues there, whose friendship, help, and tolerance I have been able to depend on during the time I was finding my way about Futurism, this study is dedicated.

J. J. W.

Kew
July 1988

Acknowledgements

I SHOULD like to thank the following individuals, publishing houses, and institutions for granting me permission to reprint/reproduce the material indicated below. The Design and Artists Copyright Society Ltd: works by Carlo Carrà and Fortunato Depero (Figs. 1.1, 1.21, 1.23, 2.2, 2.4, 2.5, 2.13, 2.14, and 5.1): © DACS 1989. Works by Gino Severini (Figs. 2.1, 2.3, 2.7, and 2.8): © ADAGP, Paris, DACS, London, 1989. Rag. Giuseppe Dal Pian: quotations from Umberto Boccioni, *Gli scritti editi e inediti*, ed. Zeno Birolli (Milan, 1971). Il fauno, Editore: material from Paolo Buzzi's *Conflagrazione* (Florence, 1963). Arnaldo Mondadori, Editore: F. T. Marinetti's works, ed. Luciano De Maria (Milan, 1963–). The Department of Special Collections, University Research Library, University of California, Los Angeles: the manuscript version of Marinetti's 'Bombardement d'Andrinople' (Fig. 3.4). Sig. Massimo Carrà: collages from Carlo Carrà, *Guerrapittura*, Milan, 1915, and Luciano Caruso (Le Brache di Gutenberg, Florence) for permission to make use of his 1978 facsimile reprint of this volume for illustrative purposes. Patrick Bridgwater: for allowing me to quote various poems in his English translation, August Stramm, *twenty two poems* (Wymondham, 1969). For their help in supplying photographs of material in their possession I should like to thank: the British Library (Fig. 1.19), the Musée National d'Art Moderne, Centre Georges Pompidou, Paris, Sce photographique du MNAM (Fig. 2.3); the Deutsches Literaturarchiv, Schiller-Nationalmuseum, Marbach am Neckar (Fig. 3.2); and the Galleria Museo Depero, Rovereto (Figs. 1.1 and 1.21). Additional photographic credits: Jacqueline Hyde, Paris (Fig. 2.3); Madame S. Delaunay (Fig. 3.2); Städelsches Kunstinstitut, Frankfurt a. M. (Fig. 2.7); and Professors Carlo and Mary Vieira Belloli (Fig. 5.1).

Contents

List of Illustrations ix
Abbreviations xi
Introduction 1
1. The 'Typographical Revolution' and Futurism's Exploration of the Iconic 8
2. The Development of Futurist Collage and the *Dipinto Parolibero* 73
3. Telegraphic Lyricism and the Cult of the Minimal 143
4. Linguistic Literature: Neology, New Syntax, Word-Games 215
5. Futurist Primitivisms 288
Glossary 359
Index 369

List of Illustrations

1.1. Fortunato Depero, *Grattacieli*, 1929 (Museo Depero, Rovereto) 13
1.2. Francesco Cangiullo, *Milano-Dimostrazione*, 1915 (Private Collection) 14
1.3. The balloon 'designed analogy' from F. T. Marinetti, *Zang tumb tumb*, Milan, 1914 16
1.4. 'Designed analogy' from Paolo Buzzi, *L'ellisse e la spirale*, Milan, 1915 17
1.5. 'Designed analogy' depicting suitcases on a station platform from Francesco Cangiullo, 'Fumatori II.', 1914 17
1.6. Final part of F. T. Marinetti's *8 anime in una bomba*, Milan, 1919 18
1.7. F. T. Marinetti, 'Corvée d'acqua sotto i forti Austriaci', 1925 19
1.8. Extract from Mario Bètuda, 'Looping the Loop', 1914 20
1.9. From F. T. Marinetti, *Zang tumb tumb*, Milan, 1914 23
1.10. Smoke 'designed analogy' from Francesco Cangiullo, 'Fumatori II.', 1914 24
1.11. Part of Francesco Cangiullo, 'Fumatori II.', 1914 24
1.12. From F. T. Marinetti, *Zang tumb tumb*, Milan, 1914 30
1.13. From F. T. Marinetti, *Zang tumb tumb*, Milan, 1914 31
1.14. From F. T. Marinetti, *Zang tumb tumb*, Milan, 1914 34
1.15. Sea 'designed analogy' from Corrado Govoni, *Rarefazioni e parole in libertà*, Milan, 1915 42
1.16. Umberto Boccioni, study for *Dinamismo di un ciclista*, 1913 50
1.17. Cesare Simonetti, 'Treno in corsa', 1925 52
1.18. Cover of F. T. Marinetti, *Zang tumb tumb*, Milan, 1914 53
1.19. Cover of the second edition of Alexei Kruchenykh, *Vzorval'*, St Petersburg, 1914 54
1.20. Cover of Auro D'Alba, *Baionette*, Milan, 1915 55
1.21. Cover of Fortunato Depero, *Liriche radiofoniche*, Milan, 1934 56
1.22. Iconic page from Alexei Kruchenykh, *Vzorval'*, St Petersburg, 1913; drawings by N. Kulbin, poetry by Kruchenykh 58
1.23. Carlo Carrà, *Sintesi futurista della guerra*, 1915 59
1.24. F. T. Marinetti, *Le soir, couchée dans son lit, elle relisait la lettre de son artilleur au front*, 1919 62
1.25. F. T. Marinetti, *Lettre d'une jolie femme à un monsieur passéiste*, 1914, used as the cover design for *Les Mots en liberté futuristes*, Milan, 1919 68
2.1. Gino Severini, *Danza serpentina*, 1914 78

2.2. Carlo Carrà, *Lacerba* version of *Festa patriottica*, 1914 79
2.3. Gino Severini, *Ritratto di Paul Fort*, 1913 (Musée National d'Art Moderne, Paris) 102
2.4. Carlo Carrà, *Sintesi circolare di oggetti*, 1914 103
2.5. Carlo Carrà, *Inseguimento*, the *Guerrapittura* version, 1915 105
2.6. Extract from Umberto Boccioni, 'UOMO + VALLATA + MONTAGNA', 1914, showing the 'compenetration' of two sizes of type 108
2.7. Gino Severini, *Cannoni in azione*, 1915 (Städelsches Kunstinstitut, Frankfurt a. M.) 118
2.8. Gino Severini, *La guerra*, 1915 (present whereabouts unknown) 120
2.9. Paolo Buzzi, *Bombardamento aereo*, broadsheet, 1915 125
2.10. F. T. Marinetti, *montagne + vallate + strade × joffre*, 1915 126
2.11. F. T. Marinetti, *Une assemblée tumultueuse. Sensibilité numérique*, 1919 128
2.12. Ardengo Soffici, *Tipografia* from *BÏF§ZF+18: Simultaneità e chimismi lirici*, Florence, 1915 129
2.13. Carlo Carrà, *Cielo di guerra*, 1915 130
2.14. Carlo Carrà, *La notte del 20 Gennaio 1915 sognai questo quadro (Angolo penetrante di Joffre sopra Marna contro 2 cubi germanici)*, 1915 132
2.15. From Paolo Buzzi, *Conflagrazione*, 1915–18 135
2.16. From Paolo Buzzi, *Conflagrazione*, 1915–18 136
2.17. Vasily Kamensky, 'Konstantinopol′', 1915 141
3.1. Eiffel Tower 'auto-illustration' from Guillaume Apollinaire, *Calligrammes. Poèmes de la paix et de la guerre*, Paris, 1925 156
3.2. Cover of Iwan Goll, *Der Eiffelturm*, Berlin, 1924. Designed by G. Salter after a drawing by Robert Delaunay 157
3.3. From F. T. Marinetti, *Zang tumb tumb*, Milan, 1914 179
3.4. F. T. Marinetti, 'Bombardement d'Andrinople', undated manu script (University Research Library, University of California, Los Angeles) 184
3.5. Card One of Mykhaylo Semenko, 'Kablepoema za okean', 1924 200
4.1. Fortunato Depero, 'Tramvai', 1927 259
4.2. Giacomo Balla, 'canzone di maggio', undated 260
4.3. Guillaume Apollinaire, from 'Quelconqueries II', 1914 262
4.4. Part of Vasily Kamensky, 'Konstantinopol′', 1915 265
4.5. Mykhaylo Semenko, 'Sil′s′kyy peyzazh', 1922 267
4.6. Beginning of Velimir Khlebnikov, 'Kakim obrazom v *so* est′ oblast′ sna', 1912–13 276
4.7. Card Two of Mykhaylo Semenko, 'Kablepoema za okean', 1924 280
5.1. Fortunato Depero, *Grattacieli a New York*, 1927 (Collection Prof. Mary Vieira Belloli, Milan-Basilea) 332

Abbreviations

THROUGHOUT the present study, the following abbreviations will be used in page-references made in parentheses after quotations.

Archivi	*Archivi del Futurismo*, ed. Maria Drudi Gambillo and Teresa Fiori (2 vols.; Rome, 1958–62)
FF	*Futurismo & Futurismi*, catalogue of the exhibition organized at the Palazzo Grassi, Venice, by Pontus Hulten, Milan, 1986
Gli scritti	Umberto Boccioni, *Gli scritti editi e inediti*, ed. Zeno Birolli (Milan, 1971)
GM	F. T. Marinetti, *La Grande Milano tradizionale e futurista: Una sensibilità italiana nata in Egitto*, ed. Luciano De Maria (Milan, 1969)
Manifesty	*Manifesty i programmy russkikh futuristov*, ed. Vladimir Markov (Slavische Propyläen, 27; Munich, 1967)
PSS	Vladimir Mayakovsky, *Polnoe sobranie sochineniy* (13 vols.; Moscow, 1955–61)
SP	Velimir Khlebnikov, *Sobranie proizvedeniy*, ed. Nikolai Stepanov (5 vols.; Leningrad, 1928–33)
TIF	F. T. Marinetti, *Teoria e invenzione futurista*, ed. Luciano De Maria (Milan, 1968)

Of course one can 'go too far' and except in directions in which we can go too far there is no interest in going at all; and only those who will risk going thus far can possibly find out just how far one can go.

T. S. Eliot

Introduction

> They say I'm a Futurist. What is a Futurist? I don't know. They don't exist. 'Mademoiselle Critique' has told you about them. I'll show her!
>
> V. Mayakovsky, 'We want meat'[1]

THE chapters that follow are not intended to furnish a full-scale introduction to the subject of Futurism in literature. With the exception of Russian Cubo-Futurism, literary Futurism has arguably suffered from too many introductions and not enough detailed, scholarly investigation of individual works. Nor does this study set out to survey systematically or in any depth the full range of characterizing features that distinguish Futurist writing (many of these have, in any case, already received the attention or the neglect they deserve). Rather, its main aim is to attempt a comparatist examination of a number of specific, and in my opinion important, Futurist themes and techniques in the writings of certain authors, coupled with an analysis of some of the movement's principal works. If the reader's picture of Futurist literature in the corporate sense does happen to become any clearer during the course of looking at my more circumscribed arguments and interpretations, so much the better. But 'Futurism'—or as the plural in the title of the 1986 Venice exhibition more accurately put it '*Futurismo* and Futurisms' (that is to say, Italian Futurism and its various foreign analogues)—is too wide-ranging and too diffuse a subject for any one commentator to attempt to do justice to it singlehandedly without occasioning a sense of growing frustration in both the writer and the reader.

The welter of books, articles, and new anthologies concerning Italian Futurism that were published in connection with the Venice exhibition—the individual surveys of Russian Futurism by Vladimir Markov, of its Polish counterpart by Bogdana Carpenter, of Ukrainian Panfuturism by Myroslava M. Mudrak, the discussions of Czech Futurism in Peter Drews's *Die slawische Avantgarde* and in Zbigniew Folejewski's useful study-cum-anthology of international Futurism (works which will be referred to in the course of the present study)—all reveal just how many-pronged and ultimately just how unique the Futurisms of the various European countries were. To talk meaningfully about a number of these movements at once adds up to more than a critical challenge: it would demand the kind of scholarly sweep that is beyond most of us. In any case, reliable information about most of the individual Futurisms (with the notable exception of Italian Futurism) is too patchy for one

[1] *Nov'* (16 Nov. 1914).

to be able to make the kind of meaningful macroscopic juxtapositions that such an approach would require.

In addition to such problems of perspective and information, there is also something inherently protean to the writings of the Futurists, a quality that militates against such an epic approach, even if it were practicable. It may be possible to think in European terms about the impact of a movement like Naturalism, but to try to do so in the case of Futurism would be to confront insuperable difficulties. 'Most statements about the Futurist movement', according to one seasoned critic in the field (and he was at the time restricting his comments to the secondary literature on Russian Futurism), 'tend to generalize from a partial selection of facts, and this is understandable, since we customarily place under the rubric Futurism such diverse and unhomogeneous phenomena as Mayakovsky and Khlebnikov, the one an urbanist, the other a Slavic primitivist.'[2] Vladimir Markov, the most discerning and by far the most impressively comprehensive of all the chroniclers of the early Russian avant-garde, came to the conclusion, at the end of his monumental study of the movement, that Russian Futurism was really no more than a 'complex conglomeration' consisting of 'elements of impressionism, expressionism, neo-primitivism, constructivism, abstractionism, dandyism, theosophy, and so forth', with the result that only very tentative generalizations could be made about any common characteristics to the movement. His proposed definition of such a complex 'ism' as the Russian version of Futurism, a period of literature comprising a whole network of continually changing, as well as often incompatible, factions and coteries, was couched in suitably cautious terms. He called it 'a postsymbolist movement in Russian poetry of 1910–1930 which, roughly, put under the same roof all avant-garde forces'[3]—a cornucopia of a definition which is not far removed from Vladimir Mayakovsky's suggestion that, in the Russian sphere, Futurism could be thought of as virtually synonymous with the term avant-garde.[4] In such a context, 'Futurism' would seem to be little more than a handy convenience-label, hardly a precise critical concept.

Once one introduces the image of Italian literary Futurism into the picture, matters become in one sense more complicated still, but in another respect clearer. Camilla Gray's remark that 'the name "Futurism" is . . . almost all that unites the Russian and Italian movements'[5] would for many commentators seem as applicable to literary Futurism as to the visual arts which she was at the time considering. Although scholarship has in recent years[6] tended to accept

[2] Edward J. Brown, 'Mayakovsky's Futurist Period', in George Gibian and H. W. Tjalsma (eds.), *Russian Modernism: Culture and the Avant-Garde, 1900–1930* (Ithaca, 1976), 109.

[3] *Russian Futurism: A History* (Berkeley, 1968), 384.

[4] Letter to Trotsky, published in *Literaturnoe nasledstvo: Novoe o Mayakovskom*, ed. V. V. Vinogradov, 65 (Moscow, 1958), 176.

[5] *The Russian Experiment in Art, 1863–1922* (London, 1962), 94.

[6] I am thinking in particular of Cesare G. De Michelis, *Il futurismo italiano in Russia, 1909–1929:*

that there was a greater influence on the Russian Futurists from Italy than was once assumed—or certainly than was at the time admitted to by the Russian writers themselves—the actual history of this interaction is still only patchily recorded and the story hardly becomes any easier to reconstruct with the passage of time. However, if one leaves aside the vexed question of lines of influence that still clouds much Futurist scholarship with nationalist interest, and one tries to concentrate on directly comparing the two main Futurisms (Italian and Russian) within a typological framework, it becomes clear that even this can be a hazardous undertaking. The establishment of a concept of 'Futurism' adequate to such a comparative task is, for a start, itself a venture fraught with obstacles. One danger is the temptation to rely too heavily on the Italians' own extensive and readily available manifesto-pronouncements about their Futurism's status as the literary expression *par excellence* of the technological twentieth century. As a result, one may be inclined to work with Milanese ideas about what epitomizes 'Futurism' and hence judge a lot of so-named Russian Futurist writing (with the possible exception of its urbanist strain[7]) as falling short of the mark of 'Futurism' proper. Italian Futurism is, as we shall see, certainly more homogeneous in its response to the modern world and in its concern for the discoveries that were in the process of creating a distinctive future than its Russian counterpart(s)—which was the point I had in mind when I tried to suggest that the matter becomes a little more sharply focused when one brings *Futurismo* into the picture. Indeed, there are areas which may undoubtedly make Italian Futurism seem more 'Futurist', in a qualitative sense, than its Russian analogues. Given such differences, one can at least indulge in the pursuit of playing off the relative homogeneity of the one group, under Marinetti's tutelage, against the sheer diversity of the other, with its cult of artistic anarchy in many quarters. Nevertheless, it would hardly seem permissible, for that reason, to derive from the one (Italian) group any hard-and-fast criteria for either evaluating—or even setting about understanding—its various Russian namesakes. This would surely scarcely mark an advance on the situation that prevailed at the time, when quarrels about anteriority led to the spokesmen of the two Futurisms judging each on the dubious assumption that whichever one had started experimenting in a particular area first had the right to become the judge of the other's activities. Precisely because of early Italian Futurism's clearer profile and the vociferous claims it repeatedly laid to being the leader of world-Futurism, one has to be wary of attributing to it the dual status of *fons et origo* and yardstick that it itself was quick to covet. Similarly, one

Temi e problemi (Bari, 1973); the section 'Marinetti et la Russie' in Jean-Claude Marcadé (ed.), *Présence de F. T. Marinetti: Actes du Colloque International tenu à l'UNESCO* (Lausanne, 1982), 177–274; and Juliette R. Stapanian, *Mayakovsky's Cubo-Futurist Vision* (Houston, Texas, 1986).

[7] Much useful material on the subject can be found in Kjeld Bjørnager Jensen, *Russian Futurism, Urbanism and Elena Guro* (Århus, 1977).

has a right to be sceptical about other aspects of Italian Futurism's own self-image. As we shall see, many things are not quite as the manifestos would have us believe, either on the question of the calibre of a lot of Futurist works or when it comes to the long-term significance of some of the movement's much-trumpeted innovations.

In any case, even Italian Futurism invariably begins to assume a less monolithic appearance, the more closely one starts to investigate it. Those who present its literature as a relatively homogeneous output usually tend to focus too readily on Marinetti himself, and then often largely on his theoretical statements rather than on the rich spectrum of his writings, which cover more than four decades.

In contrast to this often procrustean 'outside' version of the Futurist writers' achievements, here is what Giovanni Papini had to say on the subject of the diversity of Italian literary Futurism in his 1913 article 'Why I am a Futurist' (when he refers to 'Futurism', he of course means exclusively Italian Futurism):

> Futurism is not a Prussian regiment or a closed school where a single leader gives orders and is in charge, where one's duties and tasks are prescribed, and where the activities of members are limited and laid down. Marinetti is a man of talent and an innovative poet . . . but he is neither the Pope nor the leader of a school, neither a general nor a jealous prophet, he is neither a despot nor an office boss. He acknowledges what he owes to others and lets everyone think and act according to his own wishes, but is always ready to support all forms of initiative and to help with all kinds of undertaking. Everyone remains responsible for the ideas he displays and the works he creates: and the tendencies common to all (liberation from the past, the search for novelty and individuality, the use of violence, etc.) are so wide-ranging that they hardly inhibit anyone in what he does. . . . One should not equate Futurism with a few phrases in certain manifestos or the stance of certain Futurists. One must look beyond the words, the formulae, the placards, and the banners and hold on to the live concrete reality of the situation: in other words, to the FACT that twenty or so young men of great talent and great audacity are in the process of creating a new art, a new way of thinking and a new intuition of life. . . . Not everything that they do is admirable, or superior or perfect; not all of their works will stand the test of time as preparing the way for a new art or as definitive statements. But they have on their side activity, boldness, unselfishness, experimentativeness, enthusiasm; and apart from them there exists, at least in Italy, in poetry and painting only an archeological spirit, shallowness, timidity, commercialism, the continual imitation of sacred models, the unperturbed and fastidious repetition of completely worn-out clichés.[8]

Very frequently, those inclined to view Italian literary Futurism in relatively monolithic terms have a habit of restricting their vision to the achievements of early Futurism, i.e. of concentrating on what is known as the Futurism of the first 'heroic' period, rather than that of Second Futurism.[9] Once critical

[8] 'Perchè son futurista', *Lacerba* (1 Dec. 1913), 268.

[9] For information concerning this period-concept and a number of other terms used mainly in the

coverage of even Italian literary Futurism is widened to include the post-1918 period, the picture becomes far more diverse and less easy to equate with any narrowly Marinettian influence or any clearly defined 'sensibility', to use one of the movement's favourite words.

In the light of all this, it would look as if Futurism scholarship is caught between the Scylla of selectivity and the Charybdis of shapeless pluralism. Faced with so much amorphous material, it can either choose to explore a limited number of features, while pretending that they are representative, with the result, as Edward J. Brown suggests, that it is likely to be claiming too much on the basis of too little or from a partisan standpoint; or it runs the risk of surrendering to a convenient terminological permissiveness which allows for there being only loose similarities between the various avant-gardes—Futurismo, Panfuturism, Ego-Futurism, Cubo-Futurism, Komfut—that actually chose to use the word as part of their self-image (leaving aside those various coteries that have been judged quasi-'Futurist' in critical retrospect). The result is that 'Futurism' comes to mean such radically divergent things in different contexts that it can hardly appear fruitful to think *comparatively*, and that includes thinking in *contrastive* terms, about the movement as a supra-national phenomenon. The danger here is a retreat into the very myopia and parochialism to which the Futurisms of the time were often a prey.

Yet, if it is not possible, with a phenomenon as vast, as variegated, and as elusive as international literary Futurism, to see the wood for the trees, then one solution may be to stop trying to do so for a while and to examine some of the individual trees in a little more detail. Then, at least, we should know whether the wood is worth returning to and walking in again.

There is, however, an alternative way forward, and it is one that has already been taken by Marjorie Perloff in a recent book on the subject. Employing Renato Poggioli's idea of a 'futurist moment' which 'belongs to all avant-gardes',[10] she sets out, in an excellent study which even sports the title *The Futurist Moment*, to detect that vein in a whole seam of early avant-garde literature and art, some of it nominally Futurist, much of it, such as the work of Blaise Cendrars and Ezra Pound, belonging to a far more general 'rapprochement between avant-garde aesthetic, radical politics, and popular culture'.[11] Perloff's illuminating deployment of Poggioli's idea enables her, without becoming entangled in disputes about the nature of historical Futurism, to posit examples of such a 'futurist moment' in a whole series of avant-garde works; and it is, of course, consistent with her method (and Poggioli's

context of literary Futurism, the reader is referred to the Glossary at the end of the present study (pp. 359–67).

[10] *The Theory of the Avant-Garde* (Cambridge, Mass., 1968), 68 f.

[11] *The Futurist Moment: Avant-Garde, Avant Guerre, and the Language of Rupture* (Chicago, 1986), p. xvii.

underlying notion) that she should find the 'futurist moment' as much, if not more, in evidence in the work of modern writers who are *not* normally regarded as Futurists as in literary Futurism itself.

As my sub-title suggests, I shall in some respects be adopting a similarly selective, essentially typological strategy in the following chapters, but with one major difference. I intend to concern myself primarily with examples of characteristic techniques, experiments, and devices *within* literary Futurism itself. The desire of Futurist writers to make literature more 'expressive' (iconic), to find new ways of synthesizing various media so that the word might interact with non-verbal elements, to introduce an innovative 'telegraphic' brevity into the language of their writings, indeed even to create new languages, seems to me to be at the heart of both Futurism's blind rejection of the past and its attempted response to the early part of the twentieth century. These various approaches to the challenge of forging a new literature for the modern world are not only exciting in themselves; they can also be seen to reflect the new Futurist sensibility. And this is also true of the final subject I shall be looking at: the cult of the primitive in its various positive and negative manifestations. I would not want to go so far as to claim that the techniques and topics I shall be exploring are in any way peculiar to Futurism (I doubt, in fact, if any of them are), or even that I have managed to single out all the main typologically 'Futurist' manifestations, in whichever Futurist context they appear. But I would want to insist that there is more of the qualitatively Futurist to these features than one finds in many works that are conventionally ascribed to the Futurist canon, whether one is thinking of the poetry of a Palazzeschi, a Buzzi, or a Severyanin. And I would hope that my choice of these topics has not been unduly prejudiced by the Italian Futurist perspectives of Marinetti's manifestos, or the ideas so persuasively set out in Boccioni's *Pittura scultura futuriste* (*Futurist Painting and Sculpture*).

One final point about the coverage of this study and in explanation of the principle of selectivity involved. If on the whole I have chosen to approach '*Futurismo* and Futurisms' narrowly by focusing restrictively on aspects of the work of a limited number of writers, and thus foregrounding features which from our vantage-point may be taken to embody something of the essence of the Futurist literary endeavour in a way that whole sad tracts of would-be Futurist writing fail to do, there is another sense in which I have cast my net deliberately wider than might have been expected. For I have included, at a number of points in my discussion, a consideration of the work of the German Expressionist *Sturmkreis* writers—not only of the poetry of the German 'Futurist' August Stramm,[12] but of a number of the other writers in the circle around Herwarth Walden during the pre-war and war years. This is not because I share

[12] The term is used by Patrick Bridgwater in his introduction to August Stramm, *twenty two poems*, trans. Bridgwater (Wymondham, 1969), n.p.

Gottfried Benn's view that German Expressionism is in its entirety really just Futurism by another name.[13] This seems to me to over-widen the relevant category. But I do believe that there is as much 'Futurism' (though perhaps less of Poggioli's 'futurist moment' with a small 'f') in the work of many of the *Sturmkreis* writers as there is, for instance, in the majority of poetry published in Marinetti's 1912 anthology *I poeti futuristi*. By 1918, according to Friedhelm Lach, the editor of the collected works of Kurt Schwitters, who began writing within the orbit of Walden's circle, the *Sturmkreis* was permeated with Marinetti's ideas.[14] Which is more than one could say of the work of some of Milan's own self-proclaimed Futurists or of many of the Hylaea Group in Russia. For this reason, it seems sensible not to be put off by the fact that the *Sturmkreis* writers did not actually choose to call themselves Futurists, even when they were producing what was to all intents and purposes Futurist literature. Their achievements certainly belong within the purview of a study of literary Futurism. On the other hand, although I have occasionally glanced across the Channel at the activities of the English Vorticists, I am far from convinced that they belong within the framework of a study of this kind. But these remain issues for debate, no matter what strategy is adopted in the present volume; for in the study of literary Futurism, there is very little that is not still open to debate of a fairly fundamental order. 'Futurism', Papini (who was always Futurism's gadfly) remarked, 'has made people laugh, shout and spit. Let us see if it can make them think.'[15] He made the remark in 1913. Joshua Taylor quoted it almost half a century later in the catalogue to the famous 1961 New York Museum of Modern Art exhibition that was to become a milestone in international Futurism reception.[16] Yet Papini's 'let us see' has not lost any of its relevance today, when it comes to the subject of literary Futurism.

[13] 'Rede auf Marinetti', *Gesammelte Werke*, ed. Dieter Wellershoff, iv. *Essays, Reden, Aufsätze* (Wiesbaden, 1959), 243.

[14] *Der Merz Künstler Kurt Schwitters* (Cologne, 1971), 16 f.

[15] 'Il significato del Futurismo', *Lacerba* (1 Feb. 1913), 22.

[16] *Futurism* (New York, 1961), 10.

1.
The Typographic Revolution and Futurism's Exploration of the Iconic

THE Italian Futurists, like so many of the self-appointed legislators of the early avant-garde, suffered from a tendency to be prescriptive rather than permissive in spirit. The contagiously dogmatic mood of the second decade of the twentieth century is well captured in the way that the literary journal *Montjoie!*, one of the main contemporary French rivals to *Lacerba*, could flaunt the unashamedly assertive sub-title 'Organe de l'impérialisme artistique française'. In fact, a similar declaration of colonizing intent would not have been out of place beneath the title of almost every Futurist manifesto and every collection of the movement's poetry and theories. In the period leading up to the First World War, the Italian Futurists concentrated much effort on dictating the new directions which painting, literature, sculpture, music, and architecture should take, if they were to be rescued from their current ignominious state of passéism. The resultant welter of aesthetic proposals, coupled with the strenuous promotion of exemplary Futurist works, allowed little room for alternatives. Yet the movement's measurable influence on later modernism largely proved to be inversely proportional to the dictatorial nature of these attempts at artistic hegemony. Thus, when looked at from beyond the citadels of Futurism and with the advantage of hindsight, much Futurist painting pales into insignificance alongside the achievements of Cubism and Expressionism. And the exciting architectural dreams of a Sant'Elia or a Chiattone remain mere theoretical blueprints, producing nothing tangible to set alongside the work of Le Corbusier and Gropius. Experimental typography, on the other hand, was one area where the Futurists were able, over the decades, to exercise an exceptionally profound aesthetic influence—one which did leave its unmistakable mark on the future. With their innovative layouts, the Futurist poets had an impact extending well beyond the realm of adjacent avant-gardes into the world of modern advertising, visual displays, and other forms of attention-grabbing communication, as well as modern experimental poetry in general.

Writing of the way the new expressive typography of the commercial world had successfully reduced the 'appalling discrepancy' between 'technology and design development', Lázsló Moholy-Nagy credits the Futurists with a major part in this development. 'Apollinaire's ideogram and Marinetti's poems served,' he argues, 'not so much as models, but as tradition-breakers which freed experimenters to create a quick, simultaneous communication of several messages.'[1] In fact, in the eyes of Herbert Spencer, the whole '"heroic" period of modern typography could be said to have begun with Marinetti's manifesto of 1909'.[2] Not that this influence was the result of unsubstantiated manifesto-theorizing. Futurist typography had itself opened up new horizons with the striking radicality of its layouts, and, in the intervening half century or more since 1909, Marinetti's 'original use of typography has not been superseded', in B. B. Bowler's assessment.[3]

If Spencer and Bowler are anywhere near correct—and their views have been frequently echoed in the standard histories of typography—then the Futurists did no more than fulfil one of Marinetti's guiding literary ambitions. In the section of his 'Destruction of Syntax' manifesto ('Distruzione della sintassi—Immaginazione senza fili—Parole in libertà' (1913)) entitled 'Typographical Revolution' one finds him declaring his intention of initiating—not only through the impact of his theories, but by the bold example of Futurist practice—

> a typographical revolution directed against the bestial and nauseating concept of the book of passéist and D'Annunzian verse, on seventeenth-century handmade paper, bordered with helmets, Minervas, Apollos, elaborate red initials, vegetables, mythological missal ribbons, epigraphs, and roman numerals. . . . My revolution is directed against the so-called typographical harmony of the page, which is contrary to the flow and counterflow, the leaps and bursts of style that run through such pages. . . . With this typographical revolution I intend to redouble the expressive force of words.
>
> I reject the decorative, precious aesthetic of Mallarmé and his search for the rare word, the unique, indispensable, elegant, suggestive, exquisite adjective. I have no desire to suggest an idea or a sensation with passéist airs and graces. Instead, I want to grasp ideas brutally and hurl them in the reader's face. (*TIF* 67)

The sideswipe at Mallarmé is probably in part a move to play down an undeniable influence (Marinetti did, after all, translate some of the work of the French Symbolist, who was one of the main precursors of his own 'typographical revolution'[4]). In addition, it also serves to dissociate Futurist experimentation from what was at the time seen as the relatively 'static' quality of *Un coup de dés*'s layout. For promotional reasons, Marinetti was in any case committed to promising a full-scale typographical *revolution*, not some evolutionary tinkering with the possibilities of any previous non-horizontal typography.

[1] *Vision in Motion* [6] (Chicago, 1961), 306.

[2] *Pioneers of Modern Typography* (London, 1969), 13.

[3] *The Word as Image* (London, 1971), 133.

[4] Although Marinetti was the first to translate Mallarmé into Italian (with his *Versi e prose* (Milan, 1916)), *Un coup de dés* did not appear until 1920, in a translation by another Futurist, Enrico Cardile.

During the decade that followed the Paris publication of Marinetti's founding 'Manifeste du Futurisme' (*Le Figaro* (1909)), Italian Futurist poetry was to make a far more substantial contribution to the systematic exploration of possible relationships between the printed medium and its message than did Dada, that other early avant-garde movement which 'hurled' words 'in the reader's face'. Admittedly, the Futurists' claim to fame may rest more upon the systematic nature of their experiments than upon the calibre of their individual best-known works (in this respect, they stand in the shade of both the Constructivists and the Dadaists). Yet as explorers of the iconic, at least, they remain unrivalled. But, whereas the Dada experiments have received their due measure of attention in recent years,[5] much of the Italian Futurists' typographical work, although well documented and regularly anthologized, has seldom been analysed or assessed in any detail. In the case of Russian Futurism, the situation is rather more promising. Susan Compton's excellent *The World Backwards: Russian Futurist Books, 1912–1916* (London, 1978) has done much to survey the early period of Russian Futurist typography, though in historical rather than analytical terms, and Gerald Janecek's splendidly produced *The Look of Russian Literature: Avant-Garde Visual Experiments, 1900–1930* (Princeton, 1984) chronicles—and helpfully brings out the salient features of—the work of key Futurists, including Kruchenykh, Kamensky, Zdanevich ('Iliazd'), and Mayakovsky, while extending its coverage into the post-Revolution period where many of the supreme achievements of Russian Futurist and Constructivist book-design and typographical experimentation are to be found. But no comparable work exists on Italian Futurist typography, although a number of serviceable anthologies are available.[6] This is a surprising state of affairs, when one considers the far more substantial impact that these typographical experiments had on

[5] A select bibliography of research on the subject is given in Arthur Cohen, 'The Typographic Revolution: Antecedents and Legacy of Dada Graphic Design', in Stephen Foster and Rudolf Kuenzli (eds.), *Dada Spectrum: The Dialectics of Revolt* (Madison, Wisconsin, 1979), 72. Cohen assesses the Futurists' importance exclusively in terms of the 'reconstruction of sound words on the page'. But this is only one part—and an early one, at that—of their wide-ranging 'typographical revolution'.

[6] The best is *Tavole parolibere futuriste (1912–1944)*, ed. Luciano Caruso and Stelio M. Martini (2 vols.; Naples, 1974, 1977). This can be supplemented by reference to Maria Carla Papini, *L'Italia futurista (1916–1918)* (Rome, 1977). Although a chapter of Willard Bohn, *The Aesthetics of Visual Poetry, 1914–1928* (Cambridge, 1986) is devoted to Futurist experiments, and his study includes a detailed discussion of Carrà's *Festa patriottica*, the resultant picture of Futurism is unrepresentative. Only four visual poems are explored, the other three being Carrà's *Rapporto di un NOTTAMBULO milanese*, which was never published at the time; Soffici's *Bicchier d'acqua*, seen as 'primarily a cubist, not a futurist poem' (p. 39); and his *Al buffet della stazione*, seen as 'frankly pictorial' (p. 40) and therefore untypically figurative. This leaves *Festa patriottica*, the subject of a 'realistic visual interpretation' (p. 28), despite Bohn's reference (p. 17) to the Futurists being on principle opposed to mimesis. As Fausto Curi has remarked, the history of Futurist visual poetry still remains to be written ('Una stilistica della materia: F. T. Marinetti e le "parole in libertà"', in Jean-Claude Marcadé (ed.), *Présence de F. T. Marinetti: Actes du Colloque International tenu à l'UNESCO* (Lausanne, 1982), pp. 46 f.). My present chapter, with its focus on the iconic, offers no more than a small part of that history.

Western modernism, and in particular on concrete poetry, and the degree of similarity between many of the Futurists' ideas and certain influential later theories such as Jan Tschichold's writings on typography and Marshall McLuhan's ruminations on the plight of the printed word in relation to other media.[7] The work of the Italians is, I would venture to suggest, more seminal in culturally historical terms, and, in quality, more in keeping with Futurist theory than its Russian counterparts.

Despite this, the Italian Futurist poets have often been demoted to the role of mere precursors, early innovators who simply 'forged from technological means devices of assault which were later to be re-appraised and adapted by De Stijl, Bauhaus and Constructivist', as well as Dada typographers.[8] This contrast between innovative Futurist devices and their later, more successful deployment at the hands of other avant-gardists may at times result from understandable qualms about the morally or ideologically unacceptable content of much Futurist experimentation. But it can only serve to obscure the fact that, although Italian Futurist poetry was often war poetry and frequently spoke with a reactionary voice, its many typographical achievements were and remain technically of major significance for the history of modern printing. In any case, even though it may have been initiated by Marinetti, the 'typographical revolution' of the Futurist period is by no means synonymous with his own war poetry, notably in *Zang tumb tumb*, *Les Mots en liberté futuristes*, and numerous contributions to *Lacerba* and *I nuovi poeti futuristi*. Other prominent—and less bellicose—Futurists, among them Cangiullo, Carrà, Folgore, Govoni, Severini, and Soffici, also made creditable contributions to the development of a distinctively new typographical repertoire. Even the work of minor figures within or on the fringes of the movement, such as D'Alba, Bètuda, Fivé, Jannelli, Masnata, Meriano, and Steiner, contains important examples of experimentation within the newly emancipated typographical idiom. And the passage of time has not made all of this into a mere matter of now distant literary history: Marinetti and his fellow Italian Futurists remain a live and palpable source of inspiration, especially since their rediscovery during the past two or more decades. As Bowler reminds us, their 'use of letters to produce sound qualities has been seminal not only to the opto-phonetic poems of Hausmann and Schwitters', but to contemporary 'concretists such as Ernst Jandl and Bob Cobbing'.[9]

[7] The similarities and differences between McLuhan's *Understanding Media: The Extensions of Man* (London, 1964) and *Counterblast* with Harley Parker (London, 1970), on the one hand, and Marinetti's early manifestos, on the other, have been pointed to by many commentators, including, most perceptively, William C. Wees (*Vorticism and the English Avant-garde* (Manchester, 1972)). But they still deserve to be explored in more detail, as does the indirect line of influence—via *Blast*—from Italian Futurism to McLuhan's theories.

[8] Cohen, 'The Typographic Revolution', p. 83.

[9] *The Word as Image*, p. 133.

Marinetti once defined the chief purpose of his projected typographical revolution as the 'redoubling of the expressive force of words', a description that could of course be applied to Futurist innovations in many areas. In fact, one of the most striking features of literary Futurism, in both Italy and Russia, was a dissatisfaction with the existing means of expression. While in the movement's numerous manifestos this could often hover at the level of empty invective against traditional (passéist) modes of communication, in Futurist literature itself the result was a series of radical attempts at an intensification of language (both spoken and written). It was this that was to take typography in a more original direction and with more dramatic results than are commonly found even in most modern literary avant-gardes.

In the history of Italian Futurism, this quest for a renewal of language largely took the form of a campaign to strengthen the imitative (or iconic) aspects of the linguistic signs writers were using. It is a programme that was gradually to leave its mark in five main areas:

1. in a predilection for onomatopoeic effects in performed poetry,
2. in a search for typographical equivalents for them when performed poems appeared in print,
3. in the creation of innovative forms of spelling, 'free expressive orthography', to use Marinetti's phrase ('expressive' was, in general, one of the favoured terms used by the Futurists at the time to refer to the iconic),
4. in various forms of shaped typography: 'auto-illustrations' and 'designed analogies', as the Futurists called them,
5. in 'synchronic tables' and various types of poetry-influenced collage.

Ironically, at the time the Futurists (and, above all, Marinetti their leader) rapidly became, usually pejoratively, identified with what was in innovative terms the least significant of these five areas: the use of sound-effects and onomatopoeia during Futurist performance, usually for the purpose of illustrating battle poems. 'I futuristi del *bumbum*', as they were soon to be derisively nicknamed, became figures of fun associated with the 'rattatata' of machine-gun fire, so lampooned by the Russians when Marinetti visited them in 1914.[10] Sound-effects of this kind were the easiest target for ridicule; and they were also one of the few features of Futurist poetry in performance that could still be registered by someone with only a modicum of Italian or French. However, Italian Futurism's real achievement in this general area of enhanced 'expressiveness' lay in the development of typography's potential in new directions, beyond the kind of mimetic effects that had (with the exception of a few earlier works like Mallarmé's *Un coup de dés* and Morgenstern's *Fisches Nachtgesang*) characterized 'figured' poetry from Alexandrian times right through to the *calligrammes*

[10] See A. Spaini, 'Futurismo', *La Voce*, 5 (1913), 1110, and Kruchenykh's hostile reaction to Marinetti's performances in Russia (*Manifesty*, p. 71).

FIG. 1.1. Fortunato Depero, *Grattacieli*, 1929 (Museo Depero, Rovereto)

and contemporary forms of visual concrete poetry. Nevertheless, as will become plain later in our exploration of the subject, these experiments do partly have their roots in the Futurists' preoccupation with onomatopoeia.

It makes sense to start an investigation into this 'typographical revolution' with the Italian Futurists' early sallies into shaped poetry, since this is one of the more ostentatious forms which their quest for a new 'expressiveness' took and certainly the one that has most frequently suggested, not so much a revolution, as an indebtedness to the more familiar work of Guillaume Apollinaire.

Although a number of Apollinaire's *calligrammes* had been privately sent on postcards to friends before this date, very few were available in published form by 1914, the time of the major Italian experiments in expressive typography. These printed versions must, however, have been known to Marinetti and his

FIG. 1.2. Francesco Cangiullo, *Milano-Dimostrazione*, 1915 (Private Collection)

circle, yet, tellingly, few Italian poets produced autonomous *calligrammes* of the kind one associates with Apollinaire. They did, admittedly, occasionally turn their hand to self-contained word-pictures, but mainly at a later stage in the history of Futurism, and these tended to be of a very different kind from the *calligrammes*. Fortunato Depero's view of New York skyscrapers, *Grattacieli* (fig. 1.1)[11] even has a constructed quality, and certainly invites a more exclusively synoptic reception than Apollinaire's visual poems, which are usually based on a conventionally linear and metrical sequence of lines, wittily arranged to suggest the object described. Francesco Cangiullo's *Milano-Dimostrazione* (fig. 1.2; this will be discussed in Chapter 2) represents a skilful fusion of words and graphic elements, with the patriotic Interventionist slogans representing the crowds packing the Piazza del Duomo in Milan.[12] But such works are rarities: not only is their *modus operandi* distinctly dissimilar from that of Apollinaire's visual poems; they even remain untypical of the mainstream Futurist exploitation of shaped writing, which tends to be for local impact, rather than constituting a self-contained word-picture. (If anything, a more apt analogy would be with some of the local effects in Apollinaire's 'Lettre-Océan', published in

[11] Depero also planned a book-length series of word-pictures depicting the city's skyscrapers: *New York—Film vissuto*. However, the volume unfortunately never appeared, despite the existence of a series of fine individual collages on the subject.

[12] See below (pp. 136 f.) for a discussion of this work.

Les Soirées de Paris on 14 June 1914, a work which Roger Shattuck has analysed as a 'single visual unit combining several disparate parts' and probably strongly influenced by Marinetti's own *Zang tumb tumb* and Cangiullo's 'Fumatori II.' that had both appeared earlier the same year.[13])

The simple, yet widespread visual device which Marinetti (possibly in order to forestall any confusion with Apollinaire's work) called an 'auto-illustration' does not constitute an autonomous *calligramme*, but entails a restricted use of a local 'concrete' effect as part of some larger poetic whole. In his manifesto 'Geometric and Mechanical Splendour and the Numerical Sensibility' ('Lo splendore geometrico e meccanico e la sensibilità numerica' (1914)), Marinetti explains how these 'auto-illustrations', 'in a continous effort to express things with the greatest force and profundity', at a certain point in the flux of the poem 'naturally transform themselves into auto-illustrations. . . . As soon as this greater expression is reached', however, they automatically 'return to their normal flow' (*TIF* 85). At vital poetic junctures, in other words, horizontally linear sequences of Futurist poetry would find their apotheosis in a pictogram or some other form of experimental effect. If, however, one actually turns to the best-known examples of 'auto-illustrations' in the poetry of the time, it becomes difficult to reconcile the indifferent quality of much Futurist practice with Marinetti's own evident enthusiasm. Admittedly, such 'auto-illustrations' do, as Marinetti claims, only form part of much larger poetic structures. Consequently, it would be unfair to judge them in isolation, even though anthologies inevitably do tend to reproduce them in this way. But unfortunately, even making allowances for the contextual relativity of their significance, 'auto-illustrations' of this kind were on the whole less grandly a matter of 'force and profundity' than Marinetti had evidently hoped. The classic example—his own setting, in *Zang tumb tumb*, of the Italian words for 'captured Turkish balloon' in the shape of a communications-balloon (fig. 1.3)—hardly amounts to more than a static object-image, with the main dynamic aspect coming from the sounds radiating outwards as signals from the balloon. The words thus 'auto-illustrated' do not stand out in any substantial degree of poeticity from the lines that precede them and those that follow, except that their appeal will now be primarily to the aesthetic eye, rather than to our intellectual reading faculty alone. Similarly, one is still left with the sense of an underexploited discovery when one turns to the arrangement of the words 'biplane', 'triplane', and 'polyplane' on one, two, and three lines (fig. 1.4), or the undulating printing of the word 'oscillamente' (in Paolo Buzzi's *L'ellisse e la spirale*) or the

13 'Apollinaire's Great Wheel', in Roger Shattuck, *The Innocent Eye: On Modern Literature and the Arts* (New York, 1984), 244f. In his 'Histoire sommaire du tableau-poème', *XXe siècle*, 3 (1952), 21–5, Michel Seuphor suggests that it was primarily Marinetti who influenced Apollinaire; and Apollinaire himself, using his pen-name 'Gabriel Arbouin', cites a number of Futurist precedents for his work: 'Devant l'idéogramme d'Apollinaire', *Les Soirées de Paris* (July–Aug. 1914), 383–5.

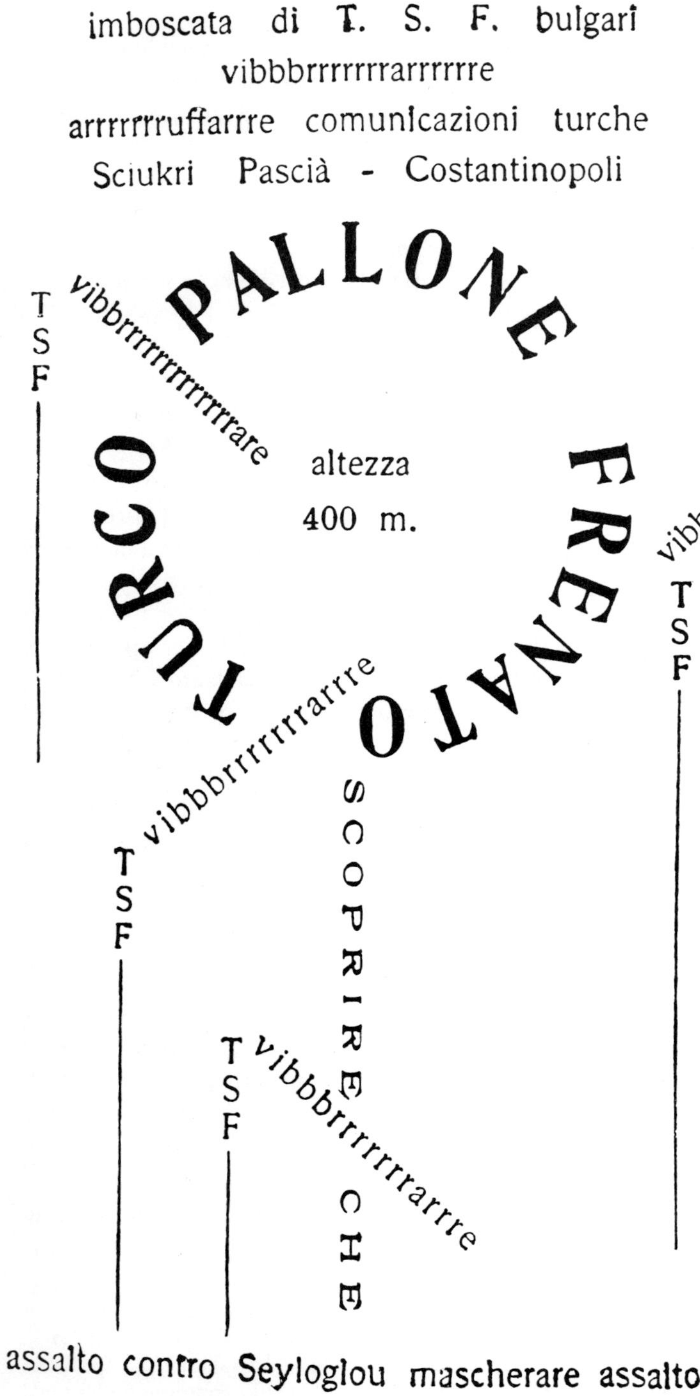
imboscata di T. S. F. bulgari
vibbbrrrrrrrarrrrrre
arrrrrrruffarrre comunicazioni turche
Sciukri Pascià - Costantinopoli

PALLONE FRENATO TURCO

vibbrrrrrrrrrrrrrare
altezza
400 m.
vibbrrr Tsarigraad

TSF
TSF
vibbbrrrrrrrarrrre
TSF
SCOPRIRE CHE
TSF
vibbbrrrrrrrarrrre
TSF

assalto contro Seyloglou mascherare assalto

FIG. 1.3. The balloon 'designed analogy' from F. T. Marinetti, *Zang tumb tumb*, Milan, 1914

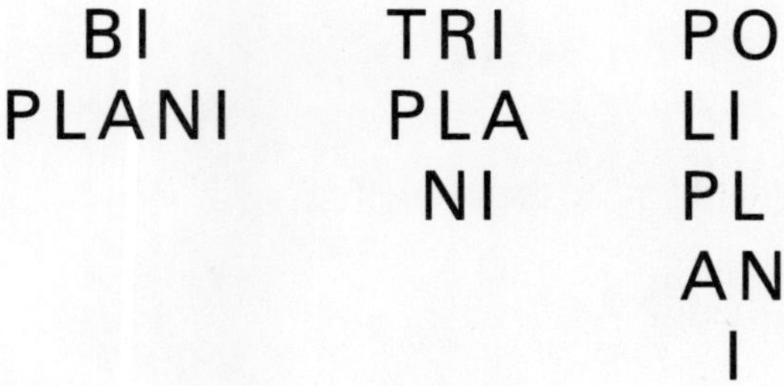

FIG. 1.4. 'Designed analogy' from Paolo Buzzi, *L'ellisse e la spirale*, Milan, 1915

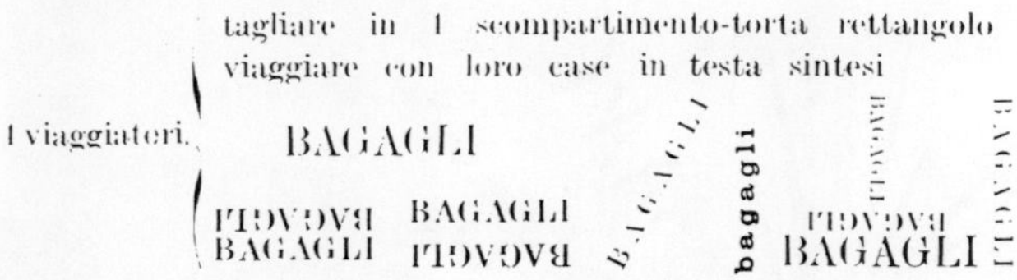

FIG. 1.5. 'Designed analogy' depicting suitcases on a station platform from Francesco Cangiullo, 'Fumatori II.', 1914

similar use of wavy lines in Corrado Govoni's poem about the sea in *Rarefazioni* (see fig. 1.15). A great deal more inventiveness comes with an effect such as Francesco Cangiullo's reproduction of the word for 'baggage' a number of times in a configuration suggesting the actual shape of a pile of luggage on a station platform (fig. 1.5); or in Armando Mazza's printing of the word 'MoNTaGNe', in his poem 'Linea Palermo–Messina', in a jagged manner to suggest the crenellations of peaks seen from a distance; or, to take another example from Govoni, the arrangement of the word 'arcobaleno' ('rainbow') in such a way as to imitate the shape of an arc in the sky. Yet not only are these examples[14] on the whole (*pace* Marinetti) relatively undynamic; they even fail to register in visual terms the actual 'expressive' highlights of the poems in which they occur. Indicating balloons, planes, waves, or suitcases in this manner is more in the nature of a Futurist five-finger-exercise than the pinnacle of visual poetic achievement that the movement's leader would have wished it to be. When Cangiullo donated a copy of his *Poesia pentagrammata* to Marinetti in 1925 and signed his name with a large letter 'C' in imitation of the shape of the Bay of Naples,[15] this whimsical cartoonist's effect was arguably no more and no less poetic than the kinds of 'auto-illustration' that the early Futurist poets soon

[14] The 'auto-illustrations' mentioned are to be found, respectively, in *TIF* 640; Paolo Buzzi, *L'ellisse e la spirale* (Milan, 1915), 222 f.; Corrado Govoni, *Rarefazioni e parole in libertà* (Milan, 1915), 58; Francesco Cangiullo, 'Fumatori II.', *Lacerba* (1 Jan. 1914), 10; Armando Mazza, *Firmamento* (Milan, 1920), 104; and Govoni, *Rarefazioni*, p. 47.

[15] For details, see Marjorie G. Wynne and Luce Marinetti Barbi, *F. T. Marinetti and Futurism: Catalogue of an Exhibition in The Beinecke Rare Book and Manuscript Library, Yale University* (New Haven, 1983), item 98.

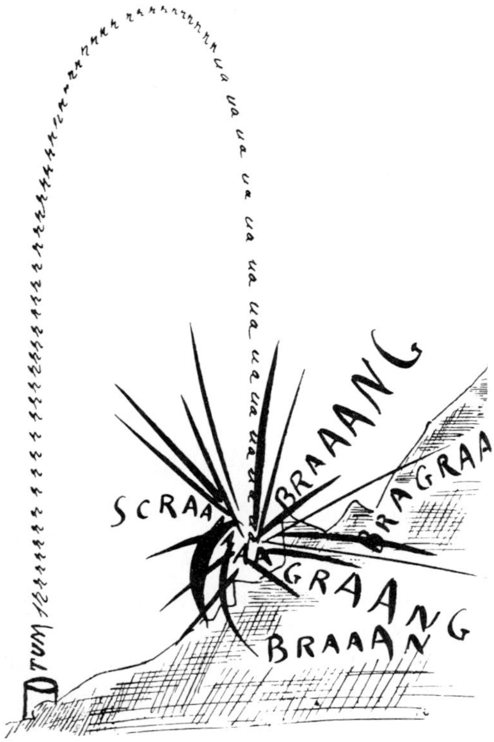

FIG. 1.6. Final part of F. T. Marinetti's *8 anime in una bomba*, Milan, 1919

began to overestimate. It scarcely compares with the subtleties of the Baroque figured poem or with the sheer wit and diversity of Apollinaire's *calligrammes*, though one has to allow for the fact that there may well be a deliberately anti-poetic element of provocation to the relative crudity of some of these effects—which avoid the decorativeness of the shaped poem as much as that of the passéist book's ornate typography.

More dynamic instances of the 'auto-illustrative' technique do, fortunately, exist. There is, for example, the final howitzer explosion (fig. 1.6) at the end of Marinetti's *8 anime in una bomba* (*8 souls in a bomb* (1919)) and the bombardment depicted in his 'Corvée d'acqua sotto i forti Austriaci' (fig. 1.7, published in *I nuovi poeti futuristi* (1925)). What these exceptions show, in contrast to the earlier balloons, suitcases, and plane shapes, is that there may be a certain tendency towards the static endemic in the 'auto-illustration', since it is not easy to suggest by mere outline shape the qualities of an object in motion. As a result, Futurist poetry is more likely to advance towards its goal of rendering dynamism typographically by using shaped writing to indicate *lines of movement*.

FIG. 1.7. F. T. Marinetti, 'Corvée d'acqua sotto i forti Austriaci', 1925

One of the earliest poems to reveal an awareness of this fact is Mario Bètuda's 'Looping the Loop', which appeared in *Lacerba* in April 1914 (fig. 1.8).[16] Bètuda's poem, reflecting a newfound sense of both poetic and aerial freedom, is no longer trapped within the kind of static and slavishly visual iconicity that involved the simplistic use of arrangements of words to suggest concrete objects via their outlines. Here, it is the flight path of the aircraft—not the silhouette of the plane itself—that becomes the focus of attention. The poem, like so many of the Futurists' subsequent incursions into expressive typography, advances beyond mere pictorial mimesis. And yet the latter is too often what the Futurists have been accused of cultivating.

Initially, Giovanni Papini must surely (and not for the only time, as we shall see, when we consider the polemics surrounding Futurist collage) bear the

[16] Although the poem is in Italian, there appears to be no Italian term for this aerial feat—hence the English title.

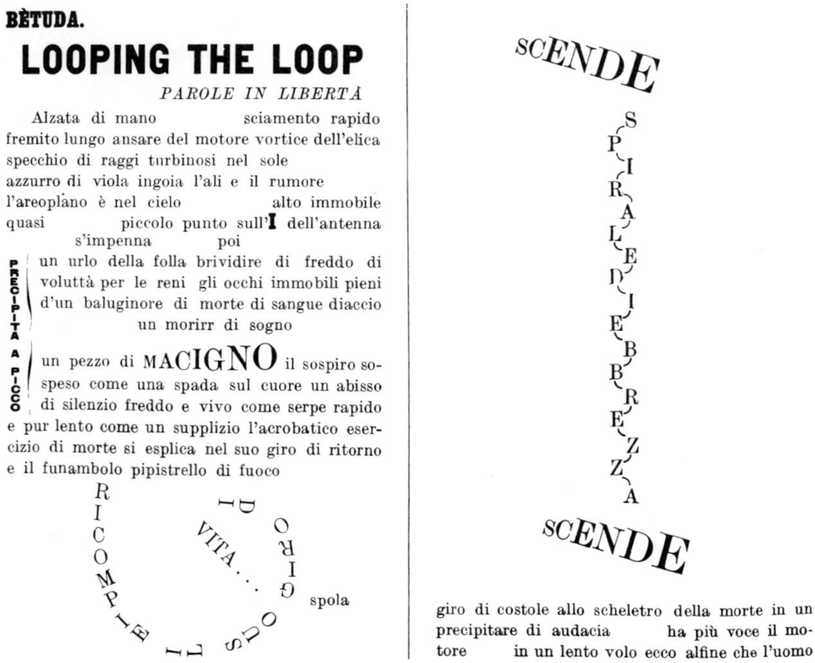
BÈTUDA.

LOOPING THE LOOP

PAROLE IN LIBERTÀ

Alzata di mano sciamento rapido
fremito lungo ansare del motore vortice dell'elica
specchio di raggi turbinosi nel sole
azzurro di viola ingoia l'ali e il rumore
l'areoplano è nel cielo alto immobile
quasi piccolo punto sull'**I** dell'antenna
s'impenna poi

PRECIPITA A PICCO

un urlo della folla brividire di freddo di
voluttà per le reni gli occhi immobili pieni
d'un baluginore di morte di sangue diaccio
un morirr di sogno
un pezzo di MACIGNO il sospiro so-
speso come una spada sul cuore un abisso
di silenzio freddo e vivo come serpe rapido
e pur lento come un supplizio l'acrobatico eser-
cizio di morte si esplica nel suo giro di ritorno
e il funambolo pipistrello di fuoco

RICOMPIE IL SUO GIRO DI VITA... spola

SCENDE

SPIRALE DI EBBREZZA

SCENDE

giro di costole allo scheletro della morte in un
precipitare di audacia ha più voce il mo-
tore in un lento volo ecco alfine che l'uomo

FIG. 1.8. Extract from Mario Bètuda, 'Looping the Loop', 1914

lion's share of the blame for the misunderstanding that arose. For, with his *Lacerba* article 'The Circle is Closing' (1914), he was, more than anyone else, instrumental in popularizing the view that Futurist typography was aiming at nothing more than pictorial mimesis.

The seemingly uncontroversial starting-point of Papini's polemic is the idea that 'Marinetti, with his words-in-freedom, relies on the visible image . . . in the form of words or phrases arranged typographically in such a way as to suggest with an ideogram the vision of the things spoken of'. But Papini quickly proceeds to the unwarranted premiss that 'if this method were accepted and rigorously taken to its ultimate conclusion'—the big 'if' and the big 'and' in the whole of Papini's argument—then 'the best poetry would be the spectacle of a battle with its sound cinema'. Papini's comment that these 'absurd prospects . . . are the direct extensions of premisses and experiments which already exist'[17] displays scant regard for the fact that the mere introduction of a controlled visual *element* into typography need not necessarily imply a criterion of creativity that must necessarily end up equating aesthetic value with the degree of imitation.

In fact, it was patently obvious, even at the time when Papini launched his

[17] 'Il cerchio si chiude', *Lacerba* (15 Feb. 1914), 49.

polemic,[18] that the general artistic goal of maximum realism was alien, not only to the spirit of Italian Futurism in the visual arts, but to the movement's aesthetic as a whole. We will 'totally invalidate all kinds of imitation', the painters had already announced in the second conclusion to their 1910 manifesto (*Archivi*, i. 64). And in the Sackville Gallery catalogue, one finds them again declaring that 'all forms of imitation must be despised'.[19] In 'Les Exposants au public' (1912) the same painters present themselves as 'destroying each day in ourselves and in our pictures the realistic forms and obvious details which have served to build a bridge of understanding between ourselves and the public' (*Archivi*, i. 108). Defining the new aesthetic in an equally anti-mimetic vein, the 1912 manifesto 'Futurist Sculpture' ('La scultura futurista') speaks dismissively of the 'figurative value' traditionally attributed to shapes and lines (*Archivi*, i. 72). In the light of such outspoken declarations, it would be more than a little perverse for Futurist typography, in particular the 'auto-illustration', to be at the very same time wilfully reintroducing on to the printed page precisely the kind of imitative and pictorial dimension being so vigorously outlawed in the visual arts. It comes as no surprise, therefore, to find Marinetti, for all his preoccupation with the new possibilities of the 'auto-illustration', pointedly warning that the Futurist poets must 'avoid any pictorial preoccupation' (*TIF* 89).

Papini's misrepresentation is, however, to some extent excusable, if one allows for the possibility of its forming part of a well-tried satirical-cum-polemical strategy, disingenuously postulating the sort of mimetic obsession that would presumably only find total satisfaction in *trompe-l'œil* of some kind. Admittedly, certain Futurists did on a later occasion (albeit in the context of a non-literary manifesto) make a glowing reference to the way shaped words-in-freedom 'destroyed the boundaries of literature and marched towards painting . . . throwing a marvellous bridge between the word and the real object'.[20] But to be marching *towards*, or throwing a bridge *between*, two things may not necessarily be the same as seeking a total congruity between sign and object, or indulging in any very pronounced or highly developed form of typographical mimesis. Had the main thrust of the Futurists' 'typographical revolution' been essentially towards pictorial mimesis, as has often been assumed,[21] the result

[18] Papini and Soffici (in '*Lacerba*—il Futurismo e *Lacerba*', *Lacerba* (1 Dec. 1914), 324) call this article, coming a year after the 'first Futurist number of *Lacerba*', the harbinger of their first 'real conflict' with the Futurists, adding that even after the final rejoinder the hostility remained 'latent'.

[19] Quoted from *Futurist Manifestos*, ed. Umbro Apollonio (London, 1973), 30.

[20] 'La cinematografia futurista', manifesto signed by Marinetti, Bruno Corra, Emilio Settimelli, Arnaldo Ginna, Giacomo Balla, and Remo Chiti (1916) (*Archivi*, i. 97). Significantly, Marinetti is the only literary signatory to this manifesto devoted to one of the more mimetic media with which the Futurists ever experimented.

[21] When Carrà uses the phrase 'we painter–poets' (*Lacerba* (1 Feb. 1914), 39), he simply means that they are painters *and* poets, in many cases. But commentators have tended to stress the pictorial aspect

would have been little more than one further—and rather belated—chapter in the long history of figured poetry. What is more, when judged as attempts at any great degree of imitative fidelity, their endeavours come through as crude by comparison with those of many of their predecessors. But the fact is that pictorial mimesis was seldom the Italian Futurists' ultimate goal. Ardengo Soffici was much nearer the mark when he chose to describe the seemingly pictorial element in Futurist poetry as 'approximately like hieroglyphic writing, *reduced to the schematic*'.[22] For the schematic—and often highly codified—element is far more integral to Futurist layouts than any straining towards faithful pictorial representation. Few Futurist poets could have used the phrase 'et moi aussi je suis peintre' in the mimetic sense in which the author of the *Calligrammes* had intended it.

Even on the very first page of Marinetti's *Zang tumb tumb*, the most influential of all Futurist typographical achievements, one finds an effect which is not a straightforward 'auto-illustration', in the sense that the words in the shape of a balloon or the piled up suitcase-words were. Here the verb *nascere* ('to be born') and the noun *Poesia* (also, and hardly coincidentally, the name of a magazine previously published by Marinetti), are printed in letters of increasing sizes, so that the layout reflects the meaning of the words (fig. 1.9). This effect (*TIF* 563) remains undeniably visual, but no one would equate it with any attempt at pronounced pictorial mimesis. In whatever sense poetry 'is born' (and 'grows', for the verb is also used in this sense of plants), this can only be metaphorically analogous to the growth in size of the letters on the page. It is by way of being a programmatic signal that this device is placed at the very

of their typography in such a way as to reduce it primarily to a passéist preoccupation with mimesis. Fausto Curi, for instance, emphasizes the 'pictorial quality' of the movement's shaped poetry in 'La "distruzione del modello lineare" e la letteratura d'avanguardia', *Lingua e stile*, 3 (1970), 450. Pär Bergman talks of the 'imitative element it contains' (*'Modernolatria' et 'Simultaneità': Recherche sur deux tendences dans l'avant-garde littéraire en Italie et en France à la veille de la première guerre mondiale* (Uppsala, 1962), 200), while Michel Seuphor injects a welcome element of abstraction by using the phrase 'poème plastique' to describe some of the works ('Histoire sommaire du tableau-poème', p. 22). Rosa Trillo Clough has no hesitation in interpreting many of her illustrations using the term 'pictorial' (*Futurism: The Story of a Modern Art Movement. A New Appraisal* (New York, 1961), 48); and Luciano De Maria expresses the view that the 'extensive introduction of designative elements' into such poetry has succeeded in shifting it substantially 'in the direction of painting' (*Per conoscere Marinetti e il Futurismo* (Milan, 1973), p. xxxiv). Most of these comments record a move in the direction of pictorialism as a matter of aesthetic programme, while often implying that this is a 'good thing'. But a condemnatory position is adopted by Zbigniew Folejewski, who castigates such an unwarrantable straying beyond the legitimate bounds of literature. He refers critically to 'the limitations of such Western attempts to introduce pictorial elements into a literary text as, for example, Marinetti's collage "Joffre après la Marne" [*sic*], Sanzin's "Tunnel" or Apollinaire's *Calligrammes*'. Khlebnikov, he announces with some relief, 'at least stayed within the sphere of verbal devices, although some of his fellow-poets indulged in the same kind of mixing of verbal and visual material' (*Futurism and its Place in the Development of Modern Poetry: A Comparative Study and Anthology* (Ottawa, 1980), 78).

[22] 'Chimismo lirico', *Primi principî di una estetica futurista* (Florence, 1920); quoted from *Archivi*, i. 586.

POESIA NASCERE

FIG. 1.9. From F. T. Marinetti, *Zang tumb tumb*, Milan, 1914

beginning of the work, for it is an example of the kind of 'expressive' effect that Marinetti referred to as a 'designed analogy'. (And unlike experimental poetry of the 'auto-illustrative' kind, it really does take shaped poetry in a direction where more ingenious analogy-creating experiments become possible.) In fact, 'designed analogies' represent a much greater breakthrough in Futurist typography and play a substantially more crucial role in the movement's various experiments with the sign-meaning relationship than 'auto-illustrations' could ever have done.

In a reference to Cangiullo's free-word *Lacerba* poem 'Fumatori II.', Marinetti praises the moment when the poet 'had the happy thought of rendering with the following *designed analogy*:

FUUUUUMARE

the long and monotonous reveries and self-expansion of the boredom-smoke of a long train journey' (*TIF* 89). In reality, the poem spells the word 'fumare' with a 'self-expanding' multiplicity of 'u's, contained within two diverging lines, as if to give some sense of the time factor in the expansion of the mood (see fig. 1.10).[23] The shape of the word 'fumare' (in either version) expands as an analogy not only to the increasing boredom of the journey, but also as a reflection of the expansion of smoke (the visual token of accumulated boredom) which the journey generates. Increasing boredom is obviously a difficult mood to suggest via visual effects, but, by hitting upon the smoke/boredom analogy, Cangiullo has managed to discover a physical objective correlative for such an atmosphere, and hence a cipher for it that can be rendered typographically.

In fact, within the actual context of 'Fumatori II.',[24] Cangiullo's effect is both more complex and decidedly more explicit than Marinetti's isolating of the

[23] The fact that Marinetti goes on in the next paragraph to cite his own balloon effect in *Zang tumb tumb* suggests that he may not always have made a clear distinction between 'auto-illustrations' and 'designed analogies'. In his discussion of the 'designed analogy' in 'Fumatori II.' and Joan Salvat-Papasseit's 'Edison'-effect (in 'Marxa nupcial'), Willard Bohn contrasts the funnel outline in the latter with Cangiullo's lack of linear containment (*The Aesthetics of Visual Poetry*, p. 130). But this is only true of Marinetti's manifesto-misrepresentation of the effect. In *Lacerba* the word is also contained within a funnel, which makes the Futurist influence on Salvat-Papasseit even more pronounced.

[24] *Lacerba* (1 Jan. 1914), 10.

FIG. 1.10. Smoke 'designed analogy' from Francesco Cangiullo, 'Fumatori II.', 1914

FIG. 1.11. Part of Francesco Cangiullo, 'Fumatori II.', 1914

single word 'fumare' would suggest. For we are offered an equation-like depiction of what the four passengers confined in a train compartment are feeling and doing (fig. 1.11). Of the people in the compartment, two are calm, whereas a third is anxious; but it is specifically the fourth one, described as being in no hurry ('senza FRETTA'), to whom the activities 'leggere . . . fuuuuumare' are attributed. Which does rather put a question mark against Marinetti's interpretation concerning 'monotonous reveries' and sheer 'boredom'. Be that as it may, the underlying point remains that what we are dealing with in the case of 'fumare' is a very different use of shaped typography from the by then traditional arrangement of words to indicate the contours of objects. It may well be that the decision to use letters of increasing (and, elsewhere, decreasing) size and in different typefaces owes something to Mallarmé's *Un coup de dés*, a work which was particularly revolutionary in extending the possibilities of typography beyond the purely pictorial (even though it does contain occasional figurative moments, such as those relating to the images of the abyss, the feather, etc.). Nevertheless, various differences between this innovation and both Mallarmé's and Apollinaire's work suggest that Marinetti's 'designed analogy' is firmly rooted in another tradition. First, there is the fact that as a genre the 'designed analogy' is not abstract in the way that Mallarmé's layout largely is, but still signals certain features of a denotatum in the real world. (I shall shortly try to show that the semiotic concept of the 'iconic sign' is the most efficient way of characterizing this relationship.) Second, the use of the 'designed analogy' at certain peak moments in a poem gives it the role of a *language-intensifier*, rather than an overall generic coding. (This much it shares with the 'auto-illustration'.) It is not that the Futurists are content to use conventional language in all but these heightened moments of expressiveness, though. Their various other syntax-orientated language reforms are calculated to create a radically innovative idiom. Yet the 'designed analogy' still stands in a more subservient relationship (as an intensifier)—and then only at certain points in a poem—to the Futurist medium than is the case with the visual effects in *Un coup de dés* or in modern shaped poetry in general.

Before we go on to examine in any detail further examples and features of Futurist typography, it will help to identify the more immediate pedigree of the 'designed analogy'—which lies in the movement's extensive concern with onomatopoeic effects in performance and in the use of innovative spelling to reflect these acoustic features on the printed page. Like 'auto-illustrations' and 'designed analogies', both of which they substantially pre-date, these are also examples of the Futurist concern with making language more 'expressive', which, in semiotic terms, invariably means 'more iconic'. In fact, to analyse this with any degree of precision, a semiotic framework becomes desirable, since only the relatively recent advances in the study of iconicity[25] can offer an adequate conceptual framework and interpretive model for the analysis of such complex sign-meaning relationships. Prior even to a detailed look at the Futurist concern with onomatopoeia, therefore, a brief excursus on the semiotic implications of the Italian Futurists' work is called for.

Ironically, at just the same time as Marinetti and his fellow Futurists were breaking new ground with various methods of language-intensification, Ferdinand de Saussure was defining language as being inherently characterized by the very features to which they were opposed. Compare, for instance, Saussure's seminal emphasis on the 'arbitrariness of the linguistic sign' with Soffici's declaration that for the Futurist the word would no longer be 'a mute symbol of convention, but a live form among live forms, one that becomes one with the material of representation'.[26] Or note the way that many of the experimental layouts of Futurist poetry represent an attempt to combat the very 'linearity' that Saussure defines as language's second principal characteristic.[27] Apollinaire once suggested that the Futurists had moved away from the tradition of 'horizontal poetry' towards that of a 'vertical poetry',[28] which is an

[25] An attempt at a methodological justification of an iconicity-based approach to Futurist typography can be found in J. J. White, 'The Argument for a Semiotic Approach to Shaped Poetry: The Case of Italian Futurist Typography', *Visible Language*, 10 (1976), 53–86. For the theoretical background to such a method of interpretation, see Paul Bouissac, Michael Herzfeld, and Roland Posner (eds.), *Iconicity* (Tübingen, 1985).

[26] *Cours de linguistique générale*, ed. Charles Bally and Albert Séchéhaye (Paris, 1949), 102–5; Soffici, 'Tipografia', in *Primi principî di una estetica futurista* (*Archivi*, i. 687).

[27] *Cours de linguistique générale*, pp. 186–90.

[28] 'Nos amis les futuristes', *Les Soirées de Paris* (15 Feb. 1914), 78 f.

understandable reaction when one thinks of the layouts of *Lacerba* and the word-columns of Palazzeschi and many of his Russian counterparts. But it would be more accurate to say that the main achievement of the Futurists in this respect, particularly in Italy, lay in the pushing of poetry in the direction of a new, more dynamic, diagonal poetry, and even towards multi-directional layouts. One Futurist response to the limitations of language's linearity was (as we shall see in Chapter 3 in the case of telegraphic writing) not to *abandon* linearity totally, but to force it to function in a more economic and (so they thought) more expressive way. Experimental poetic layouts involve a similar compromise with linearity, rather than its total rejection. But in trying at the same time to make poetic language more expressive through expanding its material potential, the Italian Futurists were setting out to create a medium with more immediacy and sensual impact. In semiotic terms, they were moving language from the *arbitrary* in the direction of the *motivated* (Saussure), or from the *symbolic* to the *iconic* (Peirce).

The printed word acts as a sign in two respects. It is a token of a set of sounds, and these in turn stand for either an object or some other kind of denotatum. Within most modern European languages, 'words, sentences, and other conventional signs are', to quote Charles Peirce, no more than 'symbols'.[29] They are unmotivated, or, as Saussure puts it less helpfully, 'arbitrary'.[30] That is to say: there is no discernible connection between the shape or other properties of the letters, words, or their arrangement within an utterance, on the one hand, and their meaning, on the other. Under normal circumstances, this unmotivated, 'symbolic' aspect of language can be thought of as a form of functional streamlining. As Massin puts it: 'carried along with the flow of the narrative from page to page, barely glanced at by hurried and impatient readers, the essential job of the letter is to be as unobtrusive as possible.'[31] (This may explain why most of the Futurists, for all their emancipated layouts, retained conventional typefaces and did not go in for the decorative effects one associates with Cangiullo, Govoni, and Zdanevich; this conventional aspect accelerates the reading process, in some respects.)

It was in *Existential Graphs* and *Elements of Logic* that Peirce first elaborated his distinction between iconic and symbolic signs (the third type of sign in his

[29] *Collected Papers*, xi. *Elements of Logic*, ed. Charles Hartshorne and Paul Weiss (Cambridge, Mass., 1932), 165.

[30] Because words are not 'arbitrary' in the sense of being chosen randomly at *each* use, Stephen Ullmann suggests that the difficulty created by the Saussurean term 'can be reduced . . . if "arbitrary" is replaced by "conventional"', since all it 'means is the absence of any intrinsic motivation or justification, any natural connection between the name and the sense' (*Principles of Semantics* (Oxford, 1957), 83). For similar reasons, the contrast 'motivated'/'unmotivated' has been preferred in other quarters. See Emile Benveniste, 'The Nature of the Linguistic Sign' in *Problems in General Linguistics* (Coral Gables, Florida, 1971), 43–8, and F. François, *Le Langage: Arbitraires et motivations en français et en allemand* (Berne, 1981).

[31] *Letter and Image* (London, 1970), 19.

triad, the index, being of less importance in the present literary context [32]). It was Peirce's disciple Charles Morris, however, who was to offer a much clearer résumé of this division of signs:

A characterizing sign characterizes that which it can denote. Such a sign may do this by exhibiting in itself the properties an object must have to be denoted by it, and in this case the characterizing sign is an *icon*; if this is not so, the characterizing sign may be called a *symbol*. A photograph, a star chart, a model, a chemical diagram are icons, while the word 'photograph', the names of the stars and of chemical elements are symbols. . . . The semantical rule for the use of icons is that they denote those objects which have the characteristics which they themselves have—or more usually a certain specified set of their characteristics.[33]

Morris also introduced into semiotic theory the helpful qualitative concept of 'iconicity' and demonstrated its importance, within the aesthetic realm, as a means of reinterpreting various forms of quasi-mimetic effect within a wider, yet more differentiating framework. Explorations of the iconicity of shaped poetry—and the iconic in the other visual arts—have, however, tended in recent years to be dominated by taxonomic concerns. Just as Peirce, in his discussion of the degeneration of signs, identified various types of iconic effect, from the high iconicity quotient possible in pictures, via diagrams, to the low iconicity of metaphors, so recent studies[34] have tended to organize their examples typologically according to a notional scale of iconicity. This will not be my concern in viewing Futurist onomatopoeia and its importance for poetic layouts within an essentially semiotic framework. Instead, I shall be availing myself of the concept of the 'iconic' in the present context mainly for two other reasons: first, because even the mimesis of the iconic sign has now been recognized as consisting of a complex mixture of motivated and conventional elements (and this is a particularly important feature in motivated poetic experiments, where an interplay between the symbolic and the iconic aspects of

[32] In Peirce's system, the index relates to its object through some resultative connection: smoke is an indexical sign for fire, clues and medical symptoms are indices. In his paper 'Il concetto semiotico di "indice" nella sua applicazione alla poesia parolibera futurista' (given at the 12th Meeting of the Assoziazione per gli studi di lingua e letteratura italiana), Christopher Wagstaff has pleaded for the relevance of indexical, as well as iconic signs, to Futurist shaped poetry.

[33] *Foundations of the Theory of Signs: International Encyclopedia of Unified Sciences*, 1, 2 (Chicago, 1938), 24. Whereas Peirce merely declares that an icon is a sign which is 'like [some]thing and used as a sign for it' (*Elements of Logic*, p. 143), Morris's concept of 'iconicity' has the additional merit of interpreting the iconic as one element within the sign-structure. The iconic has recently been presented as a special case of synecdoche; see Ewa Siemińska, 'Connotation and denotation in film art', in A. J. Greimas *et al.* (eds.), *Sign—Language—Culture* (The Hague, 1970), 416. The issue at stake, however, is how much *pars* stands *pro toto*, and under what conditions of codification. And this is something recent developments in the theory of iconicity are in a strong position to determine.

[34] See Mieczyzław Wallis, *Arts and Signs* (Bloomington, Indiana, 1975), which categorizes material on an iconicity-scale ranging from 'schemata' to 'pleromata', and the similarly taxonomic approach to the iconic element in shaped poetry in Giovanni Pozzi, *La parola dipinta* (Milan, 1981).

language is bound to be far stronger than in the visual arts in general, since the printed word is by its very nature largely conventionalized); and, second, because the theory of iconicity has in recent years been fruitfully modified, particularly by Umberto Eco and the semioticians associated with the journal *VS*, so as to take account of various underlying codes of recognition and features of the reading process which can alone do justice to free-word poetic practice.

The notion of 'iconicity' has the added advantage of bringing together, within the framework of a general theory of signs, diverse features of the Futurists' interest in onomatopoeic effects, in non-conventional spelling, and in 'expressive' free-word layouts. Above all, it helps reveal the connection between what was the starting-point for many of these innovations—the challenge of onomatopoeia—and the later, more sophisticated devices the visual poets went on to create.

When Marinetti came to England and performed his acoustically powerful war poetry, reactions ranged from Edward Marsh's damning judgement that the whole performance scarcely managed to rise above 'the level of a very good farmyard imitation—a supreme music-hall turn'[35] to Wyndham Lewis's undisguised 'astonishment [at] what Marinetti could do with his unaided voice'. According to Lewis, 'a day attack upon the Western Front, with all the "heavies" hammering together, right back to the horizon, was nothing to it. My equanimity when first subjected to the sounds of mass-bombardment in Flanders was possibly due to my marinettian preparation—it seemed "all quiet" to me in fact, by comparison.'[36]

But whether praising or condemning the Italian's onomatopoeic performances, most people were at the time agreed on at least one point: that his poetic method was at root of an imitative order, aiming at great aural accuracy. Such an assumption is echoed in Hugo Ball's view that the verbal innovations of the Futurists were essentially 'naturalistic'.[37] In England, Harold Munro captured the consensus reaction of his contemporaries with the comment: 'We admire his [i.e. Marinetti's] extraordinary inventiveness; we were enthralled by his declaration; but we do not believe that his present compositions achieve anything more than *an advanced form of verbal photography*.'[38] In other words, we are

[35] *A Number of People* (London, 1939), pp. 259f.

[36] *Blasting and Bombardiering: An Autobiography (1914–1926)* (London, 1967), 33. In 'Les "mots en liberté" de Marinetti et la "transmentalité" (zaoum) des Futuristes russes', E. E. Kovtun quotes from an unpublished manuscript in the archives of the Leningrad History Museum in which Mikhail Matyushin remarks apropos of Marinetti's declamatory performances: 'He was well capable of imitating the sound of a propeller, of explosions, and the roll of a drum, as if he were enacting the future European war' (*Présence de F. T. Marinetti*, p. 235). These onomatopoeic performances are also discussed in Matyushin's 'Russkie kubo-futuristy', *Rossiya*, 1 (1974), 174.

[37] *Briefe 1911–27* (Einsiedeln, 1957), 278.

[38] 'Futurism and ourselves', *Poetry and Drama*, 4 (1913), 389 (my italics).

back on the same familiar territory we encountered with Papini's insinuation that Futurist 'auto-illustrations' aimed at a totally imitative effect. And once more, as in that case, we shall find that the charge of out-and-out mimesis is ill-founded, that a high degree of deliberate schematization is, in fact, involved in Futurist onomatopoeia.

When translated from live performance, of the kind Lewis, Marsh, Matyushin, and Munro recalled, to the printed page, the poetic cacophonies with which Marinetti assailed his audiences were invariably invested with further iconic dimensions, as in two examples from *Zang tumb tumb* (figs. 1.12 and 1.13). Here, intensity and size of typeface become indicative of the volume and possibly even dramatic quality of the war-experience evoked. Although there is, as yet, no sign of any departure from the dominant convention of horizontally linear typography, size and boldness of typeface are clearly being exploited here (*TIF* 620f., 606f.) to indicate the intensity of gunfire (and possibly even the effect on the combatants hearing it). In fig. 1.13 the sounds are even explained in parenthesis, not only because Marinetti's readership may be unaccustomed to the noises of battle, but because even acoustic iconicity remains ill-defined and often ambiguous. (Without the help of well-established conventions, one might even wonder whether anyone could deduce the significance of 'tut tut' or 'POW'!) The point is made in other terms by Stephen Ullmann when he remarks of onomatopoeia that it is 'essentially incomplete: it merely suggests a similarity or a parallel' since 'not all sound elements of the name are concerned'. Consequently, 'motivation is by no means a perfect replica of acoustic structure, but merely a rough approximation'.[39] Nevertheless, to lampoon these Futurist effects for their inaccuracy, as Apollinaire once did,[40] is to misinterpret them. Futurist onomatopoeia, whether on the stage or on the page, is no more a botched attempt at some form of high-definition realism than the 'auto-illustration' is.

As John Lyons points out in *Semantics*, it is linguistically inevitable that 'there is some arbitrariness or conventionality, even in onomatopoeic forms'.[41] Indeed, Marinetti would appear to have established his elaborate typology of acoustic devices (in Point Eight of the manifesto 'Geometric and Mechanical Splendour

[39] *Principles of Semantics*, p. 87. The point is explored in respect of non-creative contexts by M. Grammont in 'Onomatopée et mots expressifs', *Revue des langues romanes*, 44 (1961), 97–158.

[40] In 'L'Esprit nouveau et les poètes', Apollinaire makes a thinly disguised attack on Marinetti and the Italian Futurists with his assertion that it 'would be absurd, if not dangerous, to reduce poetry to a kind of onomatopoeic harmony, *which does not even have accuracy as its excuse*' (my italics). Unfortunately, while being correct about the device's lack of accuracy, Apollinaire does not consider any other possible *raison d'être* for the device.

[41] *Semantics*, i (Cambridge, 1977), 101. Lyons's main plank for this claim is that onomatopoeic forms 'are made to conform to the phonological systems of particular languages, rather than being directly imitative of what they . . . stand for'. But other forms of conventionalization are also involved. Indeed, once one enters the realm of the creative use of onomatopoeia in literature, a variety of other work-related factors may be at stake in severely curtailing the degree of acoustic iconicity.

eccitazione dell'atmosfera 3 miliardi di vibrazioni nuove sensibilità irrrritata degli echi affaccendati sotto l'assalto dei frastuoni pesanti perforatori **zang tumb-tumb-tumb** pallore sinistro della muraglia faccia di calcinaccio dal lungo nasoooone d'acciaio scivolare oliiato delle ruote del pezzo all'interno puntare fremito nelle cellule nervose che comandano riflessi di difesa ad un tratto sternuto della batteria inondazione di frastuoni rrrrussare d'ingranaggi 2° spasimo esplosioni successive di 6 riflessi di difesa **tumb-tumb tamb-tamb tumb-tumb tumb-tumb** prurito prrrurito prurito del forte Kutch-Tépé risveglio esaltazione danza delirio pazzia di tutti i centri di difesa *600 000 000 molecole dell'atmosfera in trambusto smarrimento depressione delle masse disgregate* **tron-tron-tapata-traack** sbocciare di fiamme buco scavato da obice *(DIAMETRO 80 CM.)* turrrrbine rivoluzione di calcinacccci fumo acre acido solfidrico gaz ammoniacale odor di bruciato *schiacciarsi delle gobbe gialle della calce aprirsi di vallette liscie nei mattoni fame e smorfie di 4 cccrepacci pressione di 3 ventri dell'atmosfera per penetrare nel buco cuocersi dei loro 3 ombelichi*

limarsi degli orli del buco gorghi di granelli di sabbia nel nuovo corridoio d'aria lacerarsi di molecole fiiiltrare di 800 grammi d'idrogeno 1 m. più in alto oscillazione striiiidente d'un trave aperto a forbice sotto il cooolar della sabbia

FIG. 1.12. From F. T. Marinetti, *Zang tumb tumb*, Milan, 1914

and the Numerical Sensibility') precisely according to the amount of interplay between convention and iconicity present in the various types, although in this case the conventional element is often a matter of an idiolect based on local Futurist codes. It is significant that, within his typology, what Marinetti designates as '*direct, imitative, elementary, realistic onomatopoeia*' is but one among a whole arsenal of sound-effects construed as being available to the Futurist war poet. And, even if some of Marinetti's types of onomatopoeia[42] may strike the less adventurous non-Futurists among us as somewhat fanciful, his thinking on the subject is at least refreshingly differentiated when compared with the position of those who apodictically condemn him for a regression to out-and-out mimesis.

Marinetti's use of the—perhaps surprising—epithet 'elementary' to describe his most crassly imitative form of onomatopoeia serves as a hint that the crudity of the effect frequently forms part of a deliberately provocative anti-art strategy. In the 'Destruction of Syntax' manifesto, the reader at one stage finds Marinetti attributing an important secondary function to the device: the provocative and vitalizing effect of suggesting primitivism. Onomatopoeia is heralded as a means of 'enlivening lyricism with crude and brutal elements'.

[42] Types 3 (*abstract onomatopoeia*) and 4 (*psychic onomatopoeia*: an umbrella concept for 'the fusion of 2 or 3 abstract onomatopoeias') are unnecessarily esoteric when compared with Marinetti's poetic practice or any conceivable form of poetic efficacy. To be told that in the 'Dune' section of *Zang tumb tumb* the sound '*ran ran ran* corresponds to no natural or mechanical sound, but expresses a state of mind' raises the question of whether one is dealing with onomatopoeia in the conventional sense, or simply with some form of acoustic objective correlative.

mulino oscillare
telaio tessitura spole di fuoco seta del
rumore fucileria = stridore di mola fuga
alternata di sottili fumi bianchi *(SHRAPNELS)*
grossi fumi neri *(GRANATE)*
concorrenza industriale di gridi + fiamme
+ proiettili + sfracellamenti sotto il vento
regolatore commerciale sciopero
di rumori a 1000 m. davanti a
me 2 aratri pazieeeenza scriiiicchiollo di zolle
sassose pettini-furtivi-del-vento su 6 pecore
tosse d'un ufficiale seduto su un cassone
nella trincea risonanza dei pol-
moni tisi flussssso e riflussssso di foglie
tremanti sulla spiaggia del silenzio bulgaro
mare agonie eleganti Nizza Menton
San Remo **patapum-pluff** *(ONDA)*
fraaaaaah *(GHIAIA)*

plupumflae *(DURO PRECIPITATO LUGUBRE)*
fraaaaaah *(INDULGENTE SCONSOLATO)*
pluuuum *(CADAVERE O CANNONATA)*
sciaaaaaah *(VENTAGLIO D'ACQUA ANGOSCIE RICORDI)*
a 8000 m. a destra al di là della Maritza
valle piena di esplosioni imbavagliate *(CALMO NONCURANTE AZZUUURRO VERDE GRIGIO FRE-SSSSSSCO GIOVANE SORRIDENTE)* ingenuità di
4 fumi bianchi su quell'altura shrapnels quasi
silenziosi garbatezza di una batteria
invisibile spazzare la CrRrREsta sopra l'arram-
picarsi *(INVISIBILE POCO RUMOROSO CERTO ACCANITO)* fanteria sussulti graziosi d'acqua
bollente quella caldaia borbottare *(MITE LON-TANO DOLCE DOLCISSIIIIIMO PACIFICO IN-NOCENTE)* risposta ugualmente gar-
bata shrapnels turchi *(LEGGIERO FESTOSO*

FIG. 1.13. From F. T. Marinetti, *Zang tumb tumb*, Milan, 1914

The new 'audacious onomatopoeia' is saluted as an attempt to 'destroy and mock the garlands, the palms, the aureoles, the exquisite frames, the mantles that comprised a large part of poetry up to now' (*TIF* 66). This second function—that of acting as a sign for 'brute reality' and hence doubling as a gesture of contempt for what was conventionally considered permissible or desirable in the realm of post-Symbolist poetry—must actually have required a certain flaunting of crude effects. At one and the same time then, the sound of a bombardment could be made to represent an incursion of brute reality into the realm of modern poetry and yet, in its schematic crudity, remain provocatively unrealistic. The sign thus bears a double coding: indicating *the real world* (as part of an attempt at communicating the actual feel of the battlefield) and *the primitive* (in contrast to the decorative *and* mimetic preoccupations of those whom the Futurists opposed).

That even relatively small—unrealistically minimal—quotients of iconicity could facilitate, rather than impede, the kind of sign-function which Marinetti envisaged for his onomatopoeia readily emerges from his analysis of *analogical onomatopoeia*, in the same manifesto:

In my 'Contrabbando di Guerra', in *Zang tumb tumb*, the strident onomatopoeia ssiiiiii presents the whistle of a tug on the River Meuse and is followed by the muted onomatopoeia ffiiiii ffiiiiii, an echo from the opposite bank. The two onomatopoeias save me from having to describe the width of the river, which is defined by the contrast between the two consonants *s* and *f*. (*TIF* 90).

On balance, even Marinetti's sympathizers would probably be inclined to dismiss this as expecting too much from a change from 's' to 'f'. For even if, in the poem itself, the words *riposta dell'eco dall'altra riva* ('reply of the echo from the other bank') at least help specify the relationship between the two sounds, the idea that a single consonantal change should be adequate to quantify (or 'define') the width of the River Meuse at this point is imposing unrealistic demands on one local effect.[43] This is another of those not infrequent occasions where the hypercharged inventiveness which helped make Marinetti a redoubtable force within contemporary European avant-garde circles succeeds in getting the better of him, to the point where he seems unable to distinguish between ideal effects and what he can plausibly expect certain innovative signals to say to the reader. Yet, for all the apparent lack of pragmatism to this particular sequence, it does throw light on the essentially proto-semiotic way in which Marinetti views the Futurist deployment of onomatopoeia.

The two sounds, 'ssiiiiii' and 'ffiiiii', form a contrastive structure; in linguistic terms, they act as a 'minimal pair'. In structural linguistics, the normal function of a minimal pair is as a strategy for isolating the phonemes of a given language.[44] But here Marinetti's minimal, but semantically meaningful, contrast receives a new poetic task. Realistic details, such as the fact that a double echo would hardly consist of two almost equally long sounds or the likelihood that such a sound would, if more faithfully reproduced, involve some measure of vowel mutation, as well as a change of initial consonant, are not pertinent to the present context—the reason being that Marinetti is not primarily concerned with the accurate reproduction of real sounds. What he is offering is a linguistic sign to convey a sense of distance. The resultant contrasting pair 'ssiiiiii'/'ffiiiii' allows the second sound-cluster to function as an *indexical sign*, for the change between the one sound and its echo is an index not only of the tugboat's presence (in the same way as smoke functions indexically to denote fire), but also of the width of the River Meuse. The fact that the indexical sign-function overrides any preoccupation with acoustic fidelity in this instance is by no means untypical of Futurist onomatopoeia. At first sight, the device may look crudely unrealistic. But that is only because it is performing another, more important function than the acoustic equivalent of the pictorial.

It is, of course, conceivable that these somewhat primitive onomatopoeic

[43] In fact, Marinetti misquotes his own effect in 'Contrabbando di guerra (Rotterdam)', where the contrast is between 'siiiii-siiiii' and 'fiiiii-fiiiii' (one consonant and five vowels in each case). But when Russolo later garbles Marinetti misquoting himself, he ironically ends up with a more realistically differentiated version of the same effect: 'ssiiiii' and 'fffiiiiiiii fiiiii' ('I rumori di linguaggio (le consonanti)', *L'Italia futurista*, 1 (1 Oct. 1916)). In other words, the effect is now less codified and more superficially mimetic.

[44] See the discussion of 'minimal pairs' as a device for isolating phonemes in Eugene Nida, *Morphology: The Descriptive Analysis of Words* (Ann Arbor, Michigan, 1962), and Arthur Wagner, *Das Minimalpaar* (Hamburg, 1982).

signs on the printed page were intended first and foremost simply as cues for Marinetti *the performer*. (This might at least account for the disconcerting discrepancy between the impression of aural realism his early audiences clearly registered when listening to the poet declaim and the evident simplicity, and even schematic primitiveness, of the signs chosen to represent the various battle-sounds and noise of machines.) No doubt, an unsubtle, shorthand form of annotation would have sufficed as a signal that a certain, in actual fact highly realistic, imitation was called for from the declaiming poet. But a number of factors speak against such a notion of an encoded onomatopoeic realism, realizable only in performance:

1. The *visual* iconicity of shaped free-word poetry at this time was equally schematic and crudely unrealistic, yet here there could be no equivalent opportunity for supplementation in performance. (The mind's eye does not replace the Turkish balloon or the bi-plane effect with a more accurate picture of the denotatum while the 'auto-illustration' is being read.)
2. Marinetti's own theoretical remarks on the subject of onomatopoeia, few in number though they are, invariably focus on highly coded sign-functions, not on mimesis as an aesthetic goal in its own right. When he tells the Futurist performer to 'metallize, liquefy, vegetablize, petrify, and electrify the voice, grounding it in the vibrations of matter itself, as expressed by words-in-freedom',[45] Marinetti cannot be proposing some new form of acoustic realism; a voice used to suggest the vegetable, the liquid, or the electric is going to do so in a far more encodedly iconic way than via acoustic imitativeness. (After all, what does a vegetable sound like?)
3. The published versions of these poems using onomatopoeic effects in any case acquire a separate identity, no longer tied to their author's personal style of delivery. They then demand to be read as autonomous texts, independent of Marinetti's possible idiosyncrasies of interpretation. And received thus, they are bound to be endowed with patently non-realistic forms of acoustic iconicity.

Marinetti's onomatopoeic effects have in the past occasionally been seen as 'iconic', rather than imitative, in any faithful sense.[46] But categorizing them in this way only really becomes meaningful if one acknowledges that, whereas onomatopoeia is *always* in some respect iconic, the essential aspect of the Futurists' concern with this feature of language lies in the new directions in which it enabled them to take their poetry. Any common denominators with onomatopoeia *outside literature*, or even in earlier literature, are relatively unimportant. It is, above all, the interaction between onomatopoeia and innovative

[45] 'La declamazione dinamica e sinottica', in Francesco Cangiullo, *Piedigrotta* (Milan, 1916), p. 'e'.

[46] See Sandro Briosi, *Marinetti* (Florence, 1969), 32.

BOOOOOOOMBAAAARDAMENTO

BOOOOOMBOOOOOMBAAAAARDAAAMENTO

24 Marzo
25 Marzo } **BUM BUM BUM**
26 Marzo

FIG. 1.14. From F. T. Marinetti, *Zang tumb tumb*, Milan, 1914

orthography that effectively paved the way for the most significant achievements of the movement in trying to make the language of poetry more ‘expressive’.

The strong onomatopoeic component in Marinetti's *Zang tumb tumb* is an amalgam of some already conventionalized and a series of invented clusters of vowels and consonants to suggest noises associated with battle. The following sound-sequence was undoubtedly already available in the Italian repertoire of the time as the representation of the staccato sound of machine-gun fire, even if it did not necessarily have a single, agreed spelling:

tatatatatata
regolare le mitragliatrici
(*TIF* 650)

By contrast, the following more rhetorically Futurist words are almost certainly neologisms coined by Marinetti himself, and were invented for the very simple reason that the sounds he wished to describe were in themselves relatively new to the language of literature. The first example, however, represents an obvious, but effective deformation of a familiar phrase (express train):

treno express—express—expressssssss press—press press—press
press—press—press—press—press—press—press—press—pressssssss
(*TIF* 564)

nuvole nuvole VRRRRRRRR (*TIF* 602)

locomotiva asmatica
pfu—pfu pfu—pfu (*TIF* 669)

Yet instances of Marinetti resorting to pre-existent words with an onomatopoeic component are fairly rare in *Zang tumb tumb*. Even when he does so, as in the example in fig. 1.14, he prefers to engage in vowel- and consonant-duplication in order to register the degree of acoustic intensity to the word.

When turning to onomatopoeia, the Futurists are usually not content with what is already available in the language; it has to be made more 'expressive'. In this instance (*Zang tumb tumb* (*TIF* 642)) of onomatopoeia being strengthened by the way it is presented, we see one of the most seminal examples of early iconicity for the development of Futurist poetry, for it was in harnessing the potential of creative orthography to take it beyond the superficial function of depicting battle-sounds and engine-noises that the Italian Futurists found their way forward to more subtle forms of iconicity.

The 'liberation' of the printed word from conventional spelling was an important item on the manifesto agenda of the Italian and various Russian Futurists alike, but for very different reasons and with markedly different results. Indeed, the divergences one finds here are symptomatic of the way in which these two forms of Futurism strain in different directions, even when their starting-point is a shared rejection of certain ossified passéist conventions.

In the case of Russian Futurism, the liberation of the word from the shackles of an orthodox, consistent spelling often led to something akin to outright orthographical anarchy. Words spelt in new ways, it was claimed, would acquire a freshness: 'In the name of freedom of personal chance we reject orthography', the influential preface to *Sadok sudey* (*A Trap for Judges*) declares.[47] David Burliuk praises the slip of the tongue in similar spirit as the 'centaur of poetry', a strategy for carrying it to new heights.[48] And the revitalizing of language in this way often goes hand in hand with a certain revelling in verbal anarchy. Kruchenykh's declaration 'the more disorder . . . the better'[49] would seem to have furnished the essential guiding principle for the deliberate scattering of misprints in his own work—e.g. *Starinnaya lyubov'* (*Old-time Love* (1914))—and in much of Vasilisk Gnedov's poetry. Instead of being a means to some circumscribed end (more powerful onomatopoeia, increased iconicity, the creation of a new idiom with its own rules and codes), the act of liberation from a confining tradition here often constituted an end in itself, as well as another 'slap in the face' for public expectations—as when the foreword to Terentyev's *17 erundovykh orudiy* (*17 Nonsensical Tools*) declares that 'there are no misprints in this book', despite the fact that the work contains numerous provocative instances of misspelling.[50]

If anything, it is more at the level of the individual letter that the Russian Cubo-Futurists showed an interest in a potentially iconic dimension. As Raymond Cooke shows, 'not only . . . did Khlebnikov perceive a link between sound and meaning . . . he also wished to establish a link between the *visual form*

[47] 'Bukva kak takovaya', *Manifesty*, p. 52.

[48] 'Poeticheskie nachala', *Manifesty*, p. 80.

[49] 'Novye puti slova', *Manifesty*, p. 69.

[50] Igor Terentyev, *17 erundovykh orudiy* (Tiflis, 1919), n.p.

of a written character and its meaning'.[51] Citing Khlebnikov's view that 'in the beginning the sign of a concept was a simple sketch of this concept' (*SP* v. 219), Cooke remarks: 'He clearly had in mind the ideographs of Chinese or the hieroglyphics of ancient Egypt.' Khlebnikov's views that 'V' represents 'the rotation of one point around another' or that 'Ch' (i.e. Ч) is 'in the shape of a goblet' (*SP* v. 219), are just two of a whole host of examples of the possibilities of iconic letter-interpretation that engaged Khlebnikov's attention. In fact, just as we shall find his interest in semantics went beyond the conventional morpheme down to single letters, so his concern with language's long-lost iconicity is often at the level of the letter, rather than that of shaped clusters of words or super-signs.

If there was on the whole a cult of orthographical serendipity among the Russian Futurists, the Italians, by contrast, sought to emancipate the word from spelling conventions for a far more restrictedly goal-directed set of reasons—and by that token, with more homogeneous results. Needless to say, Marinetti's influence is very much to be felt in all of this.

The phrase coined by Marinetti to herald this codified departure from traditional spelling—'free expressive orthography' (the title given to the final section of 'Destruction of Syntax')—is an instructive one, for the formulation itself already emphasizes this more channelled form of emancipation. What Marinetti is in essence advocating is that orthography should be liberated quite specifically in order to enable it to become *more 'expressive'*. The operative part of his manifesto declaration on the subject reads:

> our lyrical intoxication should freely deform and reflesh words, cutting them short, stretching them out, reinforcing the centre or extending the extremities, augmenting or diminishing the number of vowels and consonants. Thus, we have the *new orthography* which I call *free expressive*. This instinctive deformation of words corresponds to our natural tendency towards onomatopoeia. It matters little that the deformed word becomes ambiguous. It will marry itself to the onomatopoeic harmonies or the noise-summaries, and will soon permit us to achieve *onomatopoeic psychic* harmony, the sonorous but abstract expression of an emotion or pure thought. (*TIF* 69)

Although Marinetti enumerates a whole range of potential innovative spellings, from varous forms of the additive to the subtractive, and including the deliberate distortion of vowel and consonant elements at various junctures in a word, 'free expressive orthography' soon settled down to operating with a much more restricted and rapidly conventionalized set of features than might in theory have been generated from these proposals. (The dialectic, however, is a not unfamiliar one, with what might be thought of, in Futurist terms, as the 'destruction' of the passéist order resulting by and large in no more than a new, equally conventionalized idiom, not necessarily in anarchy for its own sake.)

[51] *Velimir Khlebnikov: A Critical Study* (Cambridge, 1987), 81.

And, in spite of the Futurist poets' general cult of brevity, the device was invariably used additively, as in the 'Bombardamento' example (fig. 1.14), with the duplication of vowels and consonants tending to be used to reinforce the meaning of a familiar word.

The most significant aspect of the new spelling, from our present point of view, was the way in which it quickly emancipated itself from its original function (the intensification of onomatopoeic effects in war poetry) to become a means of registering—on the whole, synaesthetically—other forms of intensity and even other qualities, apart from the acoustic. (Marinetti appears to have been thinking in this direction in his discussion of 'free expressive orthography', when he talks of its 'psychic' potential and relates this to forms of 'abstract expression' no longer restricted to imitative onomatopoeia.) Although Marinetti was right to claim that the new orthography largely corresponded to the Futurists' 'natural tendency towards onomatopoeia', it fortunately soon developed beyond this immediate preoccupation.

The spelling *fiiiiiischiare* ('to whiiiiiistle') in 'Contrabbando di guerra' (*TIF* 670) may still be primarily intended to give the impression of a long drawn-out sound, and the later *anch'iiiiiio verroooo* ('stiiiiiiill truuuue'), while semantically no longer relating to the idea of sound, may also still be essentially acoustic in intention: one may actually hear a voice dying away as it says these words. But when, in the 'Correzione di bozze' section of *Zang tumb tumb* (*TIF* 565), Marinetti spells the Italian word for 'blue' as *azzzzzzurrrrrro*, the chosen deformation is intended to suggest an analogy for the brightness of the colour. Spelling is here no longer confined to an onomatopoeic function, but widens to reflect the intensity of some non-acoustic quality.

The following example, from Carlo Carrà's manifesto 'The Painting of Sounds, Noises and Smells' ('La pittura dei suoni, rumori e odori' (1913)), entails a comparable use of verbal intensification for a non-acoustic purpose. It makes the process of suggesting intensity of colour by means of the analogy of 'intensified' or 'elastic' spelling even more clearly an instance of *synaesthetic* technique:

THE PAINTING OF SOUNDS, NOISES AND SMELLS DESIRES:

1. Reds, rrrreds, the rrrrreddest rrrrrrrreds that shouuuuuuut.
2. Greens that can never be greener, greeeeeeeeens that screeeeeeeeeeam. . . . (*Archivi*, i. 74)

According to Cangiullo's memoirs, the Futurists felt that the word *rosso* was *more red* when written with four 'r's than when it was spelt with the conventional two.[52] By the same token, the elastically spelt word 'dolcissssimi' (*TIF*

[52] *Le serate futuriste: Romanzo storico vissuto* (Milan, 1961), 119. Boccioni's 1910 caricature of Carrà (illustrated *FF* 441) is entitled 'Carràààà!!!', with the triple exclamation marks reinforcing, but hardly clarifying the function of the three 'à's with grave accents.

607) was designed to register an even greater degree of sweetness than the superlative it already is. But this device, like many forms of iconic intensification, is not without its pitfalls and ambiguities. Carrà at the time took to signing his name 'Carrrà', with three 'r's and a triple accent. (Both the cover and title-page of *Guerrapittura* are instances of this habit.) But does the mannerism signify that he is 'more intense', 'more Futurist', 'more vital', or 'more Carrà' than when employing the orthodox spelling of his name with only two 'r's and one grave accent? Tellingly, when Ugo Tommei published a poem 'in honour of the painter' Carrà in the 22 May 1915 issue of *Lacerba*, he respectfully spelt the subject's name as 'CARrRRÀ' in his title, so what began as an onomastic device for self-promotion eventually shifted in the direction of conventionalization, as a form of tribute to the 'Futurism' of a founding Futurist.

Elastic spelling of this kind is not restricted to the Italian Futurists, however. Examples of it can also be found in Julius Evola's poems in *Arte astratta*,[53] although I know of no instances of the device occurring in the Dada writings of other countries.

In his poem 'Direttissimo', Gino Cantarelli spells the word for 'speed' (*velocità*) with three final 'à's, an effect which is clearly intended to function iconically. Christopher Wagstaff interprets the device as a 'metaphor' for the intensity of speed;[54] the Futurists themselves would no doubt have preferred the word 'analogy'. Yet the elastic spelling used here is once again open to more than one interpretation. On the analogy with *rosso*, one might conclude that it conforms to the widespread Futurist code 'duplication = intensity', hence the spelling would simply signal 'faster than fast'. But the triple 'à's at the end of the word also give it a kind of tail. Anyone familiar with contemporary Futurist paintings where a series of overlapping shapes (such as one also finds in Bragaglia's experiments with photo-dynamism) is used to suggest an object in motion might imagine a more vivid analogy here: for the word *velocità* appears, as a consequence of its new spelling, to be moving so fast that it leaves an after-impression. As these two possible readings indicate, although the iconic may well *intensify* language, this does not necessarily make utterances less ambiguous. Indeed, it can even add new competing registers of signification.

This is also true of an example of the technique in Vasto Ciclone's poem 'In auto'[55] which Noëmi Blumenkranz-Onimus analyses in the following terms. In order

> to translate the sensation of speed felt in a trip by car, [Ciclone] is content to . . . write in large thick black capital letters the word VELOCITÀ, a speed which, progressively, makes the very letters of the word disappear, with 'velocità' becoming VEL TA, and

[53] See Evola, *Arte astratta: Posizione teorica / 10 poemi / 4 composizioni* (Rome, 1920); for example 'raaaaaaaaaaaaaaaga blanca' in 'I sogni' (p. 15) and 'bleghedaaaaaaia' in '10678^{999} e le sete' (p. 22).

[54] 'Postfazione a commento', *Tavole parolibere futuriste (1912–1944)*, ii. 596.

[55] *L'Italia futurista*, 12 (1917).

ultimately V TA. A clever use of arrows indicates at the same time the car's direction and the deafening sound as it passes another car.[56]

This, like the previous examples, I would suggest, represents a marked departure from any simple introduction of supposedly pictorial or acoustically imitative emphases in Futurist poetry.

There was an interesting historical postscript to the development of 'free expressive orthography' in the direction of elastic spelling. It is an indication of the fact that the Futurists themselves were not always fully aware of the implications of their own discoveries. The very same Cangiullo, who in *Le serate futuriste* had succinctly summed up the 'expressive' rationale behind Futurist consonant- and vowel-duplication, in 1923 felt compelled to launch a further reform that ran completely counter to the established iconic nature of the device.

Such an uneconomic duplication of letters, Cangiullo argued in the introduction to his new collection of 'staved poems', *Poesia pentagrammata*, was 'anything but synthetic' (i.e. in the Futurist sense of 'abbreviating'). Hence he reasoned that it was at variance with the movement's cultivation of brevity and intensity of expression. He therefore proceeded to advocate an alternative method of rationalizing the printing of such effects created by lengthening, and suggested that a sign should be placed over the relevant vowel to represent duplication. (He appears to have had no comparable solution to the problem of consonantal proliferation.) Cangiullo proclaimed that his volume *Poesia pentagrammata* 'eliminates, with the use of the corona, the kind of primitivist repetition that is inexact and anything but synthetic, but on the contrary longwinded'. Instead of writing 'ma siiiiiiii', Cangiullo suggested simply 'ma sì̑'. And, in order to indicate the long prospect on to an avenue, which would have been written as 'viaaaaaaaaaaale', he proposed simply the re-coded word 'vià̑≥le'.[57]

The poems that follow in the volume are, in fact, set out on lined paper as musical scores (hence the title 'staved poetry'). And, as the first page of the introduction to the new collection makes clear, the adoption of length-marks is part of a general process of importing musical principles into poetry. Not that Cangiullo had initiated the experiment with this collection, even though the theoretical rationale for the device is first presented here. Various signs, such as ⌒ (pause) and > (diminuendo), are already to be found in *Piedigrotta* of 1916. The principle was even extended to his *sintesi* for the theatre—most pronouncedly in *Canzone pirotecnica*[58] of 1915 which includes a series of musical notations.

[56] *La Poésie futuriste italienne: Essai d'analyse esthétique* (Paris, 1984), 78.

[57] *Poesia pentagrammata* (Naples, 1923), 10. See in this connection the words 'dà̑..........dà̑..........' (in 'Golfo Estivo', p. 23), where the dots occupy just as much space as any vowel-duplication would have required.

[58] *Canzone pirotecnica* was first published in *La balza*, 1 (10 Apr. 1915), 14–15; it has been reprinted in Mario Verdone, *Teatro italiano d'avanguardia: Drammi e sintesi futuriste* (Rome, 1970), 53. The signs > (diminuendo) and ⌒ (pause) used here also figure in *Piedigrotta* (1916).

Not surprisingly, for all its ingenuity, this quasi-musical importation failed to arouse much enthusiasm among Cangiullo's remaining Futurist contemporaries. The poet Alberto Viviani did feel able to welcome the device as one which functioned 'more efficiently' and appeared 'more complete, more interesting' and even 'more inspired' than 'free expressive orthography',[59] but he does not appear to have followed in Cangiullo's footsteps. And, tellingly, in Oreste Marchesi's 'Un bel sogno', published two years after Cangiullo's proposals, one has the lines:

vola vvvola vvvvooooola
viiia viiiiia viiiiiiiia[60]

Given that 'viaaaaaaaaaale' was specifically one of Cangiullo's negative illustrations, the elastic spelling of 'via' here looks like a deliberate affront to the author of the *Poesia pentagrammata* preface!

The main reason for such a lack of resonance was the fact that this change of code was fundamentally at variance with the dominant principle underlying 'free expressive orthography', which was above all else a matter of (quantitatively iconic) expressiveness, and not mere fetichized verbal economy. Whenever the Italian Futurist quest for brevity—and sometimes even easy readability—came into potential conflict with the demands of material iconicity, it was invariably iconicity that was permitted to win the day. Nowhere is this axiomatic truth more strikingly in evidence than in the many varied forms of iconic typography we are considering.

It is clear, then, that 'free expressive orthography', and even many of the cruder onomatopoeic effects in Futurist poetry, are more intricate than any facile return-to-mimesis theory, in Papini's sense, would account for. They involve subtly *analogical* (or *synaesthetic*) forms of codification, and their iconicity is often relatively abstract (or 'degenerate', in Peirce's sense). It would, of course, be wrong to assume that the Futurists were fully cognisant of the nature and possible aesthetic repercussions of such experiments or that they were even systematically transferring their iconic/'expressive' principles of creation from poetic onomatopoeia and liberated orthography to non-linear, free-word layouts. But, in line with the general stress they placed on the role of intuitive processes in Futurist writing, one can safely assume that they must at least have been unconsciously preparing in these other areas for the application of iconic codes to their typographical configurations as well.

In a discussion which has a strong bearing on Futurist typography's strategies, Umberto Eco borrows from Wolfgang Köhler's *Gestalt Psychology*[61] the point

[59] Alberto Viviani, *Dal verso libero all'aeropoesia, 1905–1942* (Turin, 1942), 86.

[60] *F. T. Marinetti presenta I nuovi poeti futuristi* (Rome, 1925), 262.

[61] *Gestalt Psychology* (New York, 1959), 106.

that objects in real life do not have dividing lines or possess linear outlines, and yet throughout the ages graphic conventions have enabled us to represent them thus. 'Suppose', Eco explains,

that I draw the outline of a horse on a sheet of paper by one continuous and elementary line. The sole property that my horse possesses (one continuous black line) is precisely the property that a real horse *does not* possess. My drawing has defined by that line the space inside the horse separating it from the space outside the horse, whereas the actual horse is in fact a body within or against a space.

Thus, according to Eco, 'a *graphic convention* allows one to transform, on paper, the elements of a *schematic conceptual or perceptual convention* which has motivated the sign'. To cover such contingencies, Eco redefines Peirce's concept of the icon as a sign which 'reproduces a few conditions of perception, but only when these have been selected on the basis of codes of recognition and explained on the basis of graphic conventions'.[62]

Even simple iconic effects in shaped poetry, e.g. devices like the minimal 'auto-illustration', have to be reassessed in the light of such a Gestaltist perspective. For they are outline-conceived; hence this is a matter of further modification of a pre-existent graphic convention, not an actual property common to the shaped print and the object signified. In fact, there is often a further complication at the typographical level, for iconic effects such as Marinetti's balloon or Buzzi's bi-plane can do no more than gesture towards the continuity of an outline. Just as with Apollinaire's *calligrammes*, and unlike many Baroque 'filled' figured poems, we are called upon to translate the lines of print into the conceptual outline which is itself only the poetic equivalent of Eco's schematically drawn horse.

Let us compare what happens in the 'auto-illustration', where, as we have seen, the effect is less dependent on actual shared features and more on graphic codes and patterns of recognition, and the *modus operandi* of the 'designed analogy'.

First, a simple example of 'auto-illustration'—an image of the sea from Govoni's *Rarefazioni* (fig. 1.15).[63] It could be objected that without the explanatory words 'il mare' Govoni's image might be of dunes in the Sahara, the flight of the bumble-bee, or a couple of nude sunbathers at Saint Tropez. As it is, even with the aid of the word for 'sea', we need the added graphic convention of undulating lines standing for the waves, in a way in which they are seldom found in reality. And we need mentally to join up the implied line formed by the individual 'srrrrrrrrrrrrrrrrr . . .' lines in order to prepare for the availability of this graphic code. If this is iconicity, then it is more conventionalized and more governed by codes of recognition than the usual pictorial thesis would imply.

[62] *A Theory of Semiotics* (Bloomington, Indiana, 1976), 193 f.

[63] Govoni, *Rarefazioni*, p. 48.

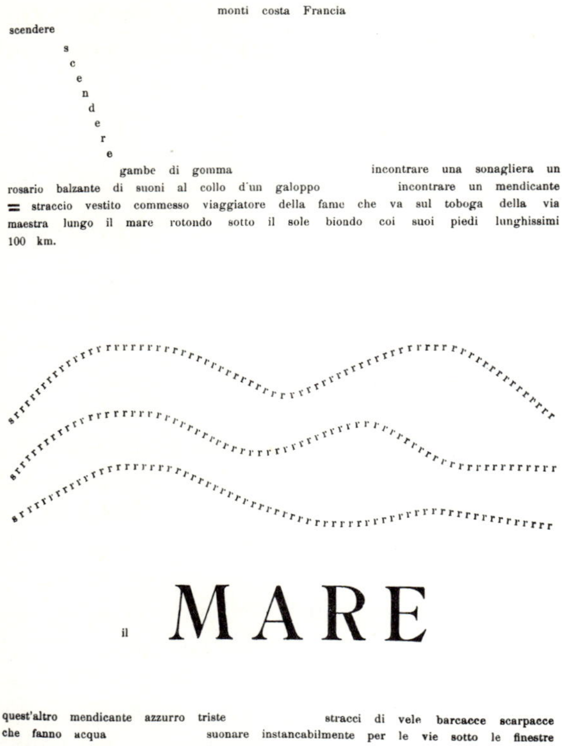

FIG. 1.15. Sea 'designed analogy' from Corrado Govoni, *Rarefazioni e parole in libertà*, Milan, 1915

Turning to the 'designed analogy', we encounter even more highly conventionalized forms of the iconic. Papini, of all people, concludes the first section of his 'Marcia del coraggio',[64] with the word *coraggio* printed six times in ever-increasing typefaces. The device is primarily analogical, for *coraggio* is an abstract concept, whereas the visually iconic effect involves a very concrete increase in size each time the word is repeated. Since the piece is entitled 'March of Courage', it might just be feasible, given the penchant for allegory among certain Futurists, that this increase in size is meant to depict the actual dimensions of Courage (personified) as he/she approaches. However, a more likely reading is that the growth in size of the repeated word is being presented as an analogy to the upsurge in courage Papini expects of his readers at this momentous period in history—or possibly even hopes to detect in them as a result of reading his article. An added acoustic feature lies in the fact that it is difficult to read these final lines, even silently, without an increasingly declamatory tone setting in.

Such a 'designed analogy' is as dependent on conventions (this time typo-

[64] *Lacerba* (1 Nov. 1913), 237.

graphical ones relating size to importance, and possibly even acoustically to volume) as Marinetti's balloon 'auto-illustration' is. The main difference in the way they work is that the outline convention is a longstanding one, familiar from many periods of drawing, whereas the 'coraggio'-effect depends on a new code exploiting analogies which has probably been established within the Futurist period, but has even possibly had to be generated by the individual work itself. In the case of outline-based effects, we may not even be aware, without the benefit of Eco's thoughts on the subject, that we are not literally dealing with a shared feature. With the *coraggio* example, the problem is of a different order, relating to the question of how the link between the typography and the semantics of the device can in retrospect be conceptualized, even if its gist is grasped intuitively by the reader at the time of reading.

Peirce's semiotic theory had already, it is true, recognized the importance of the psychology of perception for an understanding of the way in which iconic signs operate. In *Elements of Logic* the icon is defined as a sign which displays qualities that 'resemble those of its denotatum and *excite analogous sensations in the mind* for which it is a likeness'.[65] In recent years, however, Eco and Krampen have not only reinforced the point that iconicity 'must be defined in connection with the process of perception';[66] they have also demonstrated the advantages of conceiving of perception in modified Gestaltist terms and pointed to the need for recognizing the place of graphic, as well as perceptual, factors within the process of semiosis. Peirce's idea that an icon excites 'analogous sensations' to its object is one way (a 'pragmatic' way, in semiotic terms) of looking at the relationship between 'designed analogies' and their denotata. For a number of them are not easy to account for in terms of *shared characteristics*. When a polemic by Giuseppe Prezzolini prints the words 'Risposta a Carrà Futurista' with the last two words becoming increasingly smaller, and also sets Marinetti's name in ever-diminishing typefaces, he is using an analogical effect to signal his disregard for these two founding fathers of Futurism.[67] When Carrà, in turn, prints the word *rispettare* with each letter larger than its predecessor,[68] or Guglielmo Jannelli prints the word *passato* ('past') with a crack running through it (as if the edifice of the Past were crumbling) and uses ever-increasing letters to print the word for the 'future' (*avvenire*),[69] it is not hard to appreciate that they are trying to stimulate attitudes of sympathy or rejection in their readers via an inventive form of iconicity. The way in which *passato* is printed is, in

[65] *Elements of Logic*, p. 168.

[66] *La struttura assente: Introduzione alle ricerce semiologica* (Milan, 1968), pp. 113f. Cf. Martin Krampen, 'The Role of Signs in Different Sign Processes: Towards a Basis of Generative Semiotics', *VS*, 4 (1973), 116.

[67] 'Risposta a Carrà Futurista', *Lacerba* (17 Jan. 1915), 23.

[68] '1900–1913. Bilancio', *Lacerba* (1 Feb. 1914), 38f.

[69] 'Messina', *Lacerba* (15 Feb. 1914), 60.

Wayne Booth's sense, a form of typographical *rhetoric*: a literary icon calculated to arouse specific evaluative reactions in the reader.

In his discussion of Apollinaire's *Calligrammes*, Theo Hermans interjects a fundamental *a priori* objection which might, in theory at least, also apply to the majority of Futurist 'auto-illustrations', and possibly even to 'designed analogies', as well. 'Essentially', he suggests, 'the calligram is a compound sign with a structure similar to that of a pun, but whereas the pun, through the coincidence of different signifiers, points to several signifieds at once, the calligram contains an obvious tautological element in that its two signifying levels, one linguistic, the other iconic, carry the same semantic load.'[70] This is to some degree also true of other forms of iconic typography, although with the examples we have just been looking at, it would be difficult to conclude that any pronounced degree of tautology was in operation.

Hermans's second stricture, regarding the actual nature of the tautologies, may hold in the case of Apollinaire, but it does not apply to the majority of Futurist experiments with 'designed analogies'. Hermans complains that 'it is difficult to see what the genuine advance with the calligrammatic experiment consists in. With its limited semantic scope, its obvious tautologies and its crude mimeticizing (the poem as picture is never more than a crude cliché), the technique ultimately appears to be a self-defeating over-simplification.'[71]

A related objection, this time in terms of 'redundancy', can be found in Franco Gisuti's *Italian Futurism as a Sociolinguistic Phenomenon*. Here the charge is that 'the facile mimicry of graphic poetry diverts the text from the pursuit of a cognitive goal and represents the surrender to an infantile zest for redundant expression'; hence, 'redundancy is the only contribution made by this operation: the image of the referent is sketched twice'.[72]

In the case of the kind of 'designed analogies' we were examining earlier, neither tautology nor redundancy is the main outcome. On the contrary, the analogy-based technique is often quite subtle and hardly amounts to evidence of an 'infantile zest for redundant expression'. Obviously there is a sense in which any well-executed *calligrammes* and 'auto-illustrations' are bound to give the lie to such rejection *en principe* of a genre. But, in any case, 'over-determinism' can never be held to be axiomatically a literary sin; it is present in all aesthetic structures, as scholars from Tynyanov onwards have amply

[70] *The Structure of Modernist Poetry* (London, 1982), 81 f.

[71] Ibid. 82. The idea that the calligram remains no more than a tautological device has been given the stamp of authority in recent years by Michel Foucault, *Ceci n'est pas une pipe* (Montpellier, 1973), 21 f. For a counter-argument, see Bohn, *The Aesthetics of Visual Poetry*, p. 46 and Chap. Four.

[72] Ann Arbor, Michigan, 1981, 141. Gisuti is equally critical of Futurist onomatopoeia, condemning it as 'superficial' on the grounds that 'it reproduces in the text a moment of hearing perception which is obvious to the addressee and therefore not informative, since it does not add anything to the appraisal of that element of reality' (p. 104). For this reason, the device is dismissed as 'ornamental'.

demonstrated. Or, as Peter Mayer puts it (more discerningly than the device's opponents have), such seemingly tautological arrangements 'reflect and reinforce the meaning of what is written on the page'[73]—which is surely an acceptable function for a literary device, and one which becomes even more defensible when the over-determining structure is a local technique and not the central strategy of an entire poem.

In theory, of course, a charge of iconic tautology could also be laid against many of the visual effects used in the world of mass communication. The kind of advertising techniques that A. A. Roback has analysed as 'simulates' during the course of his study of motivated typography[74] (letters dripping blood in horror-film advertisements or the use of Gothic print to suggest things Teutonic) have much in common with Futurist 'designed analogies'. And, just as reinforcement of an iconic kind, as well as various forms of subliminal value-suggestion via 'simulates', have been deemed anything but redundant in the language-strategies of a ruthless advertising world, so too in literary Futurism this departure from an exclusive reliance on the arbitrary sign only serves to enrich the medium and reinforce its expressivity and impact. It is far from being a reduction of language to the status of a playground for tautological over-simplifications.

Compared with the pronounced enthusiasm for exploring diverse forms of iconicity among the Italian Futurists, one finds little interest in anything resembling 'auto-illustrations' or 'designed analogies' among the Russian Futurists. In 1913, in 'The Word as such' ('Slovo kak takovoye') Kruchenykh does, admittedly, suggest the replacement of the normal word for the lily (*liliya*) by the neologism 'еуы' in which the extended stem of the letter 'у' is meant iconically to resemble the flower and its stem. And in *Tango s korovami* (*Tango with Cows*) the poem 'Polet Vasi Kamenskogo na aeroplane v Varshave' ('Vasya Kamensky's Aeroplane Flight in Warsaw') by Vasily Kamensky has to be read from the bottom line upwards, so that the reader soars into the sky alongside the pilot, just as the reader of Bètuda's 'Looping the Loop' has to perform the same manœuvre while reading the poem. But these are rare effects and only incidental to the main preoccupations of the Russian Futurists.

In *The Look of Russian Literature* Gerald Janecek sets out to account for this

[73] 'Framed and Shaped Writing', *Studio International* (Sept. 1968), 114. Antonella Ansani argues, in similarly positive terms, within a framework of *reinforcement*, commenting that in 'Golfo Estivo' (*Poesia pentagrammata*, p. 23) 'Cangiullo shows his delight in visual and verbal puns. The letter V of the word "vela" (sail) is turned upside down so that it resembles a sail. Thus the depicted word reinforces its own meaning' ('Words-in-Freedom and Cangiullo's Dancing Letters', in Anne Coffin Hanson (ed.), *The Futurist Imagination: Word + Image in Italian Futurist Painting, Drawing, Collage and Free-Word Poetry* (New Haven, Conn., 1983), 58).

[74] *Destiny and Motivation in Language: Studies in Psycholinguistics and Glossodynamics* (Cambridge, Mass., 1952), 414–22.

lack of concern for typographical iconicity by approaching the situation historically. Arguing that 'the genre ... never caught on' in Russia, he presents Simeon Polotsky's figured poems of the seventeenth century (in heart- and star-shapes) as 'a purely imported product', derived from Polish and Latin practices. The further examples noted (by A. Rzhevsky in the eighteenth century, and more recently by I. Rukavishnikov, E. Martov, and V. Bryusov[75]) are so rare, in Janecek's opinion, that they merely corroborate his claim that this foreign plant never really managed to take root in Russian soil.

Nevertheless, one has to be cautious when assessing the merits of a historical argument of this kind. On the surface, the contrast between the Russian and the Italian contexts would seem to be a clear one. Italy, as G. Pozzi's *La parola dipinta* shows, had a strong and richly inventive native tradition of figured poetry, above all during the Humanist and Baroque periods.[76] Russia, by contrast, does not appear to be similarly endowed. Yet, even though the Italian tradition was there, it is none the less clear that most Italian Futurists were blissfully ignorant of it. (Which is hardly surprising for the time; the rediscovery of the figured poem is largely a result of mid-twentieth-century scholarship, a development that happened very much in the wake of modern poetry's exploration of this field of possibilities.) In practice, then, the significance of any posited contrast between the Italians' possession of a strong tradition of shaped poetry and the Russians' predicament of having virtually no such native tradition to work from—and on the whole little sympathy with the few imported foreign models—is negligible. It is not *tradition* (whether rich and potentially seminal or practically non-existent) which accounts for the divergences between Italian and Russian Futurist typography. It is the largely different aims of the two sets of writers which explains the diametrically opposite directions their various liberations of the printed word took. The Italian concern with enhanced expressivity, with literary dynamism, and with propaganda impact, as well as the avant-garde desire to create some demonstrably anti-passéist mode of printing and to engage in a provocatively anti-intellectual sensory appeal, all combine to take Italian typography in an iconic direction. Since Russian Futurist writers, with one or two exceptions, did not share these goals, the typographical work of poets like Zdanevich, Kamensky, Mayakovsky, Kruchenykh, and Khlebnikov contributes greatly to the overall art of book-design and emancipates the printed word in a series of equally fruitful directions, but seldom in that of the iconic.

However, there is one exception to this overall difference between the two groups' attitude to the possibilities of the iconic. In 'Poetic Principles' ('Poeticheskie nachala' (1914)) Nikolai Burliuk argues that 'the word has no

[75] *The Look of Russian Literature: Avant-Garde Visual Experiments, 1900–1930* (Princeton, New Jersey, 1984), 8 f. In each case, Janecek gives illustrations of the kinds of effects these poets created.

[76] Pozzi, *La parola dipinta*, especially Part Three.

way of transmitting an object other than the degree to which it represents at least part of that object's qualities' (*Manifesty*, p. 78), which is about as forthright an alternative formulation of the principle of iconicity as one could wish to find. The manifesto goes on to instance various (foreign) examples of shaped poetry from the past, including the no longer preserved writings of Apollon of Rhodes and Callimachus 'in the form of a lyre, a vase, a sword etc.'. But these are clearly only intended as parallels; the aim does not appear to be to encourage the production of more shaped poems of the same kind, but to extend the principle of revitalized typography in other directions.

The manifesto also points out that the Hylaea Group had published handwritten books and experimented with the use of written elements within typographical layouts so as to exploit a certain equivalent to the iconicity of shaped poems. Since 'poetic language is tangible', the argument runs, and 'changes qualities according to whether it is written by hand or printed or thought',[77] the use of hectographed handwritten forms could introduce a certain subjective dimension into contemporary poetry. For this reason, the title-page of the book *Troe* (*Three Men* (St Petersburg, 1913)) bears the handwritten names of Khlebnikov, Kruchenykh, and Guro. Kruchenykh's *Mirskontsa* (*Theworldbackwards* (Moscow, 1912)) also uses handwritten words, alongside printed ones. Unlike the use of handwriting in Govoni's *Rarefazioni* (where there appears to be no iconic principle at work), we have here moved beyond a mere gesture rejecting the tyranny of the conventional typeface towards an attempt at expressing subjectivity through an iconic effect. But the individualizing possibilities of the device remain underexploited. The stamp of personality that handwriting brings with it involves little more than a simple binary contrast between print and handwriting. There is less differentiation when it comes to styles of handwriting in the ensuing works than one finds when different typefaces are used at different stages in modern texts; and there is certainly no local differentiation in respect of content. In any case, this predilection for handwritten layouts remains one of the few areas in which a form of iconic effect was favoured by the Russian Futurists. Other forms of expressiveness are marginal to the movement's aims. Kamensky may have imitated a mosque's façade in one of his poems[78] and an occasional iconic element appears in the design of Russian Futurist book-covers,[79] but such forays into iconicity are rare. As both Susan Compton and Gerald Janecek have amply demonstrated, Russian Futurist typography was richly inventive and in many ways superior to

[77] *Manifesty*, pp. 77 f. Bohn considers the related issue of the significance of handwritten forms of Apollinaire's poetry (*The Aesthetics of Visual Poetry*, p. 50).

[78] See Janecek's reading of part of Kamensky's 'Konstantinopol′' ('Constantinople') from *Tango s korovami*, 1914; *The Look of Russian Literature*, p. 130.

[79] See the covers of *Vzorval′* (*Explodity*), and Kruchenykh's and Khlebnikov's *Bukh lesinnyy* (*A Forestly Rapid*), ill. in Susan Compton, *The World Backwards: Russian Futurist Books, 1912–1916* (London, 1978), 103.

the work of the Italian Futurists; but it seldom moved in the direction of iconic effects.

'We know', Massin once observed, 'that ideograms are not necessarily restricted to representing objects, that certain types of ideogram denote action, quantity or movement, while others combine a generic meaning with phonetic elements, and yet others call on analogy or alternatively are combined as groups of ideas.'[80] As can readily be seen, the Futurist poets were active in all of these areas; indeed, they were the ones who first took modern experimental typography in many of these fertile directions. But one of their major achievements, one which has not been looked at so far, is the creation of a new dynamic form of typographical layout.

In an earlier part of his declaration in 'Destruction of Syntax' (not quoted above) concerning the nature of the Futurist 'Typographical Revolution', Marinetti writes:

> I shall combat Mallarmé's static ideal with this typographical revolution, allowing me to impress on words (already liberated, dynamic and torpedo-like) all the velocity of the stars, the clouds, aeroplanes, trains, waves, explosives, globules of seafoam, molecules, and atoms.
>
> In this way I realize the fourth principle of my First Futurist Manifesto (20 February 1909): 'We affirm that the world's beauty is enriched by a new beauty: the beauty of speed.' (*TIF* 83)

Buried in among the bombast and rather precious comparisons, there is a key reference here to the purpose of much Italian Futurist typography. The creation of dynamic layouts was, in fact, of major concern to Marinetti and his fellow Futurists, although the results we shall be looking at would not have been possible in their literary works if the Futurist painters had not by this time already explored the *force-lines* which they claimed characterized objects. For it is the application of equivalents of these force-lines to the arrangement of poetry on the printed page that was to help set the Italian typography of the time in motion.

The Futurist painters explained what they meant by force-lines in the catalogue preface to the 1912 exhibition of their work at the Gallery Bernheim-Jeune in Paris. From their argument there, it is evident that they conceived of these force-lines partly as a property of *all* objects, then as a particular quality of objects *in motion* (as experienced), and, furthermore, as a product of the aesthetic codification of the sensation of movement, a way of communicating the dynamism of all objects in the visual arts. 'All objects', it is claimed,

> stretch out towards infinity by means of their *force-lines*, whose continuity is measured by our intuition. It is these *force-lines* which we must draw, to lead the work of art back

[80] *Letter and Image*, p. 174.

to true painting. We *interpret nature* by depicting on the canvas these lines *as the continuations of rhythms which the objects themselves impress upon our sensibility*. (*Archivi*, i. 105, my italics)[81]

This means that, inasmuch as force-lines coincide with the 'rhythms which the objects themselves impress upon our sensibility', they must be essentially iconic (in fact, the phrase used here bears a marked resemblance to Peirce's idea of 'exciting analogous sensations'). But the point about Futurist painting embodying 'continuations' of these structures implies that an additional measure of stylization, or codification, must be involved. This can be seen most simply and effectively in the Futurists' predilection for triangular structures, both in painting and in their typographical arrangements.[82]

'Every rapidly moving object', according to Boccioni's *Pittura scultura futuriste*,

—a train, a car, a bicycle—generates in pure sensation an emotional milieu which takes the form of horizontal penetration at an acute angle. . . . A slowly shifting crowd awakens in us the inert milieu of perpendicular directions, whereas the same crowd starting off at a run appears in our dynamic consciousness as a maze of acute angles, obliques and aggressive zig-zags. (*Gli scritti*, p. 199)

(The remarks read like a gloss on paintings like Russolo's *La rivolta* (*Rebellion*) and Boccioni's *Rissa in galleria* (*Riot in the Gallery*).) In Carrà's manifesto 'Plastic Planes as Spherical Expansion in Space' ('Piani plastici come espansione sferica nello spazio' (1913)) one comes across a similar statement to the effect that a 'pictorial composition constructed of right angles does not go beyond what in music is known as *plainchant*. The acute angle, by contrast, is passionate and reveals volition and aggressive onslaught.'[83] Balla also coins the revealing term 'speed-lines' ('linee di velocità' (*Archivi*, i. 126)), which can be found not only in his theoretical writings, but also in the titles of a number of his works: *Linea di velocità + spazio* (*Speed Line + Space*) and *Linea di velocità + paesaggio* (*Speed Line + Landscape*), for instance. (And if the singular sounds strange—especially since

[81] This contrast between the force-lines of an object in motion and those of any object stems from Henri Bergson's distinction between 'absolute' and 'relative' motion (*The Creative Mind* (New York, 1946), 21). Hence, as Christine Poggi points out, Boccioni can refer to force-lines revealing the absolute motion of even his *Still Life with Glass and Siphon* in 'Les Exposants au public' (see 'Marinetti's *Parole in libertà* and the Futurist Collage', *The Futurist Imagination*, p. 11). In 'Moto assoluto + Moto relativo = Dinamismo' (part of *Pittura scultura futuriste* pre-published in *Lacerba* on 15 Mar. 1914, 90–3), Boccioni argues that 'there is no such thing as an immobile object in our modern perception of life. . . . It is a matter of degrees of movement, and above all a matter of tempo' (p. 90). Tellingly, this theoretical discussion comes immediately after illustrations of Boccioni's *Voglio fissare le forme umane in movimento* (p. 88) and *Voglio sintetizzare le forme uniche della continuità nello spazio* (p. 89).

[82] Maurizio Calvesi, *Il Futurismo: La Fusione della vita nell'arte* (Milan, 1967), the only commentary to explore this concern with triangularity in any detail, posits various influences on Balla's more abstract preoccupation with the shape—including Jugendstil and Kandinsky's *Über das Geistige in der Kunst*, influences which are more likely to be relevant to Balla personally than to the other Futurists.

[83] 'Piani plastici come espansione sferica nello spazio', *Lacerba* (15 Mar. 1913), 53 f.

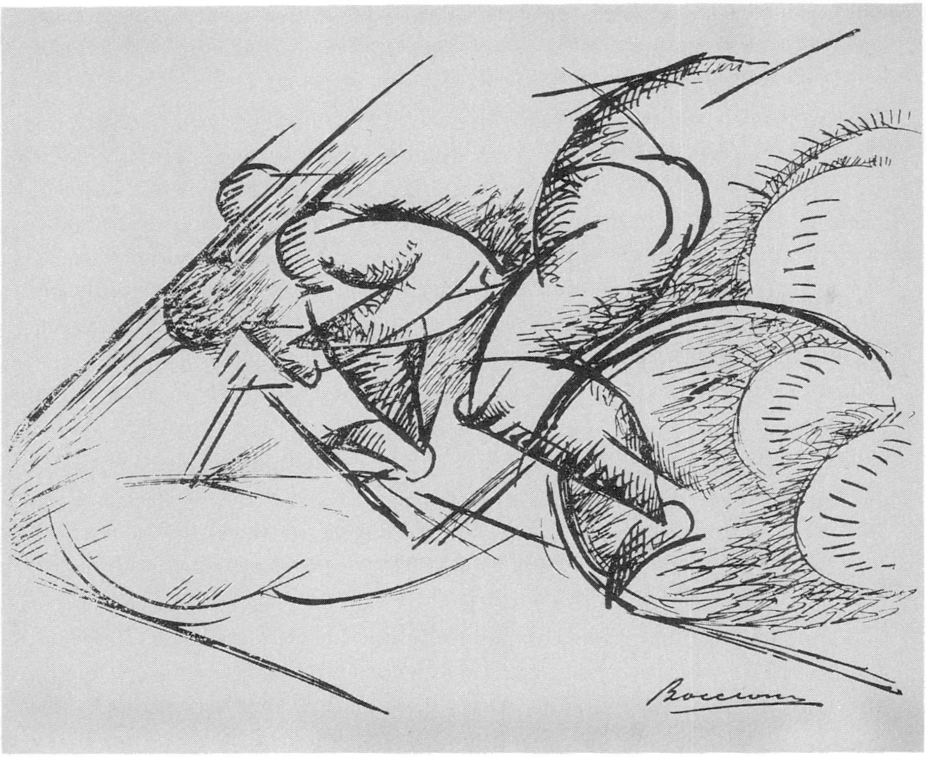

FIG. 1.16. Umberto Boccioni, study for *Dinamismo di un ciclista*, 1913

he opts for the more usual plural in his various *linee forza* titles (e.g. *Archivi*, ii, ills. 276a, 280, 285, 291c)—this is corrected in no. 291c, when he entitles a new work *Linee di velocità + paesaggio*.)

The paradigmatic vector of force-lines Boccioni calls the aesthetic 'representation of the movement of matter along the trajectory determined by the structure of the object and its actions' (*Gli scritti*, p. 156), which tends to mean, more particularly in most cases, objects *in motion*. 'Movement is triangular', as the Futurist painters put it in one of their manifestos (*Archivi*, i. 78). Indeed, in Boccioni's various preparatory studies for the picture *Dinamismo di un ciclista* (*Dynamism of a Cyclist*) (an example of which is given in fig. 1.16), one can observe a process of stylization increasing from study to study, as perceived features of the cyclist in motion gradually give way to a highlighting of radically schematized force-lines (as 'extensions' of the property of the object in motion). In my estimation, Russolo's painting *Dinamismo di una automobile* (*Dynamism of a Car*) is one of the finest examples of Futurist painting's exploitation of force-lines, with none of the jokey effects—echoing the current fad for

Muybridge's chrono-photography or Bragaglia's photo-dynamism—that one has in Balla's earlier *Guinzaglio in moto* (*Dynamism of a Dog on a Leash*) or *Ritmi dell'archetto* (*Rhythms of the Violinist*), both produced in 1912.[84]

A simple application of the dynamic, triangular force-line structure to Futurist poetry can be found in Cesare Simonetti's poem 'Treno in corsa' ('Speeding Train', a rare early combination of visual and sound concrete poetry). But there is a certain conflict in this layout (fig. 1.17) between the lines of movement represented by the overall shape of the train in motion and the to-and-fro movement of the eye as the text itself is read. On the other hand, the cover of Marinetti's *Zang tumb tumb* (fig. 1.18), one of the most effective yet simple examples of dynamic typography, has successfully managed to overcome this problem.

By dint of being an onomatopoeic title, *Zang tumb tumb* already offers an iconic signification of the sounds of war, and in this respect (like various other Futurist covers, including that of *Vzorval'*, *Baionette*, and *Liriche radiofoniche* (figs. 1.19, 1.20, 1.21)), bears out Pierre Guiraud's point that the 'title of a work refers to the code adopted much more often than to the content of the message'[85]—something which is as true of the typographical code(s) of a work as it is of its other linguistic and cultural codes.

The layout of Marinetti's *Zang tumb tumb* cover shares with the radiating noises of battle a certain centrifugality and a diminution of volume, as we move away from the focus of attention. Although the indication of the author's name and the work's publication details are so positioned as to reinforce this pattern, in itself the design is remarkably simple (as befits the cover to one of the movement's canonical texts). And yet, despite—or even because of—its simplicity, it can stand as a model of the *modus operandi* of much more complex Futurist layouts based on the new force-line principle.

Marinetti's effect consists of a small number of lines of diverging print and displays an element of duplication presumably intended to convey the impression of battle-sounds reverberating and dying away. (The parallel use of the phrase 'Zang-Tumb-Tuumb' in Paolo Buzzi's poem 'L'artiglieria' ('Artillery')[86] reinforces the supposition that the acoustic effect signifies an artillery bombardment.) It is as if 'zang' were the sound of a shrapnel shell, or some other military projectile, whereas 'tumb' represented either an echo or the noise

[84] The method of suggesting dynamism in such works is in many respects reminiscent of what are known as 'speed lines' in the world of strip cartoons. See Wolfgang J. Fuchs and Reinhold C. Reitberger, *Comics: Anatomie eines Massenmediums* (Reinbek, 1973), 32f.

[85] *Semiology* (London, 1975), p. 9.

[86] Paolo Buzzi, *Popolo, canta così! Canzoni d'arti e mestieri del popolo italiano* (Milan, 1920), 27. In the MS version (reproduced in Buzzi, *Futurismo, scritti, carteggi, testimonianze*, ed. Mario Morini and Giampaolo Pignatari, ii (Milan, 1983), 240), Buzzi even added a further 'Tuumb' to make the last line of his poem begin 'Zang-Tumb-Tuumb-Tuumb', whereas the printed version looks more like a quotation of Marinetti's original acoustic effect.

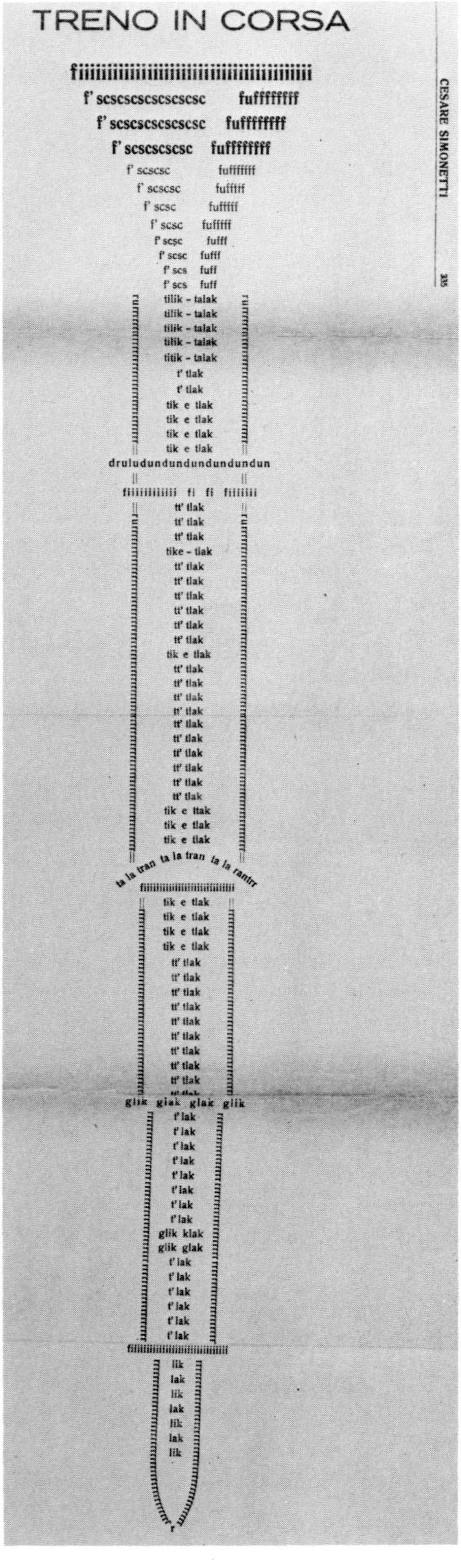

FIG. 1.17. Cesare Simonetti, 'Treno in corsa', 1925

FIG. 1.18. Cover of F. T. Marinetti, *Zang tumb tumb*, Milan, 1914

of its exploding, followed by a dull impacting sound ('tuuum tuuum'). The shield-like arc of the words 'parole in libertà' would seem to conflict with this force-line pattern and give the impression of being either some defensive ground-position assailed by the sound or the radiation of further battle-sounds from the other side of the front.

A further important feature of the layout is the way in which the basic acoustic pattern is organized iconically, so as not to conflict with readability. This was, after all, the title of the movement's major poetic work at that stage, and, as such, it needed to avoid impairing its functionality.

In *Art and Visual Perception* Rudolf Arnheim discusses what he calls our 'general tendency to read visual patterns from left to right',[87] a habit which is exploited in different ways by Futurist painting, on the one hand, and free-word typography, on the other. A study of any representative selection of Futurist paintings or the perusal of any typical corpus of the collages of the period would soon reveal that the vast majority of dynamic objects in Futurist works are depicted as moving towards the left of the canvas. This is true, to cite

[87] *Art and Visual Perception: A Psychology of the Creative Eye* (London, 1954), 18.

FIG. 1.19. Cover of the second edition of Alexei Kruchenykh, *Vzorval'*, St Petersburg, 1914

some of the more famous examples, of Russolo's *Treno in velocità* (*Speeding Train*), *Automobile in corsa* (*Racing Car*), and *La rivolta* (*Rebellion*), and it holds (as was already implied) for Boccioni's various *Dinamismo di un ciclista* works. The movement of the viewer's eyes from left to right will consequently endow the depicted object with a sense of impetus in the opposite direction. By contrast, the sounds of *Zang tumb tumb* radiate outwards from left to right (rather than seeming to move from right to left, because of the scanning process). Fundamentally, the reading direction remains the same for both cover-design and the tendency in Futurist painting, but the direction of the (super-)sign's motion has been deliberately reversed in *Zang tumb tumb*.

There are, in fact, good reasons for this fundamental difference between the two kinds of dynamic structure, and they stem from the nature of the reading process. As Jan Tschichold has pointed out, since 'our writing runs from left to right . . . , our eyes naturally return at the end of each line to the place where they started'.[88] Thus, while the contrast between radiating and converging lines may, on the whole, be appropriate to the difference between a painting of an

[88] *Asymmetrical Typography* (London, 1967), 44.

FIG. 1.20. Cover of Auro D'Alba, *Baionette*, Milan, 1915

object in motion (moving from right to left across the canvas), and sounds emanating, in the case of *Zang tumb tumb*, from a particular source (off-left), the two configurations are also skilfully accommodated to different modes of readability. And, because more than one line of writing/print will be needed in the case of Marinetti's book-cover, both to supply the full message and to create a viable typographical configuration, and given the fact that we are expected to take in each line separately *and to understand it*, certain kinetic structures are inevitably preferred. As one can see, Marinetti is here reluctant to depart too boldly from the kind of reading conditions we are accustomed to, because then the layout would risk becoming self-defeating on a book-cover. In principle, Marshall McLuhan may be right when he complains that the 'linear, eyedropper flow of printed words [has] greatly weakened the plastic power and iconic perception' in the modern book;[89] but the answer, as Marinetti shows in this case, may once more lie in some modification of the 'eyedropper' convention, not necessarily in its total rejection.

[89] *Counterblast*, p. 71.

FIG. 1.21. Cover of Fortunato Depero, *Liriche radiofoniche*, Milan, 1934

The layout of Marinetti's arrangement still tends to *relate*, though not rigidly conform, to the pressures and the codes of traditional horizontality, just as Italian 'free expressive orthography' does not lapse into spelling anarchy, but simply becomes an intensification or a modification of conventional practice. The resultant new structure will not involve a single sweep of the eyes from left to right, as Arnheim describes in the case of the reception of most paintings, but a number of repeated 'eyedropper' motions of this kind. Nevertheless, the layout's endeavour to mediate between linearity and a synoptic effect has led to much confusion about how even a simple configuration of this kind should be read. Is it *Zang tumb tuuum*, *Zang tumb tumb*, or even *Zang tumb tuuumB*?[90]

[90] Details of the various readings of the title are given in De Maria's introduction to *TIF*, p. cxvi. The book is advertised in *Lacerba* (15 Apr. 1914), 120, as *Zang Tumb Tuum*. However, Marinetti refers to the work as *Zang tumb tumb* in his introduction to Armando Mazza's *Firmamento* (Milan, 1920, 12). Without doubt, the most eccentric interpretation of the configuration comes in Moholy-Nagy's reference to *Zang-tumb-tu-uumb* (*Vision in Motion*, p. 304). A picture by Rosai entitled *Zang tumb tumb* (ill. *FF* 202) and dated 1913–14, suggests that this may have already been the consensus reading by then. Giusi Baldissone's point that the title is referred to 'indifferentemente' by Marinetti holds equally for those around him. See *Filippo Tommaso Marinetti* (Milan, 1986), 161 f.

Marinetti himself was inconsistent in his references to the work, having at one stage abandoned the, to my mind, preferable version *Zang tumb tuuum* (which nicely captures the way the sound dies away and also gestures towards each separate line of the layout) in favour of the undifferentiated form *Zang tumb tumb* which has now, regrettably, established itself as the standard form—and which has, as a consequence, been adopted, reluctantly, in the present study. (Surprisingly, only the late-1980s record company has seen fit to preserve the older 'Zang tumb tuuum' version of the phrase in its commercial title.)

An attempted compromise between linearity, which tends to produce an unambiguous reading, and the synoptic, which can permit various decoding possibilities, in the case of Marinetti's cover leads to a situation where commentators may be in a quandary as to how even to cite the title of one of the key works of Italian literary Futurism.

In examples such as this there is a sense in which the relative poverty of the medium can be exploited to distinct advantage. For, as Tschichold has pointed out, 'in special cases lines may be set obliquely, which is more eye-catching.' But he cautions that 'it can be very effective but only when done sparingly. If it is used, single lines are more effective than short words or groups of short lines, because then the oblique position is not so easily noticed.'[91] In setting out the cover of his *Zang tumb tumb*, Marinetti would appear to have shown a remarkable degree of intuitive understanding of the operative design factors which Jan Tschichold was later to elucidate.

No sign, not even an icon, is totally motivated, however. As we have seen in the cases of onomatopoeia and innovative orthography, as well as Futurist typography, any attempt at moving language in the direction of iconicity can only remain a matter of degree. Onomatopoeic effects must, as John Lyons demonstrates, still be accommodated to the sound-system of a particular language's structure. Distorted spellings also ultimately have to remain distortions of the recognizable, that is, if they are to stay within the realm of the comprehensible. And for similar reasons, much Futurist topography, as we have just observed, walks a tightrope between would-be radical liberation from linear–horizontal layouts and the demands imposed by our conventional patterning habits of reading. But the Italian Futurist free-word poets' harnessing of force-lines to their typographical experiments entails another feature of iconcity which modern semioticians have often commented on, and that is the corollary of the process of iconization—namely the reverse pattern of de-iconization.[92] In fact, during the first Futurist decade one can discern a distinct process whereby force-lines began life as an iconic device reflecting the characteristics of dynamism inherent in all objects. Then, gradually, certain triangular

[91] *Asymmetrical Typography*, p. 62.

[92] See Frans Plank, 'Ikonisierung und De-Ikonisierung als Prinzipien des Sprachwandels', *Sprachwissenschaft*, 4 (1979), 121–58.

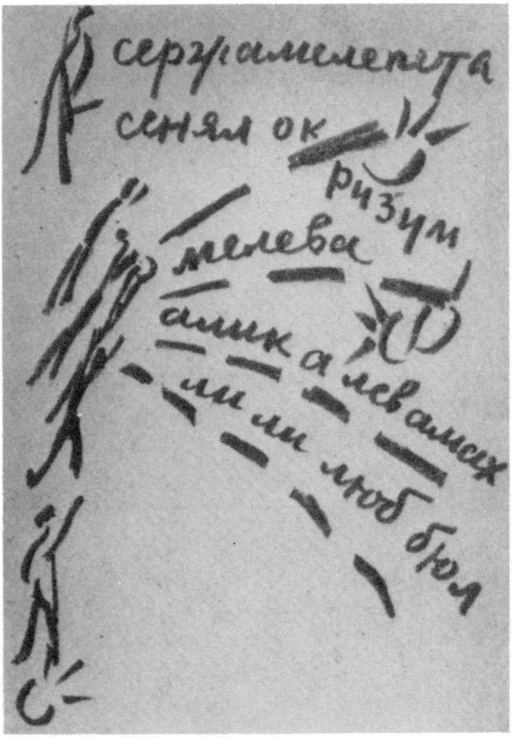

Fig. 1.22. Iconic page from Alexei Kruchenykh, *Vzorval'*, St Petersburg, 1913; drawings by N. Kulbin, poetry by Kruchenykh

shapes became abstracted from their objects in order to become conventionalized signs of movement in themselves. Compare, for example, an iconic page from Kruchenykh's *Vzorval'* (*Explodity*) (fig. 1.22) (1913)) and D'Alba's *Baionette* (*Bayonettes* (1915)), on the one hand, and Carrà's *Sintesi futurista della guerra* (*Futurist Synthesis of War*) (fig. 1.23),[93] on the other. Kruchenykh's centrifugal layout suggests an explosion, just as the configuration of sharp-pointed letters on D'Alba's cover gives the impression, not only of bayonettes, but—exploiting the dynamic potential of the acute angle—of bayonettes held by swiftly advancing soldiers. Both of these works, then, like *Zang tumb tumb*, are using the force-line principle in a familiar way to iconize the printed word. Carrà's *Sintesi futurista*, however, is not an iconic reference to anything outside itself in any simple visual sense. The use of the primitive wedge containing the word 'FUTURISMO', with the preposition 'contro' appropriately at the sharp end, and 'passatismo' printed in a much smaller typeface on the receiving end of

[93] Carrà's 'manifesto-sintesi' appears as an unpaginated fold-out at the end of *Guerrapittura* (1914), a work whose cover also betrays a diagonally organized structure. For effect, this 'sintesi' configuration compares very favourably with Apollinaire's earlier 'L'antitradizione futurista. Manifesto-sintesi', *Lacerba* (15 Sept. 1913), 202 f.

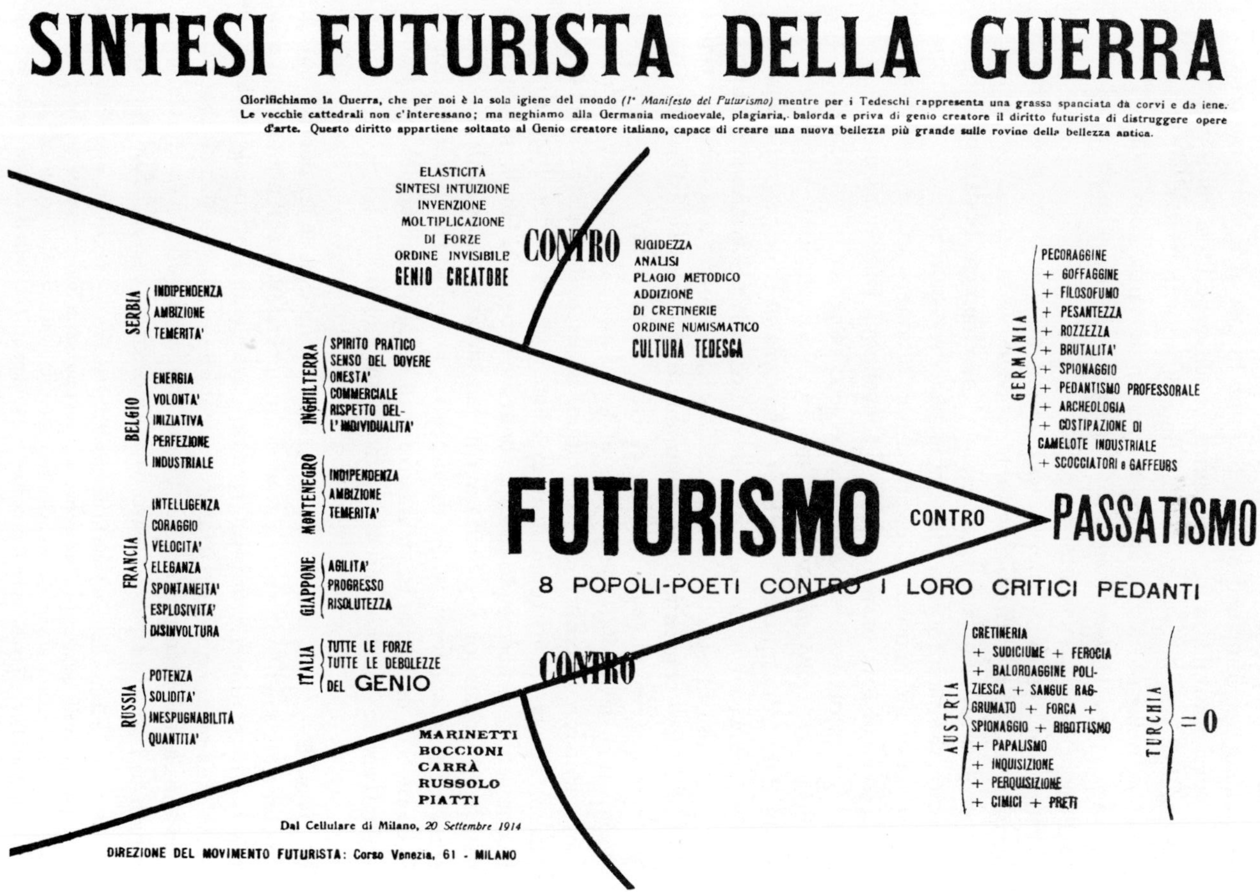

FIG. 1.23. Carlo Carrà, *Sintesi futurista della guerra*, 1915

the projectile, is a very effective structure around which to organize the rest of the manifesto's points. The wedge-shape itself, it is important to note, is deliberately not composed of lines of print. In fact, if it were, the effect of Carrà's design would, in dynamic terms, be self-cancelling. For it is clear that the result of reading two converging lines of print from left to right would inevitably make the wedge appear to radiate outwards from its apex (in the manner of the *Zang tumb tumb* layout), instead of spearheading a verbal attack on the conceptual associations of passéism.

Patterns of iconization and, to a lesser extent, de-iconization or 'conventionalization' are both distinct, observable processes within the rather breathless history of early Italian Futurist typography's development. Both in painting and literature, the quest for new forms of iconicity was quite rapidly transformed, as a result of the unusually accelerated pace of avant-garde experimentation, into a series of new, movement-based conventions. Force-lines in typography began life as the posited property of objects (whether in motion or in any condition is largely immaterial to the present discussion). They thereafter acquired the 'conventional'[94] status of 'extensions' of those properties and, thus codified, they were ultimately, as in the case of Balla's *Velocità astratta* project,[95] divorced from all aspects of referential iconicity in order to assume the function of a new convention for communicating dynamism. The process is, however, not evolutionary in any strict sense. That is to say: one cannot find a clear period devoted to the process of iconization, followed by a phase of movement-based codified de-iconization. As even the present selection of examples shows, many fundamentally iconic and 'conventionalized' forms of layout were in fact being produced at about the same time. This is one respect in which the process of de-iconization in the literary context differs radically from the linguistic form, where the process is more gradual but more pervasive in its effects—until, that is, some Futurist comes along and singlehandedly sets about reversing the due process of philological evolution.

In 'The New Spirit and the Poets' ('L'Esprit nouveau et les poètes' (1917)) Apollinaire proudly informs his readers: 'in general, you will not encounter in France any of those "words-in-freedom", exaggerated offspring of the new spirit, to which limits the Futurists' outbidding of one another, in both Italy and Russia, has been pushed, for France dislikes disorder. We are willing to return to first principles, but we have a horror of chaos.'[96] And yet, as can

[96] Whereas in Italian Futurism one finds a tendency towards de-iconization, with the result that force-lines gradually become detached from objects, to function as conventions for signifying abstract dynamism, the opposite process occurs in Russian Futurism, where the abstractions of Rayism in due course had an effect on typographical structures. See Compton, *The World Backwards*, pp. 88–90.

[95] On Balla's 'lines of velocity', see Enrico Crispolti, *Il mito della macchina e altri temi del futurismo* (Trapani, 1969), 138, 178.

[96] Paris, 1948, 5. See also Apollinaire's *Anecdotiques* (Paris, 1926), 215–21.

readily be appreciated, most of the signal achievements of Italian Futurist typography and visual poetry lie in an area of great control and of highly codified schematization. They are anything but the products of chaos and disorder. In contrast to the predominantly primitivist and even anarchic quality of much Russian Futurist typography, in Italy the abandonment of one set of (passéist) typographical conventions by and large led to the establishment of a replacement series of dynamic, 'expressive' code-systems. As Apollinaire himself acknowledged, the Italian Futurists hardly ever abandoned the use of conventional black ink (even though Marinetti's manifestos occasionally gesture towards the possibilities of multi-coloured layouts; and, as has been stressed in retrospect by one of their number, their practice was only to vary the size, but seldom the actual fount of type involved.[97] Nevertheless, the examples we have briefly examined so far may give the false impression that all Futurist layouts retained a relatively faithful relationship to linearity and discursive readability. To counter this danger, I should like to conclude the present chapter by turning to one of the most accomplished of all of Marinetti's iconic literary works: the dynamic free-word poem *Le soir, couchée dans son lit, elle relisait la lettre de son artilleur au front* (*In the Evening, lying in her Bed, she reread the Letter from her Gunner at the Front*) (fig. 1.24). Here, we are confronted with multi-directional verbal fragments and exploding shapes, rather than the conservatively horizontal telegraphic writing of much Futurist poetry. And Marinetti's material is organized in a way that makes noticeably fewer concessions to linearity and readability than the layouts of *Zang tumb tumb*, where one suspects that economic, and possibly technical, factors may well have kept a brake on the pace of innovation.

Together with *Une assemblée tumultueuse*, *Après la Marne, Joffre visita le front en auto*, and *Bataille à 9 étages du Mont Altissimo*, Marinetti's *Le Soir* was one of a number of powerful visual poems made famous by the anthology *Les Mots en liberté futuristes* of 1919. When compared with the relatively straightforward 'auto-illustrations' and 'designed analogies' of his earlier poetry, these pieces—and above all, *Le Soir*—reveal a new organizational complexity and visually dramatic impact. They stand out, even among the numerous achievements of this volume, and finally help put Marinetti on a par with Severini and Carrà, as a leading creator of Futurist visual collage-poetry. What is of primary interest for the present discussion, though, is the way in which this work successfully harnesses typography to the re-iconization of language and explores various forms of codification.

Although frequently dated 1919 on account of the printing in *Les Mots en liberté futuristes*, *Le Soir* was in fact first published two years earlier in the

[97] 'Simultanisme-librettisme', *Les Soirées de Paris* (15 June 1914), 325. Alberto Viviani remarks that the Italian Futurists show, at most, a predilection for 'characters that differ in size but not in typeface' (*Il poeta Marinetti e il futurismo* (Turin, 1940), 94).

FIG. 1.24. F. T. Marinetti, *Le soir, couchée dans son lit, elle relisait la lettre de son artilleur au front*, 1919

periodical *L'Italia futurista*. It there bore the resplendently Futurist title *Morbidezze in agguato + bombarde italiane* and was equipped with the generic heading *Pubblicità parolibera*.[98] Despite the fact that Marinetti borrowed the first part of this caption, 'morbidezze in agguato', from the title of a contemporary novel by Irma Valeria, it seems to have been the phrase alone that attracted him, for there is no further take-up from Valeria's novel in the collage.[99] *Morbidezze* (the plural, a rarity, of *morbidezza*, meaning 'softness' or 'effeminacy') is lying *in agguato* ('in ambush'). The title, in other words, must be referring to the figure at the bottom of the picture, while the rest of the wording '+ bombarde italiane' can only be pointing to the typographical explosion at its centre. When read in conjunction with the word 'Isonzo'[1] on the mid-left-hand side of the work, the apparently allegorical image of 'softness lying in ambush' is open to a number of historically more specific readings. One possibility is a reference to cowardly Austrian troops lying in ambush during the Italian offensive, the Austrians thus showing their military deviousness by stooping to conceal themselves in wait for the Italian enemy. But the figure at the bottom of the picture hardly looks as if it is 'lying in ambush'. Rather, it seems to be cowering with its hands over its ears at the noise of the Italian onslaught.

In *The Appeal of Fascism*, Alastair Hamilton points out that the word *imboscato*, meaning 'literally, someone who lies in ambush . . . during the Great War [also] came to mean a "shirker", a man who evaded military service',[2] a connotation which may not be irrelevant to the use of 'in agguato' here. Given that the outline is so feminine, albeit grotesquely so, in shape, a second possible reading is that the *morbidezza* figure is a caricaturing allegory of effeminate civilians back home, unable to do more than hide in fear, while patriotic Italian troops fight courageously at the front. As yet, the collage suffers from an inadequately defined relationship between its various parts, though. A central ambiguity still looms in relation to the question of whose *morbidezza* is being referred to or personified. Marinetti seems to have sensed this problem, for two years later the collage resurfaced in *Les Mots en liberté futuristes*, now bearing a new title.

It may at first seem surprising that the only substantive change (apart from removing the old heading) that Marinetti needed to make to 'rescue' his work was to furnish it with a new caption. But this fact tells us something about the

[98] *L'Italia futurista*, 2 (1917), 28. This version of the work is reproduced in Maria Carla Papini (ed.), *L'Italia futurista (1916–1918)* (Rome, 1977), 334.

[99] Irma Valeria's 'I gufi' had appeared in no. 38 of *L'Italia futurista* in 1917, and it may well be that the acoustic dimension of her depiction of hooting owls held a particular appeal for Marinetti.

[1] In the autumn of 1917 the region was the scene of a major Italian offensive against the Austrians, in fact one of the few occasions during the war when Italy successfully took the initiative. This perhaps explains the reference to the Isonzo Front in a work of *pubblicità parolibera*.

[2] *The Appeal of Fascism: A Study of Intellectuals and Fascism, 1919–1945* (London, 1971), 17.

nature of its iconicity and about iconicity in general. The iconic element in a sign is invariably so limited that, even if one is confronted with an iconic pattern of words, it may not be possible—without the interpretive guidance of a title or an explicit convention—to infer what the particular sign-configuration stands for. Many of the signs we encounter in the everyday world (road-signs, symbols on maps, codified diagrams) still depend, for all their iconicity, on well-established conventions to make them meaningful. And the same principle operates here.

Apollinaire once remarked caustically that the Futurists had come to Paris to 'teach us with their titles, not their works', meaning that there was often more dynamism and modernolatria, and by that token more Futurism, in these than in the paintings themselves.[3] But with the work under discussion the title does furnish a crucial indication of how the other signs must be read. That is to say: we often need symbols to tell us how to read icons. Without a title, or, rather, without an adequate title (as was the case with the *Italia futurista* version), the work has something of the ambiguity of a Rorschach Test ink-blot about it. Take, for instance, the silhouette at the bottom of the layout. The Italian Post Office thought so little of this figure that they actually removed it from the work when they reproduced it on a 1976 stamp issued to commemorate Marinetti's hundredth birthday. (No doubt political as well as aesthetic factors played their part in determining how the poem was to be modified for philatelic use, since the phrase *Guerra ai tedescofili!* ('War to the Germanophiles!') also needed to be deleted for diplomatic reasons.) Yet both of the work's titles accentuate just how important this figure is within the layout's overall strategy and how the configuration becomes practically meaningless without her.

The title which Marinetti chose for the reprint of his work in *Les Mots en liberté futuristes* (this time in French, as befitted the new context) still refers to the figure at the bottom of the work—*Le soir, couchée dans son lit, elle relisait la lettre de son artilleur au front*. Compared with the titles of other leading free-word creations, such as Carrà's *Festa patriottica*, Soffici's *Tipografia*, or Marinetti's own *Une assemblée tumultueuse*, this new caption is exceptionally long. I am shortening it to *Le Soir*, for convenience, and there have even been moves to call it *SCRABrrRrraaNNG* (a caption certainly worthy of the author of *Zang tumb tumb*).[4] But the title nevertheless remains uncompromisingly wordy. For the leader of a movement often associated with clipped, telegraphic writing, this second title seems more than a little out of character. Moreover, it has a deliber-

[3] *Anecdotiques*, 49.

[4] In the catalogue of the exhibition *Futurismo 1909–1919* (Newcastle, 1972), 77, and in Caroline Tisdall's and Angelo Bozzolla's volume in the influential Thames and Hudson 'World of Art Library' series: *Futurism* (London, 1977), 97 and 213, the work is referred to as 'SCRABrrRrraaNNG', without any indication that this is not the title of either the first or the second version of the work.

ately 'literary' ring to it. Although it has never been attributed to a documented source, it would not be surprising to find that Marinetti had in fact 'lifted' the sentence from some passéist novel. The demonstrative 'le soir' and the strange use of the past tense to refer to a visually immediate image, thus establishing a complex time-scheme, are as uncommon in Futurist works as in captions to pictures and visual poems in general. Yet 'le soir . . . elle *re*lisait' serves to inform us that she has already read the letter once (presumably on receipt) and is now doing it justice by rereading it fondly at her leisure. Even the pronoun 'elle' suggests extraction from a longer prose passage which would have already revealed who 'she' and her 'gunner' were. (The only other possible reading is that she is no more than the archetypal soldier's girl back home.)

The result of this eccentrically verbose titling is a pointed contrast between the conventional, tired, non-iconic prose of the caption and the free-word explosion itself. Indeed, the word 'esplosione' which occurs in the layout could as well be a comment on the aesthetic mode of the composition as a reference to the battle itself. The centre of the poem/collage is 'explosive'; the caption, by contrast, is statically passéist, and exaggeratedly so. Take away the Futurist visual effects, in fact, and one is left with an insipid, inconsequential piece of prose. Add the iconic collage, and the lifeless phrase 'au front' immediately becomes more important, as it explodes into life before one.

This surprise effect is strengthened in *Les Mots en liberté futuristes* by the fact that, having read the title on p. 103, the reader then unfolds the large-format visual poem which has hitherto remained unseen. The title in fact receives a page to itself, and is set out on three lines, as if to slow down the pace of its reading.

There is some contextual evidence to suggest that this is a deliberate contrast between traditional and Futurist, symbolic and iconic, modes of expression. For in this same volume of 'mots en liberté futuristes' Marinetti included another work which had originally, according to Mario Diacono,[5] appeared in 1915 in the form of a broadsheet with the snappily Futurist title *montagne + vallate + strade × joffre* (*mountains + valleys + roads × joffre*). Two years later it bore the more conventionally passéist caption *Après la Marne, Joffre visita le front en auto* (*After the Battle of the Marne, Joffre visited the Front by Car*). (The title is again translated into French from the original Italian, but this may not simply be in order to accommodate the work to the context of a French anthology: it has the added effect of associating French more with the language of passéism, whereas Italian becomes the idiom of telegraphic Futurism.) This change adds a measure of specificity not present in the original 'mathematical' title, but also—and

[5] 'L'oggettipografia di Marinetti', *Tavole parolibere futuriste (1912–1944)*, i. 11. Two of the more important features of the new title are the retention of the word 'Joffre' and the reference to the Marne. It was French resistance to the Germans at the Battle of the Marne that galvanized the Interventionist cause.

more importantly, from our point of view—it establishes the same kind of forceful juxtaposition of Futurist poetry and passéist prose that we have in *Le Soir*. In other words, we are dealing with a similar case of apparent regression from the language of Futurism to convention. The original title of *Le Soir* was also relatively telegraphic and mathematical (*Morbidezze in agguato + bombarde italiane*). It would have been out of character for Marinetti to retreat from radical Futurist methods to the conventional, without good cause. There is certainly no evidence of him going soft as a Futurist experimenter as the years went by. Quite on the contrary. Nor do I think he is simply making concessions to what he might have regarded as an unenlightened or relatively passéist French reading public (although commentators have suggested that the titles of his manifestos in Italian tend to be more polemical than when they are published in French). This is probably not a factor here. The reason for this old-fashioned title is rather to bring out a contrast between two kinds of literature.

Indeed, the whole work is built on a series of pivotal contrasts: between the language of the title and that of the central word-explosion; between the figure at the bottom of the page and the word-picture of battle that dominates the main part of the work; and, by extension, between the experiences conjured up by the work and our own predicament as readers.

Marinetti was shortly to return to the same juxtaposing structure with his *Simultaneità di guerra volutta* (*Simultaneity of Voluptuous War*), a *sintesi* of 1921. Here, at the same time as Marinetti read his *Battaglia nella nebbia* (*Battle in the Fog*), a well-dressed couple danced a 'languid tango' around him. But, according to Michael Kirby, the envisaged effect was not one of purely contrastive juxtaposition; on the contrary, 'the two unrelated actions were intended to "interpenetrate" and combine "bellicose furore and voluptuous nostalgia"'.[6]

In his insistence on the importance of the iconic sign within communication, Peirce once went so far as to claim that 'the only way of directly communicating an idea is by no means of an icon': 'every indirect method of communicating must depend for its establishment upon the use of an icon. Hence, every assertion must contain an icon or set of icons, or else must contain signs whose meaning is only explicable by icons.'[7] Even if the word *tree* is non-iconic, it will create the image of a tree in the mind of the person hearing the word, and that image will be iconic. (Kaja Silverman has recently pointed to the link between this and Freud's distinction between 'thing-presentation' and 'word-presentation'.[8]) Whatever one thinks about the status of Peirce's claim as an axiom concerning *all* communication, it offers a neat gloss to the poetic *modus operandi* of *Le Soir*.

What, though, is the word-explosion that dominates this collage? It could be

[6] *Futurist Performance* (New York, 1971), 68.

[7] *Elements of Logic*, p. 158.

[8] *The Subject of Semiotics* (Oxford, 1983), 21.

that her 'gunner at the front' wrote her a letter in conventional non-motivated language describing his experiences in combat. In that case, the free-word pattern could represent the iconic image (in Peirce's sense) that his letter created in her mind's eye. Or perhaps he is a Futurist and has written her a free-word *lettre-poème*, *à l'Apollinaire*, in which case what we see before us might well be his actual poem, not the image of reported battle created in her civilian imagination.[9] But there is a third possibility, especially attractive in the light of Marinetti's later *Simultaneità di guerra volutta*. The word-explosion could be the collage's own representation of the actual battle, a picture for the reader alone, and one contrasting then with the words she reads as she lies in bed that evening. Given the Futurist interest in simultaneity, this is by no means an implausible interpretation: while she lies there simply reading about battle, we, as readers, are privileged—by courtesy of Futurist words-in-freedom—to receive some more direct sense of the battle; we witness it happen, as it were, before our very eyes.

The epistolary motif in the 'lettre de son artilleur au front' idea also echoes the cover of Marinetti's entire anthology, which is based on a poem known as *Lettre d'une jolie femme à un monsieur passéiste* (fig. 1.25). According to the catalogue description in *The Futurist Imagination*:

> Marinetti typeset several versions of this letter. . . . All . . . play on the similar sound of the words *chère* (dear) [it should really be *cher*, since she is addressing *him*!] and *chair* (flesh). The words in the different versions vary, but they all make clear that the pretty woman is interested in a large number of valuable jewels. The words 'vitesses élégantes' appear repeatedly in Futurist literature. Their placement here [on a proof sheet] suggests that they may have originally been intended as the title for *Les Mots en liberté futuristes*.[10]

Marinetti's poem, then, takes up the contrast between a passéist and a Futurist figure, already encountered on the cover of the anthology in which the work appears.

The possibility that we are being offered a juxtaposition of actual battle with her lying reading might explain the fact that the girl has her back to the images which assail us from the centre of the work. In any case, how one interprets the relationship between the word-explosion and the shape at the bottom of the

[9] Bowler interprets the word-explosion in this way, as 'a poem, in the form of a letter from a soldier to his sweetheart' (*The Word as Image*, p. 132). For Moholy-Nagy, 'Marinetti attempts to show typographically the sweetheart reading a letter from her soldier friend who describes his dangerous experiences in the trenches. These experiences, though accumulated over a period of time, can be seen here simultaneously' (*Vision in Motion*, p. 306). I am unclear what the evidence is for this last statement, unless it is Boccioni's idea that *stati d'animo* works invariably involve a synthesis of past and present perspectives (something which I shall return to in the next chapter's discussion of collage).

[10] 'Catalogue of the Exhibition', *The Futurist Imagination*, p. 109.

FIG. 1.25. F. T. Marinetti, *Lettre d'une jolie femme à un monsieur passéiste*, 1914, used as the cover design for *Les Mots en liberté futuristes*, Milan, 1919

work—the focal point of both the titles[11]—may well depend on which features one decides are iconic and which are not. Carlo Belloli once described *Le Soir*, in terms that would surely have appealed to its creator, as a work which 'attempted to exhaust all the possibilities that typography has of reproducing an experience, in order to open up new paths for it in the future'.[12] Similar claims can, I believe, be advanced for the work, in semiotic terms. For, just as the poem is an exploration of the sheer diversity and potential of typography, so too it plays creatively with various possibilities of semiosis, allowing above all for a range of iconic possibilities.

Clearly, when the title of Mayakovsky's poem 'Skazka o Krasnoy Shapochke' ('The Tale of Little Red Riding-Hood') (*PSS* i. 142) has the word for 'red' printed in red ink, we assume that an iconic effect must have been intended. However, when it comes to the use of black ink in *Le Soir*, matters become more

[11] For Blumenkranz-Onimus, it is not so much the titles that drew attention to the figure as the fact that the reader 'often has to begin reading a work from the bottom of the page' (*La Poésie futuriste italienne*, p. 201).

[12] 'Italian Pioneers of Graphic Design', *Neue Grafik—New Graphic Design—Graphisme Actuel*, 3 (1959), 9.

complicated. Arthur Burks, Peirce's editor, once expressed the view that the philosopher's suggestion that a sign was an icon 'merely if it possesses or exhibits the quality or relation it signifies' is not without its pitfalls, when it comes to typographical signs. For, he complains, 'on this criterion, any token of "black" printed in black ink is iconic, though the reader is unaware of the fact that it is displaying the quality that it represents'.[13] Such an objection is wrong-headed, in my opinion. If the reader posited by Burks is 'unaware' of an iconic shared characteristic, then that is simply another way of saying that he is not reading the sign in an iconic way. The whole point about the process of semiosis is that something only functions as a sign when the reader decides to interpret it in a semiotic way. (Burks's intentionalist view of signs as 'transmitted signals' in any case seems more appropriate to Saussurian semiotics and later Gallic derivatives thereof than to Peirce's.) But, once one enters the aesthetic realm, the possibilities of iconicity come much more naturally into play, and as a consequence intentionalism becomes even more of a non-runner. And this is certainly the case with *Le Soir*.

Here, there is a great deal of free play about what is or is not iconic, and this enriches the work's interpretive potential. Bearing in mind that the collage represents an evening scene, it would be possible to view the girl's shape as a silhouette—possibly even a shadow cast on a bedroom wall as she lies reading her gunner's letter. On the other hand, this is only one possible decoding of the image. Instead of the blackness itself being iconic, it may be that only the outline or shape of the figure is so (as is the case with 'filled' road-signs). How one interprets the role of *blackness* in the icon of the girl may well have a decisive influence on how one deciphers other parts of the collage (or, indeed, the principle may work in the reverse direction). If the girl's *shape* is presumed to be the only motivated element in that part of the work, then black will tend to be taken to be non-iconic elsewhere, too. But if blackness is interpreted iconically, then the presence of blackness at the centre of the work may also be read in the same way: as a sign of the smoke and pall of battle.

To compare this with an effect in *Zang tumb tumb* (*TIF* 644): in the illustration of a flag on a pole in that work, using the words *bandiera bianca su* ('white flag on'), the shape of the 'auto-illustration' is clearly iconic, whereas the colour is not, since black letters are used to stand for a white flag.

BANDIERA
BIANCA
S
U

Similar choices about what is, or is not, iconic also have to be made when it comes to the question of relative size. The collage has a very small picture of a

[13] 'Icon, Index and Symbol', *Philosophy and Phenomenological Research*, 9 (1948–9), 575 f.

girl at the bottom of a large battle poem. This could be an indication of her relative unimportance compared with the content of 'her gunner's letter'. (The adulating-cum-erotic position that she adopts while reading his letter is, of course, a further indication of the fact that she is to some extent a strategem for self-glorification on the part of the male Futurist poet.) But if size, or rather smallness, is in this case assumed to be iconic, then it would be an instructive exercise to transfer such a code to the actual verbal element in the rest of the poem. The sounds of war, written so much larger than the slogans of battle ('Death to the Germanophiles!', i.e. the Austrians), might well be a reflection of the fact that it is the invigorating experience of battle *per se*—no matter what the cause being fought for—that is the ultimate concern of the Futurist poet. An alternative, more realistic interpretation would simply relate this feature to the probability that the sound of the explosion would practically drown out the noise of the soldiers' battle-cries. On the whole, there can be no clearcut answer to the question of which aspects (size or colour) are iconic within the overall layout.

As well as this range of semiotic variables, *Le Soir* also rejoices in other forms of richness—not only in the exploration, as Belloli notes, of the full range of typographical possibilities (from handwriting, printed letters, hand-drawn ones, shapes, and fragments) but also in its semantic spectrum. To take the most striking example: the work is a complex mixture of historical and aesthetic references. The key identifying word 'Isonzo' seems to locate the poem in a specific period of history, even though this reference will have been read differently in 1919 than it was in 1917.[14] But the work also contains, over and above the associations of the gunner's battle context, a whole series of aesthetic references. Three of its main words (tellingly, none of them fragmented, this time) are drawn from the realm of artistic self-commentary, for 'futurista', 'esplosione',[15] and 'simultaneità' describe the work itself, and not just the gunner's experiences at the front. Since war, as Marinetti was to note in 'The Futurist Synthetic Theatre' ('Il teatro futurista sintetico' (1915)), was 'essentially no more than Futurism intensified' (*TIF* 98), battle can be presented both in its own right and at the same time as the great paradigmatic 'explosive', 'simultaneous', 'Futurist' experience. In this sense, this work in its entirety is a

[14] The first version was published on 9 Sept. 1917, well before the major offensive of 17 Oct., hence it could only be focusing attention on a centre of Italian hopes for the future. The victories of 1917 give the 1919 version a far more patriotic dimension, leaving aside the fact that Italy and her allies had by then won the war.

[15] While, of these three, the word 'esplosione' might seem the most likely to be simply a reference to battle, it was also one of the commonest Futurist metaphors for their centrifugally dynamic word-patterns. Thus, the manifesto 'Typographical Art of the War- and Post-War Periods' ('L'arte tipografica di guerra e dopoguerra') speaks of Futurist free-word typography as something which 'explodes with typographical characters in all directions' (signed by Marinetti, Alfredo Trimarco, Luigi Scrivo, and Piero Bellanova, *Sintesi del futurismo*, ed. Scrivo (Rome, 1968), 12 f.).

'designed analogy' for battle; but, looked at the other way around, the battle it evokes is a 'designed analogy' for the Futurist sensibility or the new conception of the work of art.

It would restrict the work's implications unduly, I feel, to assume that the girl depicted is Irma Valeria or that the 'gunner at the front' is none other than Marinetti (even if the handwritten thanks at the top of the letter, even signed 'FTM', did originate as an expression of gratitude for the receipt of *Morbidezze in agguato*). But if this is part of the gunner's actual letter, he is thanking the girl for a book at the same time as returning the compliment by sending his account of the battle. Moreover, since his actual letter conceivably *includes* a picture of her, she becomes his own means of self-congratulation, and hence the relationship between sender and addressee is far more complex than the identification of a real Marinetti and a real Valeria would allow for. In any case, all Futurist woman is permitted to do under such circumstances is lie at home adopting erotic poses in the face of a series of unambiguously phallic battle images. But she at least has the modest privilege of receiving a letter from the front, and she clearly knows how to respond, although one suspects that Marinetti's ideal reader would be wanting to go out and join the soldier at the front, rather than just receive letters from him.

Reading and images of battle exist side-by-side in *Le Soir*. The concentration with which the girl appears to be engrossed in the letter, the distribution of focus (with the battle itself at the centre and the reading girl marginalized), and the subordination of conventional language to the exploding typography of battle all suggest the importance of what the girl *and we* are reading. Ultimately, in other words, within a framework that leaves the reader quite a measure of latitude in deciding how individual elements in the layout are to be interpreted, one is none the less left in no doubt at all about the final judgement to be passed on the letter's content. The work thematizes its own reading process in such a way as to indicate the requisite attitude on our part. Whatever else may be ambiguous in this complex structure, the receiver's role is not.

Le Soir has over the years received its fair share of bouquets in the histories of visual poetry, experimental typography, and collage. Luigi Ballerini has called it one of the 'three ... most important verbo-typographical compositions ever produced' (his other two candidates being *Après la Marne* and *Une assemblée tumultueuse*), his reason being that 'here Marinetti goes ... beyond free-word composition proper and engulfs verbal symbols in a wider circulation of typographical signs'.[16] For Mario Diacono, too, the work represents 'the ideological and creative apex of Marinetti's literary exploration' of the full potential of free-word typography.[17] Nevertheless, praise for such innovative iconic

[16] 'Sixty Years of Visual Writing', *Italian Visual Poetry, 1912–1972*, ed. Luigi Ballerini (Finch College Museum, New York, 1973), 6.

[17] Ibid.

exercises needs to be kept within reasonable bounds, if it is to be meaningful.[18] Futurist 'auto-illustrations', 'designed analogies', and other non-linear creations have an essentially *experimental* value in most cases. And the caveat that Robert Scholes once sounded in connection with the modern experimental novel is also pertinent here:

> Since the scientist experiments, then the experimental novelist, if he is to compete, must experiment also. . . . But in the sciences, if the experiment does not work, the savant does not insist on publishing his results in the scientific journals. It is only in literature that all experiments are deemed worthy of publication.[19]

To make language more iconic, in whatever degree and by whatever means, is, contrary to what some of the Futurists themselves may have thought, not *per se* a pinnacle of aesthetic achievement. The results may not be experiments that failed, but all versions of the experiment are not of equal value. The exercise is merely a matter of reconstituting the medium, whereas value can only arise out of what is done within the medium. In any case, no literary technique is good or bad as such. Some forays into this field are of little more than anecdotal interest. Others, like *Le Soir*, are of value, not merely for extending the repertoire but for the accomplished and powerful way in which they have done so. As Tisdall and Bozzolla so aptly put it, 'belief in innovation was both the strength and the weakness of Futurism'.[20]

There is also surely a difference, in this area, between the cognitive—or generally 'modernist'—value of the experimental for its own sake and what might be thought of as a genuine aesthetic achievement, something that can rank alongside the canon of conventional masterpieces. Sometimes the Futurists, in their quest for the innovative, are too easily content with equating the ingenious with the expressive when it comes to experiments in making language more iconic. This is truer in the case of 'auto-illustrations' than 'designed analogies', where some very impressive achievements are created. Here, and, as we shall see in the next chapter, again with their collages, they rise to heights which were seldom reached again, even during the Dada period.

[18] Compare Roger Shattuck's recent extravagant claim that Apollinaire's 'Lettre-Océan'—surely inferior to *Le Soir*—'deserves a place beside *Les Demoiselles d'Avignon* and *The Rites of Spring*' (*The Innocent Eye*, p. 244).

[19] 'The Limits of Metafiction', *Fabulation and Metafiction* (Urbana, Illinois, 1979), 136.

[20] *Futurism*, p. 108.

2.
The Development of Futurist Collage and the *Dipinto Parolibero*

As Marinetti's *Le Soir* suggests, some of the greatest triumphs of the new Futurist iconic free-word poetry could equally well be thought of as examples of typographical collage. The arrangement of letters, figures, and supporting drawn material in various collage-like layouts, often arrived at without resorting to any actual gluing-on process, means that 'free-word tables' and Futurist collage are closely related. Marinetti's *Le Soir*, *Une assemblée tumultueuse*, and *Après la Marne*, Soffici's alphabet-mosaics (in *BÏF§ZF+18*), and many of Cangiullo's animated letter-pictures (in *Palpavoce* and *Caffèconcerto: Alfabeto a sorpresa*) mark out a rich and highly fertile territory somewhere on the border between free-word shaped typography and collage proper. And the Futurists' experiments in the realm of collage itself were no less remarkable. Works like Carrà's *Festa patriottica* (also known as *Manifestazione interventista*), Severini's *Danza serpentina*, and Buzzi's *Conflagrazioni*-cycle are important milestones in the development of early collage, individually as accomplished as many of the better-known Cubist works in the field. And yet Futurist collage remains one of scholarship's minefields, both because of the vexed issue of its relationship to Cubist discoveries, to which it so often plays Cinderella, and on account of the embarrassingly reactionary content of many of its most representative works.

The whole complicated story of collage's evolution in the early twentieth century has been bedevilled by assertions and counter-claims concerning priority, by disputes about who discovered certain techniques first, and—perhaps inevitably, given the strong nationalism of the period leading up to the First World War—by the contentious issue of which country's avant-garde was to be credited with having effectively influenced what was happening elsewhere in Europe. ('We are obliged jealously to defend the primacy of our discoveries . . . because we live and work in Italy, a country . . . considered up to now by those abroad not to exist, as far as painting, sculpture, architecture, music and literature are concerned,' to quote Boccioni's defensive words (*Gli scritti*, p. 175).) The Italian Futurists, in any case not the greatest respecters of historical accuracy when it came to promoting their own cause, were thus given to subterfuge, at times even redating their creations in order to suggest a greater independence from the Cubists, or at least to avoid the ignominy of appearing to lag too far behind what was happening in Paris.[1] Subsequent discussions of early collage have unfortunately all too often fallen into the same trap of becoming obsessed with the question of *anteriorità*, of who influenced whom, and who did what first.

At the time, priority was, understandably, a matter of national honour. Avant-garde movements need to be seen to have discovered fresh art-forms; and a movement so self-consciously associated with the future and the new had an added burden to bear in this respect. Collage, one of the major artistic innovations of the period, was also of vital interest to the Futurists as an example of the breaking down of the life/art distinction and of the synthesizing of the various arts which they favoured as part of their new aesthetic. Nevertheless, the early collagists were often the victims of their own machinations and self-perpetuating legends. Delaunay accused the Italian Futurists of outright plagiarism,[2] Marinetti responded with similar sentiments concerning Delaunay and his circle,[3] and the Cubists and their literary spokesman, Apollinaire, were generally disposed to regard much of the Italians' work as overly derivative of their own discoveries, while in their turn the Futurists preferred to view Cubism and Orphism as Futurism plagiarized. In such a climate, confusion concerning specific lines of influence was bound to arise; indeed it even suited some people's purposes that it should. It comes as no surprise to find a contemporary observer, the poet Raoul Hausmann, confidently declaring in 1916 that the Dadaists 'knew that Picasso, *following the example of the Futurists*, used real materials, cuttings from newspapers, hair, wood, pieces of glass—and this already in 1912—and that the same Picasso . . . made still-lifes out of pieces of coloured paper, a process that is called "collage"'.[4] But we find Severini seeing the situation rather differently, from his Parisian vantage-point. Apollinaire, according to him, 'was the one who suggested the invention of *papiers-collés* to [him] in 1912'.[5] According to Severini, Apollinaire also proposed the idea at about the same time to both Braque and Picasso.

[1] To give but one example of retro-dating in the case of a work which will be considered later in this chapter: the tenth war collage in Carrà's *Guerrapittura* series, *Angolo penetrante di Joffre sopra Marna contro 2 cubi germanici* (*Joffre's Angle of Penetration on the Marne against Two German Cubes*), is prefixed by the caption *La notte del 20 Gennaio 1915 sognai questo quadro* (*On the Night of 20 January, 1915, I Dreamed this Picture*). By contrast, the original of this work (in the Winston–Malbin Collection, New York) bears the date 1914. Herta Wescher (*Die Collage: Geschichte eines künstlerischen Ausdrucksmittels* (Cologne, 1968), 62) concludes that the date, which does not appear in *Guerrapittura*, was added in retrospect, which arouses the suspicion that Carrà redated the work to create the impression that it was an even earlier example of the genre.

[2] *Du Cubisme à l'art abstrait* (Paris, 1957), 123.

[3] See Marinetti's letter to Soffici of 21 July 1914 (*Archivi*, i. 342).

[4] *Am Anfang war Dada* (Steinbach, 1972), 45. My italics.

[5] *Témoignages: 50 ans de réflexion*, with a preface by Georges Borgeaud (Rome, 1963), 216.

Writing to the art-historian Raffaele Carrieri in the 1960s, Severini speaks at some length of the early days of collage, claiming that

> *papiers-collés* . . . were born in 1912 in . . . Montmartre. As I remember it, Apollinaire suggested the idea to me after having spoken of it to Picasso, who immediately painted a small still-life on to which he applied a small piece of waxed paper (the type that was used for the table-cloths in the bistros of Paris). I tried to glue some *paillettes* and multi-coloured sequins onto forms of ballerinas in movement. I next saw a collage by Braque, perhaps the first, made of what seemed to be wood and large sheets of white paper on which he had sketched to a large extent with black crayon. During my trip to Italy in August of 1912, I naturally spoke about the technique to Boccioni and he, in turn, to Carrà. During 1913 the first Futurist experiments in this field saw the light of day. The reasons why Apollinaire gave us these suggestions, as I remember them from conversations of a later date, were: in the first place, the need at that time to comprehend the sense of a more profound and secret inner reality which would have been born from the *contrast* of materials employed directly as things and placed in *juxtaposition* to lyrical elements created by our artistic sensitivity. I believe that in the final analysis it was this first idea which persuaded Picasso and Braque to make use of stamped numbers and letters.[6]

Most standard accounts of the development of collage, including those by Janis and Blesh, Jürgens-Kirchhoff, Wescher, and Wolfram, likewise tend to assume that the really decisive influence was, as Severini claims, in the French–Italian direction. So exclusively so, in fact, that they habitually underrate the marked degree of cross-fertilization that took place *in both directions*. But even to acknowledge this substantial measure of interaction is still to miss the real point. For Cubist and Futurist collages are for the most part so different that any comparison based on an undifferentiating concept of influence yields little more than misleading results. While there is undeniably a Cubist strain to the collages of Carrà, Severini (who was, after all, largely Paris-based), Soffici, and some of the Turin Second Futurists, it remains less pronounced than historical studies suppose. And, if Herta Wescher is correct in suggesting that, by comparison with their Cubist counterparts, Italian Futurist collages have not received the scholarly attention they merit for the simple reason that 'it was clear . . . that the Futurists were in this case not the real inventors, but . . . merely the inheritors of French Cubist discoveries',[7] then this is a topsy-turvy state of affairs. For it means that detailed investigation of a complex and largely heterogeneous phenomenon has been held up by a set of misconceptions which the requisite research would have revealed as such. In any case, 'invention' of such a diverse and richly exploitable technique as collage is by no means comparable to having copyright on the method. Even inheritors, to change the metaphor,

[6] Raffaele Carrieri, *Futurism* (Milan, 1963), 117.

[7] *Die Collage*, p. 52. Clearly, numerous surveys have appeared since Wescher made this statement, but on the whole Futurist collage still plays only a minor role in most of them.

can make as good a use of legacies as their original owners; and the issue of probate is in any case still open to dispute, when it comes to modernist collage.

As a corrective, the following exploration of Futurist collage is principally intended to demonstrate three things: first, that most Italian Futurist collages operate in a very different—dynamic—mode from that of Cubist still-life experiments, hence too much concern with questions of priority has the effect of levelling substantial qualitative differences; second, that the aesthetic premisses which gave rise to Futurist collage differ in virtually all respects from those which determined the Cubists' interest in the genre; and, third, that within the impressive spectrum of its collage exploration[8]—and Futurist collage is in some respects richer and frequently more inventive than its Cubist counterpart—Italian Futurism was not slow to give birth to an independent and highly effective sub-genre: that of a word-orientated form of collage.[9] Alternatively, to borrow Robert Rosenblum's useful terminology, one could argue that, within the general sphere of collage/assemblage (which can involve the use of a vast repertoire of glued-on material), the Futurists were particularly interested in the semantic potential of certain kinds of carefully chosen *papiers-collés*.[10] My aim in foregrounding these facts is not to question or minimize the radically innovative achievements of their Cubist contemporaries, but merely to emphasize the essentially very different work of the Italian Futurists, and, to a limited extent, of their Russian counterparts, particularly in their verbal collage-work.

On 11 July 1914 Carrà wrote to Severini:

In the last issue of *Lacerba* I saw a drawing by you which from a certain point of view interested me very much because I have recently completed a work ... entitled *festa*

[8] There is a useful chapter on 'The Invention of Collage' in Marjorie Perloff, *The Futurist Moment: Avant-Garde, Avant Guerre, and the Language of Rupture* (Chicago, 1986), 44–79. However, with the exception of Boccioni's *Fusione di una testa e di una fenestra* (*Fusion of a Head and a Window*) and *Bottiglia nello spazio* (*Bottle in Space*) and Carrà's *Manifestazione interventista* (or *Festa patriottica*), the survey is little concerned with Italian Futurist examples of the genre. Perloff has much of value to say, though, about the genre's early history, the relationship between collage and free-word poetry, and the interplay between alterity and referentiality.

[9] Eddie Wolfram hints at this characteristic when he charges Futurist collage with being 'idiosyncratic' in its 'attempt to bring topicality and documentary realism into art, expressing the day-to-day world, the here and now ... not found in Cubism' (*History of Collage: An Anthology of Collage, Assemblage and Event Structures* (London, 1975), 42). Since what is being described is largely a matter of the politicization of collage, the inevitable result is substantially more stress on the verbal element than in Cubist collage. 'Politics, the main subject of newspaper reporting, is the one missing field' in the Cubist use of newspaper-fragments, according to Marjorie Perloff (*The Futurist Moment*, p. 50). In Cubism, such *papiers-collés* function 'as compositional rather than referential entities', whereas the opposite is usually the case with Futurist literary collage.

[10] *Der Kubismus und die Kunst des zwanzigsten Jahrhunderts* (Stuttgart, 1960), 69. Rosenblum defines the *papiers-collés* method as 'specifically ... the use of fragments of paper in contrast to the wider possibilities of collage'.

patriottica—poema pittorico . . . which has many points of contact with your own work. In it, I have abolished any representation of the human form because I wanted to depict the plastic abstraction of a civic tumult. Two days ago I took it to the photographer to have it reproduced, knowing that it would not retain much of its true quality, a quality which depends on a suggestion of poetic images and colours obtained by coloured cards. (*Archivi*, i. 341)

A monochrome version of *Festa patriottica* was duly reproduced in *Lacerba* on 1 August 1914, although, as Carrà had feared, it conveys only an indistinct and drab indication of the riot of images, the exploding colour-fragments, and the bold word-patterns of the original collage (now in the Collection Gianni Mattioli, Milan).

One of the chief 'points of contact' between Severini's *Danza serpentina* (fig. 2.1) and Carrà's *Festa patriottica* (fig. 2.2) is the tendency of both to break with mimetic representation—to such a degree that, in the case of Carrà's work, there has been a diverse range of critical opinion about the exact nature of the collage's subject, syntax, and semantics. Writing in the same year as he created *Danza serpentina*, Severini speaks disparagingly of the tendency towards too slavish a cult of the mimetic that had, in his opinion, been recently imported into Futurist typography by Marinetti and his free-word imitators:

No matter what material I am working with, I do not dream of pictorial systems; one should not confuse images with aspects, the representation of ideas with the representation of shapes.

On the contrary, it seems to me that there is in Cangiullo's 'FUMARE' device an Impressionist, pictorial element, or at least a tendency towards a pictorial form of impression. . . . *the realism we ought to attain continues to detach itself from external reality* to the point where it has nothing to do with surface reality . . .[11]

In singling out Cangiullo's 'FUMARE' device, of all examples from Futurist typography, Severini is above all reacting to the quasi-mimetic theory of the 'designed analogy' advanced by Marinetti in 'Geometric and Mechanical Splendour and the Numerical Sensibility' ('Lo splendore geometrico e meccanico e la sensibilità numerica' (1914))[12] rather than to the shortcomings of any specific concrete example.

Both *Danza serpentina* and *Festa patriottica* depart from the 'representation of shapes' attributed by Severini to the 'designed analogy' (although, as we have seen, the typographical technique was more codified in its iconicity and generally less mimetic than this contemporary view allowed for). Instead, we are offered a 'representation of ideas', embodied in a collage of stylized patterns and explanatory words (Wolfram calls *Danza serpentina* a '*scheme* for colours and for concepts'[13]). That Carrà is here developing Futurist collage in a

[11] *Témoignages*, pp. 34, 37.

[12] Published in *Lacerba* on 15 Mar. 1914, and using Cangiullo's 'designed analogy' (see fig. 1.11) as an illustration.

[13] *History of Collage*, p. 42.

FIG. 2.1. Gino Severini, *Danza serpentina*, 1914

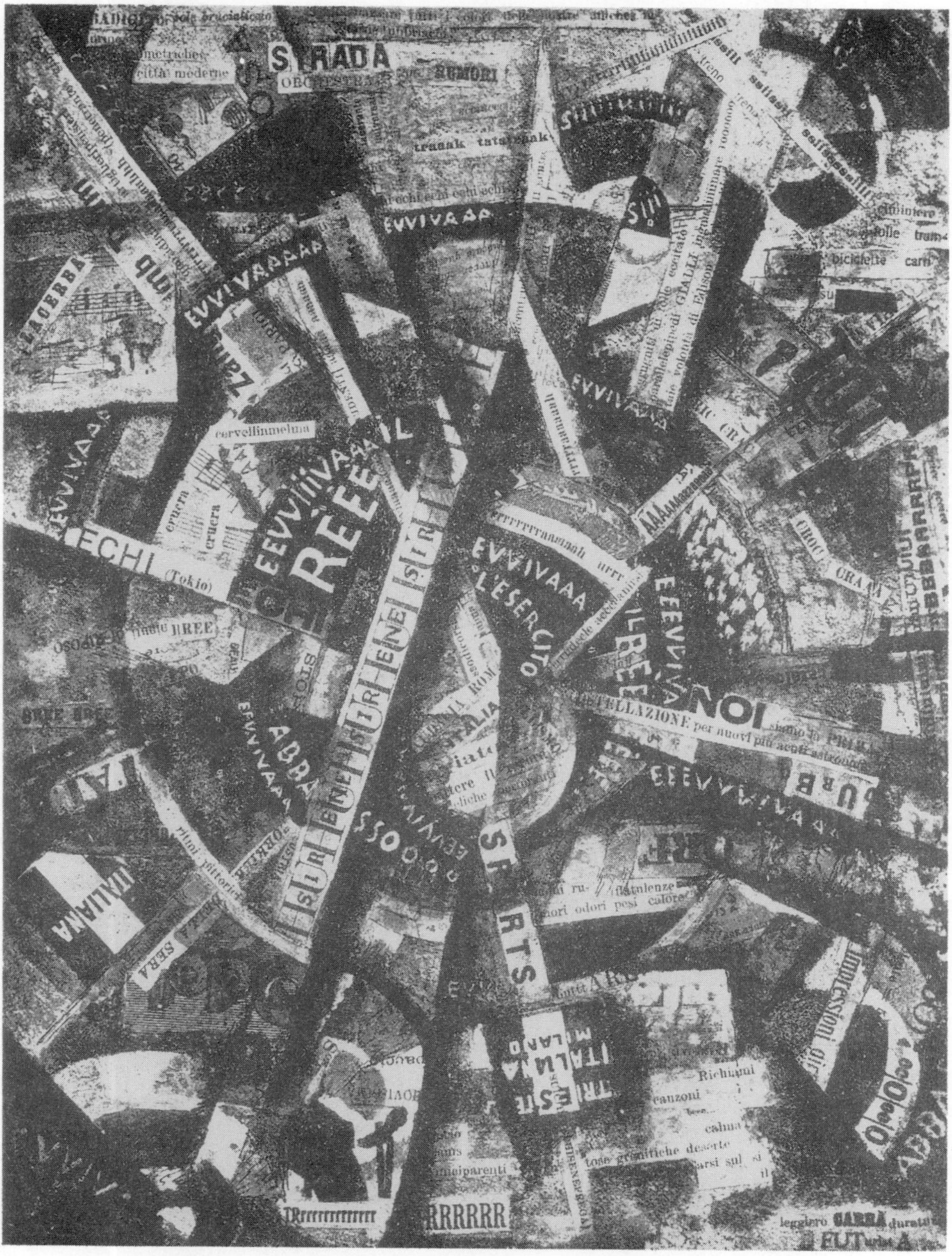

FIG. 2.2. Carlo Carrà, *Lacerba* version of *Festa patriottica*, 1914

markedly different direction from the one in which Marinetti had been leading it is clear even from the title of his work. Connotations of 'celebration' and 'patriotism' are much more closely related to the realm of feelings and ideas than much mainstream Futurist shaped typography is, with its emphasis on the visual object.

Another salient 'point of contact' between *Danza serpentina* and *Festa patriottica* is also signalled in the title of Carrà's work. Although, in his letter to Severini, Carrà gives the piece the generic sub-title 'poema pittorico', the *Lacerba* version places greater emphasis on the generic element than on the theme itself, even to the point of having the upper-case caption 'free-word picture' printed first: 'DIPINTO PAROLIBERO (Festa patriottica)'. It is clear from the background correspondence that Marinetti was instrumental in this formulation. In his letter to Soffici of 21 July 1914 Marinetti explains that he needs the photograph of Carrà's collage as soon as possible, so that it can be run in an August number of *Lacerba*. 'The urgency', he explains (perhaps disingenuously, since he may well have seen the issue of *Les Soirées de Paris* of 14 June containing Apollinaire's 'Lettre-Océan'), 'is due to our desire to establish from the outset that Carrà and Severini are the first to have achieved this absolutely original fusion of plastic dynamism and free-word poetry,' adding that this should prevent Delaunay and his ilk from poaching the technique and then claiming the discovery for Cubism/Orphism (*Archivi*, i. 341 f.).[14] 'Don't forget, I beg you,' he goes on, 'to put beneath the photograph . . . precisely the following caption: CARRÀ—DIPINTO PAROLIBERO (FESTA PATRIOTTICA).' Which he did, although he actually went so far as to place the second part of the title in lower case, so as to render the thematic component even less important. 'We also agree with Severini', Marinetti adds, 'that we should always advertise such fusions of Futurist painting and words-in-freedom in such a way as to avoid confusions and to influence the trend in this direction.'[15] One can see from this how he views the new 'literary' collage.

Despite Marinetti's general enthusiasm for overstating the innovativeness of Futurist discoveries, the works by Severini and Carrà *do* represent something new in the field of mixed-media experimentation, a fusion of free-word poetry and the kind of non-verbal collage already familiar by this time. And this innovative dimension—of intrinsically *literary*, word-orientated collage—is reflected in such hybrid terms as 'dipinto parolibero' and (a less helpful notion)

[14] The rivalry on the issue of priority was no mere figment of Marinetti's imagination. 'I was the one who began this kind of poetry', Apollinaire was to claim later (*Œuvres complètes*, ed. Michel Décaudin (Paris, 1966), i. 778).

[15] *Archivi*, i. 342. It is hard to find much evidence that Marinetti and his fellow Futurists adhered to this promotional resolution. It was, after all, Marinetti who had replaced Severini's title 'Composizione di parole e forme (Danseuse = mer)' by the words *Danza serpentina* (see below, p. 83). And significantly, when *Danza serpentina* was reproduced in Severini's *Témoignages* in 1963, it was given an exclusively generic title: 'Recherche d'une expressivité en unissant formes et mots (1914)'.

'poema pittorico'. The generic label is instructive in both of these cases, for the works are much more to be *read*, like poems, than is the case with most Cubist collages.[16] As the following interpretations will show, the constituent words are extremely important—which may possibly explain why Carrà, though mindful of the loss such reproduction would entail, was nevertheless willing to have *Festa patriottica* photographed for printing in *Lacerba*—for the resultant impoverishment could weaken the collage's propagandistic impact only to a negligible degree. What is more, since monochrome reproduction often has the effect of highlighting the words, it virtually converts *Festa patriottica* into the kind of black-and-white 'experiment in expressiveness uniting shapes and words' (as Severini was later to call it[17]) that one has with *Danza serpentina*.

Danza serpentina and *Festa patriottica* have so far been presented as if they were by and large similar examples of some new Futurist type of collage. It is now time to consider the very different possibilities inherent in the genre which they also represent.[18]

It was *Danza serpentina*, above all, that showed the way to the expressive potential inherent in Futurist collage's new 'unification of words and shapes', to use Severini's own phrase; and the most detailed and informative commentary on this work remains that by Severini himself in his 'Futurist Ideography' ('Idéographie futuriste' (1916)). Since this article has received little attention in discussions of Futurist collage, I shall quote from it in some detail.

Speaking at first in general and occasionally eccentric terms, Severini explains that 'the way of suggesting the intense realism of our emotions with new techniques of synthetic lyricism was suggested to me by my free-word drawing [*Danza serpentina*], in which I freely harmonized acute forms and acute angles with particular words, round forms and obtuse angles with other words'.

16 As the following commentary on Braque's *Bottle, Newspaper, Pipe and Glass* (1913) indicates, there are occasions when the printed word is meant to be read and to contribute to a Cubist collage's meaning: 'A trace of macabre humour is perhaps identifiable in the piece of newspaper at the bottom on the left where, framed between advertisements for Houbigant's perfume "Fougère Royale" in the form of soap, and a brand of "Violette de Parme" perfume, is a reportage on "Le Parricide de Montreuil"' (Douglas Cooper and Gary Tinterow, *The Essential Cubism. Braque, Picasso & their Friends* (London, 1983), 98). But the more typical speculation in the same volume about whether the extract from *Le Figaro* in Picasso's *Guitar, Newspaper, Glass and Bottle* (1913) is there 'because it carried some special significance for him, or perhaps simply because it had turned an appealing colour with age' (ibid. 290) is revealing: even when a printed component *can* be read, it is invariably less thematically central to a Cubist work. Thus, in Picasso's *Siphon, Glass, Newspaper and Violin* (1912), the newspaper component, cut to form the shape of the siphon, glass (and a lamp?), is semantically less important than the single word 'Journal' (which plays with the contrast between the word signifying 'Newspaper' and the actual newspaper-fragments standing for other objects).

17 Severini's unease about Marinetti's caption is suggested by the fact that nowhere in *Témoignages* does he mention the work by the title under which it is best known.

18 Some of the differences between the two works are no doubt functions of genre: *Danza serpentina* appears to have been conceived specifically for reproduction in the pages of *Lacerba*, whereas *Festa patriottica* began life as a unique multi-coloured work of art.

From this unique experiment Severini draws a conclusion destined to govern the form of a number of his later collages: 'Just as in painting we have colours and shapes which, for example, synthesize spring in its entirety, or simply joy, or heat . . . so one can equally find in literature synthetic means of uniting two groups of lyrical elements, or dynamic literary continuities.' Clearly, Severini sees his collages as belonging as much to literature as to the visual arts.

Of particular interest is the supporting analysis of *Danza serpentina*:

in the 'free-word' drawing in question, the sensations of luminous penetration coming from two electric reflectors, to which are attached, by analogy, the sensations of a flashing globe rending the air with its noise, are rendered by the contrast of two acute angles meeting at their apexes.

All the analogical sensations attached to this idea of brutal penetration can be rendered by two acute angles.

In this same part of the drawing, for the onomatopoeic szszsz . . . expressing the noise of the hissing flashing globe and the rending of the air by the sheaf of violent light, I wrote the letters with strokes of the pen and set in continuous writing. For me these lines had an intensity value (a quantitative one), just as I obtain the plastic intensity by brush-strokes using analogical colours within a shape. If one takes this intuition further, it becomes necessary to construct completely with signs expressing quantities in qualitative terms. In the end, these signs will become conventionalized, in order to allow the work to be understood, but then, are not numbers, letters, the sub-divisions of time and space also conventional?

Severini concludes by presenting *Danza serpentina* as the precursor of a new kind of literary/visual art (the status any self-respecting Futurist would no doubt gladly ascribe to his discoveries):

It will evidently be impossible to read the works produced by this new lyricism, but are we able to read our paintings, can the musician read his symphony, can one read the song of the birds?

The same lyrical work, with the images it proposes, will speak to eyes informed by universal ideas and not by sound belonging to spoken language. We will leave the latter to the lawyers (the art of oratory) and to comedians.

Actually, with the laws of simultaneity that we have discovered in ourselves, the word *sound* no longer means anything. While the particular sensations provoked in us by these hammer-blows are able to create in our *sensibility* groups of analogical *ideas* (and in the case of painters, groups of aspects) these have nothing to do with the sound or with the boom-boom, the impressionist expression of a sound. It is thus that I envisage the possibility of creating a new ideography which would be the graphic, geometrical expression of universal ideas.[19]

Severini returns to *Danza serpentina* in his later article 'Symbolism in the Plastic Arts and Literary Symbolism' ('Symbolisme plastique et symbolisme littéraire' (1916)), though many of the points he makes here are dependent on

[19] *Témoignages*, pp. 38 f.

the fact that the collage is now referred to by its original title of *Danseuse–mer*. (In 1914 Marinetti had pressed him to abandon this formulation in favour of *Danza serpentina*,[20] but by 1916—when his article first appeared in the *Merçure de France*—he had obviously decided to revert to the original caption.)

Writing now of the 'ideographical' problems associated with rendering 'the idea–image of a female dancer', Severini cites his *Danseuse – mer* as an illustration of a new (quasi-literary) analogical method—a transference of the workings of the analogy-generating 'wireless imagination' from free-word poetry to collage:

The contrasts of shapes and colours belonging to the perception of the *danseuse*-reality can be recaptured, by means of affinity or contrast, in the spiral flight of an aeroplane or the rolling of the sea.

These are the two points of the comparison, *Dancer = sea*,[21] comprising a maximum amount of universal life.

The analogy is not some generalization based on abstraction, but an element of realistic and specific intensification.

It is employed systematically in the 'synthetic lyricism' of Marinetti and the Italian Futurist poets. . . . We are technically able to achieve a corresponding aesthetic in the plastic arts, and can express, by means of a square shape and the colour blue, an idea–sensation produced in us by a reality that appears to us in the form of a round shape and the colour yellow. But the possibilities of our plastic realization require the two terms of the comparison, the point of departure and the arrival-point of the idea–sensation.

The unity of time and space being definitively destroyed in the tableau, the picture will be able to capture those realities which do not belong either to the same atmospheric milieu, or to the same moment of perception, or to any visual logic. These distant or contradictory realities will be joined together solely by our thoughts and our sensibility.[22]

Severini might have added that they may also largely be the product of a certain kind of formulaic title,[23] rather than simply the evocative interrelating of words and shapes on the canvas.

[20] In *La vita di un pittore* (Milan, 1965), 210, Severini states that it was Marinetti who chose the present title, the poet's own suggestion being at the time 'Composizione di parole e forme (Danseuse = mer)'. Marianne Martin suggests that with *Danza serpentina* 'perhaps Marinetti alluded to Loie Fuller's famous dance of that name' (*Futurist Art and Theory, 1909–1915* (Oxford, 1968), 198). Certainly, the two titles have caused their share of confusion; for example, the reproduction in Wescher (*Die Collage*, p. 60) bears the caption *Danseuse = mer*, although the work is discussed in the body of the text as *Danza serpentina*.

[21] Severini's substitution of an equal sign for the original dash is an unfortunate move, for equation now replaces the associative distant analogy of the 'telegraphic imagination'.

[22] *Témoignages*, p. 52.

[23] On the relationship between montage/collage and allegory, see Annegret Jürgens-Kirchhoff, *Technik und Tendenz der Montage in der bildenden Kunst des 20. Jahrhunderts: Ein Essay* (Giessen, 1978), 215. Both Walter Benjamin (in *Der Ursprung des deutschen Trauerspiels* (Frankfurt a. M., 1928)) and Peter Bürger (*Theorie der Avantgarde* (Frankfurt a. M., 1974)) have emphasized the parallel between the use of heterogeneous materials with conceptual connections in both allegory and modernist collocations.

What Severini is pinpointing in this further discussion of *Danza serpentina* might be regarded as a secondary stratum of the work's literariness: i.e. the use of an analogical descriptive method familiar from the practice of the Futurist poets. Yet, despite his emphatic claims to the contrary, the equation 'dancer = sea' is hardly the wished-for 'element of realist intensification', but rather a 'generalization based on abstraction', the kind of allegorizing analogy-obsession from which Marinetti's poetry often suffers. So the alternative title, *Danza serpentina*, is not without its merits. It denotes on one level a specific contemporary dance; yet, because that dance's title is itself analogical (dance : serpent's movements), it can readily mediate the 'idea–sensation' that Severini wishes his collage to convey.

It could of course be objected that, strictly speaking, *Danza serpentina* is not a collage at all, despite the fact that it usually figures in accounts of the Futurist contribution to the genre. Nothing is actually *stuck* to the canvas. Instead, the work is, in Wolfram's words, 'a lettered drawing on which the scheme for colours and for concepts such as "warmth", "fragrance", "profundity", and "silence" were all written out in curves in a centrifugal movement, and these forms were accompanied by the onomatopoeic noise of "TTATTATTA"'.[24] Herta Wescher for this reason categorizes *Danza serpentina* somewhat anachronistically as a 'lettrist *drawing*', even though it figures prominently in her historical survey of the collage-genre.[25] In his autobiography, Severini singles out this work as a particularly successful example of the use of superimposed words in a strikingly different way from that in his paintings *Ritratto di Paul Fort* (*Portrait of Paul Fort*) and *Ritratto di Marinetti* (*Portrait of Marinetti*). Words are now employed, he says, highlighting a very important feature, 'not only pictorially, for their line and tone values, but also for their essential meaning [by contrast, that is, with Cubist collage's use of word-fragments], so that a sort of drawn poem was created, in which I used Marinettian words-in-freedom'.[26]

Welcoming *Danza serpentina* at the time, Apollinaire (according to Severini) drew the painter's attention to significant precursors in shaped poetry of the seventeenth century. And Severini himself, in his autobiography, sees the work in the context of Mallarmé's *Un coup de dés* and Apollinaire's *Calligrammes*, which, he stresses, were only published some time later. (This may be true of the majority of the *calligrammes*, but conveniently ignores the appearance of 'Lettre-Océan' in *Les Soirées de Paris* by that time.) But Futurist applications of *papiers-collés* were not predominantly mimetic in the spirit of the calligrammatic, shaped-poetry tradition—nor in the way that the 'designed analogy' appeared to be to most contemporary readers. Futurist collages are constructed

[24] *History of Collage*, p. 42.

[25] *Die Collage*, p. 59. For historical details of Marinetti's involvement in the retitling of the work, see Severini, *La vita di un pittore*, p. 169.

[26] *La vita di un pittore*, p. 170.

according to a principle of *simultaneity*; they exploit the art of a-perspectival juxtaposition (a point which will be developed in more detail later). This is not true of the overwhelming majority of figured poems, which still work with the syntax of perspective (or what Severini calls 'aspect').

When reproduced in a journal like *Lacerba*, *Danza serpentina* ends up all but indistinguishable from collages, in the strict sense of the term, such as *Festa patriottica*, which are also illustrated there. By and large, photographic reproduction has the effect of blurring any qualitative difference between the use of actual *papiers-collés* and an impression of superimposed fragments of shapes and words created through graphic effects. But there are more compelling reasons than this for deciding in favour of a permissive use of the term 'collage' in the context of literary Futurism, one wide enough, that is, to embrace the techniques of both of these types of works.

First, there is the general historical fact that the term 'collage' has always possessed two connotations: a narrow (literal) and a wider (metaphorical) one. Although William C. Seitz has pleaded in favour of the narrow sense only, proposing that the word should be reserved exclusively for 'paper collages of the cubist movement',[27] his is a lone voice. Ulrich Weisstein's *Forschungsbericht* on the subject demonstrates incontrovertibly that it is not, and cannot be, meaningful to delimit the term thus—either historically just to the Cubist period or technically to any actual gluing process. In reality, 'collage' can refer either literally to the product of a gluing technique, regardless of the particular material(s) applied, or it can enjoy an umbrella-connotation signifying 'an assemblage of diverse fragments drawn from reality and juxtaposed on an essentially flat surface',[28] or, what is more important, a semblance thereof, irrespective of the specific method employed to create such an impression. (*Danza serpentina* is a prime instance of the juxtaposition of heterogeneous elements, words, and drawn shapes, on a flat surface, but it is not an example of glued-on collage material, in the literal—etymological—sense of the word.)

Louis Aragon wrote in retrospect about 'what is, by way of simplification, referred to as collage, even though the use of glue [*colle*] is not one of the characteristics of this operation, and is even not an essential feature'.[29] Similarly, Max Ernst recalls: 'in the majority of my collages, glue was not utilized at all.'[30] Weisstein assumes that this displacement of the literal

[27] *The Art of Assemblage* (New York, 1961), 150.

[28] 'Collage, Montage, and Related Terms: Their Literal and Figurative Use in and Applications to Techniques and Forms in Various Arts', *Comparative Literature Studies*, 15 (1978), 126.

[29] 'La Peinture au défi', *Les Collages* (Paris, 1965), 42.

[30] 'Quelle est la technique du collage?', in 'Au-delà de la peinture', *Écritures* (Paris, 1970), 257. Perloff quotes Max Ernst's 'Si ce sont les plumes qui font le plumage, ce n'est pas la colle qui fait le collage' ('Au-delà de la peinture', p. 256), her point being that such use of glue is not new, but that, in contrast to 'the "paste-ups" of the nineteenth century', 'collage always involves the transfer of materials from one context to another' (*The Futurist Moment*, p. 47). But a work like *Danza serpentina*

(etymological) connotation of 'collage' by the figurative one took place 'on the part of Max Ernst and his fellow Dadaists/Surrealists'.[31] But, as is abundantly clear from the present material, a blurring of any meaningful distinction between the two main forms of the genre had already come about—in both theory and practice—substantially earlier, during the first (so-called 'heroic') Futurist period. Thus, when Carrà wrote to Severini in 1914 claiming to have created something very similar in essence to *Danza serpentina*, he took no account of the fact that his own work was literally a collage using the gluing technique, whereas Severini's was not. Evidently, it is the *structure*, the simultanéiste principle underlying the works' image-juxtapositions and their use of heterogeneous materials, that interested him, not the technical means by which such an effect had been created.

The other principal reason for not confining oneself to a strictly etymological concept of 'collage'—or seeing it in terms of Weisstein's in many ways attractive binary opposition of 'art' and 'life'[32]—lies in the nature of the various Italian experiments. Since the Futurists were trying to enlarge the repertoire of techniques available in the arts, even to the point of transcending conventional generic and art-form distinctions, it would go very much against the spirit of these adventurous mixed-media activities to put them back into too procrustean a conceptual framework, be it binary or etymological. Futurist literary collage was able in practice to embrace such diverse forms as the fusion of actual *papiers-collés* and tempera (as in *Festa patriottica*), a mixture of drawn shapes and words (*Danza serpentina*), the use of painted letters and real ones (Soffici's *Tipografia*), the juxtaposition of fragments of print with other literally stuck-on objects (*Ritratto di Paul Fort*), or the combination of hand-writing and oils (such as one finds in Carrà's *Ritratto di Marinetti*).[33] And, even though I am

can also reinforce Ernst's point in another respect—by showing that it is possible to have a collage-effect without the literal application of glue.

[31] Weisstein, 'Collage, Montage, and Related Terms', p. 126.

[32] Having diagnosed the confusion surrounding the terms 'collage', 'montage', 'assemblage', 'objet trouvé', and 'ready-made', Weisstein outlines a typology based on a notional axis between art and reality: 'The criteria for assigning a given art form or technique its place along the spectrum will . . . depend on the way in which that dichotomy is maintained, modified, or dialectically resolved. . . . the *collage* in which art and life literally intersect and inter-penetrate, is located closest to the center of the imaginary spectrum, whose polar extremes are occupied by raw nature/brute reality palmed off as art, on the one hand, and by art which claims to be nothing but art, on the other' ('Collage, Montage, and Related Terms', p. 132). Collage, then, occupies the middle ground within such a range of possibilities. However, establishing such a spectrum according to the portions of a work 'which are derived from an external *reality*' (p. 135) is not the most useful approach to Futurist collage. For here, in the synthesizing of the arts, the dichotomy is often in terms of the elements which are derived from one genre (e.g. painting) and another (often words-in-freedom). It would be misleading to suggest that the excerpts from *Lacerba* which appear in Carrà's collage or the newspaper cuttings which are inserted into Boccioni's *Carica dei lancieri* (discussed below) are simply taken from something arbitrarily labelled 'reality'.

[33] The Futurists experimented in all the main areas of collage discussed in Helen Hutton's useful

trying to argue that a readable component plays an important part in many Futurist literary collages, there are other works in a more Cubist vein, such as Rosai's *Scomposizione di una strada* (*Street Composition*) and Sironi's *Motociclista* (*Man on a Motorcycle*), where newspaper elements are incorporated but positioned in such a way as to make their contents thematically unimportant. Manifestly, there is no hard-and-fast criterion for distinguishing unequivocally between a *dipinto parolibero* or *literary* collage, on the one hand, and a *painting* whose effect has been enhanced by the substantial addition of collaged written/printed material, on the other. My guiding principles in the present exploration of Futurist literary collage are twofold: first, the written material must be inserted to be read (and not simply encountered as a piece of alien print applied for painterly effect to a work of art[34]); and, second, the content of the written material must substantially contribute to the interpretation and the impact of the collage as a whole. No doubt these two interrelated criteria will not suffice to solve all forms of demarcation-dispute which can arise. But then, typological-demarcation problems of this kind were hardly of the essence of most Futurist synthesizings and 'fusions', as they liked to call them. However, such guidelines should enable us to focus on the particularly literary (or free-word) tendency within the Italian Futurists' use of *papiers-collés*.

Carrà's *Festa patriottica* is the most famous example of the new literary dimension to Futurist collage. Its verbal element is self-evidently there to be read; indeed the work is widely assumed to be part of a Futurist propaganda campaign—even if the extent to which it is an expression of Interventionist sentiment has been the subject of some disagreement.

The collage, which has been acclaimed as 'the most Futurist of Carrà's works',[35] is both visually and semantically a rousing piece, one that, in one critic's words, 'seems to go off like a siren'.[36] With its 'packed symbolism' and patterns of 'centrifugal expansion', it may give an initial impression of being 'more explosion than structure',[37] but, as Marianne Martin amply demonstrates in her now classic analysis, such an effect is actually achieved by means of

survey *The Techniques of Collage* (London, 1968), including assemblage, déchirage, photograms, photomontage, and *papiers-collés*. Too often, though, the art of assemblage is so fully associated with Dada that the Futurists are not credited with having explored this form as well. Tellingly, the prospectus-description of Kruchenykh's and Alyagrov's (i.e. Roman Jakobson's) *Zaumnaya gniga* (Moscow, 1916) on microfilm describes it as having a 'Dadaist cover' because there is a 'real button attached' (Chadwyck–Healey, *Russian Futurism, 1910–1916: Poetry and Manifestoes* (Cambridge, n.d.), 4). Yet Futurist experiments in this idiom had already occurred by 1916.

[34] This is close to being a paraphrase of Severini's distinction (in his autobiography) between words of a collage functioning 'pittoricamente' according to their 'line and tone values' and what he, less satisfactorily, refers to as their 'essential significance'.

[35] Martin, *Futurist Art and Theory*, p. 194.

[36] Joshua C. Taylor, *Futurism* (New York, 1961), 110.

[37] Harriet Janis and Rudi Blesh, *Collage: Personalities, Concepts, Techniques* (Philadelphia, 1967), 38 f.

'a controlled and carefully considered pattern and series of sub-patterns'.[38] These are all well accounted for by her. What she does not explore, however, is the feature that so distinguishes this collage from its Cubist equivalents: its literary, or, to be more specific, its political element.

Surprisingly, given the work's acknowledged calibre, the only detailed interpretation of it appeared relatively recently, in Willard Bohn's 'Circular Poem-Paintings by Apollinaire and Carrà', which is primarily an attempt at demonstrating the influence of 'Lettre-Océan' on *Festa patriottica*, although it also has an impressive number of informative, and at times contentious, things to say *en route* about the nature of Carrà's collage.

Bohn's approach emphatically denies *Festa patriottica* any immediate political relevance, arguing that the collage's alternative title (*Manifestazione interventista*) is espoused only 'by those who think the work represents a demonstration in favor of Italy's intervention in World War I, thus pre-supposing a date of August or later' for its composition. Bohn, however, claims that 'detailed analysis fails to reveal a single reference to the war' in the work, or 'for that matter to violence of any kind'. (Presumably, this must mean with the exception of the second, Interventionist title, which is not some wilful fabrication on the part of later commentators, but simply another example of the Futurists' penchant for retitling their works to suit changed circumstances.)[39] On these grounds, Bohn concludes that 'the subject of the celebration is never specified',[40] though we do know that it centres on Italian patriotic fervour.

There are various aspects to this account of the collage's historical significance, or lack of it, which remain open to question: i.e. the status of the second title, the relationship of any political content to the date of composition, the pressure of contemporary context on the interpretation of some of the written elements in the work, and the extent to which the collage represents a visual picture of an actual event or not.

To claim, as Bohn does, that the title *Manifestazione interventista* is employed exclusively by those who have decided *a priori* that 'the painting represents a demonstration in favour of Italy's intervention' in the war, as if this were some impermissible attempt at retrospectively politicizing Carrà, is misleading.

[38] *Futurist Art and Theory*, p. 194. Wescher describes the work as 'an abstract collage of a completely different kind', but she expresses the surprising opinion that the words employed in it are completely 'meaningless' and hence the work is ultimately 'a genuine sound poem [*Lautgedicht*]' (*Die Collage*, p. 61).

[39] According to Enrico Crispolti (*Il mito della macchina e altri temi del futurismo* (Trapani, 1969), 551), Balla painted a cycle entitled *Dimostrazioni interventiste* in 1915. In addition to the obvious political concerns of the moment, this may well have influenced Carrà in his title-change.

[40] 'Circular Poem-Paintings by Apollinaire and Carrà', *Comparative Literature*, 31 (1979), 246–71. Bohn includes much of this interpretation in his generally excellent *The Aesthetics of Visual Poetry, 1914–1928* (Cambridge, 1986), 24–8. The only other detailed discussion is in Alan Windsor, 'Apollinaire, Marinetti and Carrà's "Dipinto parolibero"', *Gazette des beaux-arts*, 89 (1977), 145–52.

Joshua Taylor is certainly right to point out that it is largely in recent years that the work has come to be known under the Interventionist title.[41] But this is the only discernible pattern to the use of the two titles, apart from the obvious fact that Carrà first called it *Festa patriottica* (saying in his letter to Severini that its subject was 'the plastic abstraction of a civic tumult') and only subsequently renamed it *Manifestazione interventista*.[42] To opt for the second title may be no more than a practice comparable to the time-honoured scholarly principle of attributing paramount significance to the *Ausgabe letzter Hand* (the final edition that an author saw through the press). Many with no particular political axe to grind have presumably opted for the second Interventionist title in that spirit (and this includes Massimo Carrà, in his capacity as editor of his father's collected works: *Tutti gli scritti* (Milan, 1978)). Bohn's examples (René Juillian and Max Kozloff) are speciously selective, in this respect. The most that one can conclude is that it is historically quite plausible for the collage to have acquired a far more explicitly political connotation in the war period. I do not think one has, in any case, to choose between the two titles; it is possible to see the work existing in two forms, though I shall nevertheless want to plead that the collage is more political than Bohn's reading allows for, whichever of the two titles it bears.

To assume that the political sentiments that had crystallized around Interventionism did not come into being until the actual beginning of the war in August 1914 is a historical oversimplification. The fact that the collage can be assigned, as Bohn shows beyond dispute, to 'the period June 15—July 9, 1914'[43] proves little as far as possible patriotic warmongering associations are concerned. The longstanding Irredentist cause, to which Marinetti had often lent his voice, together with the patriotic manifestations celebrating the Cinquantennio throughout 1911,[44] had generated a highly charged political atmosphere tinged with anti-Austrian sentiments and much sabre-rattling because Austro-Hungary had recently annexed Bosnia and Herzegovina. And, in general, this was a period when strident nationalism, displays of anti-Teutonic feelings, and pro-war sentiments (whether the war be in Libya, the Balkans, or the whole of Europe) would have been extremely difficult to disentangle. The Interventionism of 1914 was the culmination of this atmosphere, not some separate phenomenon. Admittedly, the Futurists' own directly Interventionist activities did not commence until September 1914. On 16 September Boccioni writes to his family that he has begun demonstrating for Italy's participation in

[41] Taylor, *Futurism*, p. 145.

[42] The collage was made in July 1914 and published in *Lacerba* (1 Aug. 1914, 233), just two weeks before the beginning of the war. I have been unable to trace the exact point at which Carrà's work was retitled.

[43] 'Circular Poem-Paintings', 255.

[44] For details, see Richard Bosworth, *Italy and the Approach of the First World War* (London, 1983), 71.

the war, despite the fact that she is nominally still a member of the Triple Alliance (*Archivi*, i. 346). And, in another letter of 19 September to his mother, he refers to the first Futurist 'Interventionist manifestation' of 15–16 of that month, organized, as he proudly reports in a later letter, 'by Marinetti and myself'.[45] Nevertheless, to see Interventionism only in terms of the Futurist organization of public demonstrations, as Marinetti himself occasionally does,[46] or to assume that political rallies involving an element of campaigning for Italian involvement only occurred, thanks to the Futurists, in September, is to misrepresent the historical situation. The signal date for Interventionism is not August 1914, but July of that year, when Italy proclaimed her neutrality[47]—and this, possibly not coincidentally, is the time when Carrà began his collage. In fact, it is the nine months between this declaration of neutrality and May 1915, when Sidney Sonnino signed a treaty with the Entente powers undertaking to enter the war within a month, that mark the true duration of the Interventionist cause and the Futurists' involvement in it.

In any case, the underlying assumption that the role of a Futurist work of art is merely to document retrospectively events happening outside, in the political arena, falsifies the future-orientation of Futurism. As one sees from 'La Guerre électrique' onwards, the Futurists saw themselves as the creators and galvanizers of the world around them, not simply slavish chroniclers of what was already happening in it. Hence Carrà's collage is itself, in deed and in spirit, a form of political 'manifestation', and may well have preceded the sort of public event which its second title more explicitly identifies.

Actually, in spite of any claims to the contrary, there are a number of allusions in the collage itself—regardless of which title it bears—which fully support a political reading of the work. The inclusion of the newspaper name *Corriera della sera*, for example, is not fortuitous, in this political context. For, under the tutelage of its then editor Luigi Albertini, a leading Interventionist,[48] the paper had rapidly established itself as one of the major organs disseminating propaganda in favour of Italy's entry into the war. Therefore, Max Kozloff's

[45] Letter of 22 Sept. 1914 (*Archivi*, i. 347). According to 'I futuristi primi interventisti' (*TIF* 519), various demonstrations took place in Milan, at the Teatro Dal Verme and on the Piazza del Duomo, and were followed, according to *L'Italia* of 17 Sept., by a number of arrests. The most detailed description of these events, from a Futurist perspective, is in Cangiullo, *Le serate futuriste* (Milan, 1961), 145–8. See also Pratella, 'Il Futurismo e la Guerra' (*TIF* 482–9, esp. 486 f.), and, for the wider background, L. Pivano, *L'Interventismo 1914–15* (Milan, 1965).

[46] In his introduction to *Les Mots en liberté futuristes* (Milan, 1919), Marinetti asserts that 'the Futurists are proud to have organized the first two popular demonstrations against Austria, on 15 Sept. 1914, in Milan, in the midst of Italian neutrality' (p. 10).

[47] For details, see Giuliano Procacci, *History of the Italian People* (London, 1970), 331.

[48] See Procacci (ibid. 331) and Bosworth (*Italy and the Approach of the First World War*, p. 32). Perloff (*The Futurist Moment*, p. 63) suggests that 'the collage ironically juxtaposes the Futurists' radical newspaper *Lacerba* to the bourgeois *Corrière della Sera* [*sic*]'; but there are good Interventionist reasons for the placing together of these otherwise unlikely bedfellows.

conclusion that Carrà's collage 'was expressly designed to exacerbate the volatile political climate in an Italy internally vacillating between the adversaries of a war known to be most imminent', that it is a piece of propaganda 'in favor of intervention on the side of the Allies',[49] is not as distortingly anachronistic a reading as Bohn would have us believe, although he is quite right to take issue with certain political readings of the work, notably René Juillian's unwarranted claim that the collage is political by dint of actually being made of 'fragments of posters and of proclamations in favour of Italy's intervention into the War'.[50] As we shall see, this is demonstrably not the case.

Bohn's point that there is no mention of the Allies and that the subject is, according to Carrà's letter to Severini, merely a 'civic tumult'[51]—whatever in the Italy of the period that might be, in non-political terms—need not prevent the occasion depicted from having very pointed political overtones, especially given the belligerent climate of the pre-war years and various other internal pointers which a careful reading of the collage furnishes.

In Bohn's view, 'the closest we get to politics is with the small, upside-down Italian flag at the bottom inscribed "Trieste Italiana Milano"'. 'This', he explains, 'is a reference to the Irredentist party, which first became prominent in 1878, advocating the incorporation into Italy of neighbouring areas with large Italian populations, such as Trieste.'[52] In fact, the inscription is much larger than most of the writing around it, and, what is more, it is written on the Italian flag—a strange place for a patriot to bury an allusion! Moreover, being one of the few inverted inscriptions in this part of the collage, it is—as many modern advertisers have evidently also decided—more likely to attract the beholder's attention than lose it.[53] In any case, if by 1914 it was generally no longer easy to draw any hard-and-fast distinction between the kind of patriotic emotions generated by the longstanding Irredentist cause and the more immediate concerns of the Interventionists, this was especially true in the case of the Futurists, who had their fingers well and truly implanted in both political pies.

As part of a nowadays fashionable process of sanitizing Futurism and making

[49] *Cubism/Futurism* (New York, 1973), 207–9; cf. Bohn, 'Circular Poem-Paintings', p. 266.

[50] *Le Futurisme et la peinture française* (Paris, 1966), 101.

[51] Bohn, 'Circular Poem-Paintings', pp. 266 f.

[52] Ibid. 267.

[53] Since Bohn elsewhere in his reading argues that 'we are viewing the spectacle *through a pilot's eyes*' as he circles above the celebration and that the words 'can only be read by someone inside the plane's spiral'—i.e. the spectator is put, in archetypally Futurist fashion, at the centre of the picture—this dismissal on the grounds of an element being upside-down seems too passéist a counsel of despair for his generally ingenious reading of the work's structure. In addition, it is clear from Marinetti's description of the first Interventionist rally of September 1914 (quoted in Cangiullo, *Le serate futuriste*, pp. 145 ff.) that 'Evviva L'Esercito', 'Abbasso L'Austria', and variants of the same Irredentist sentiments that had been in circulation for decades, were by this time doubling up as the rallying cries of Interventionism.

it generally respectable for scholarship,[54] the a-political reading denies the collage any suggestion of warmongering either by misconstruing or making metaphorical a number of other elements in the work. Thus, Bohn reasons that 'Evviva L'Esercito' is 'really not war-oriented', but 'simply reflecting general patriotic feelings'[55]—as if people were not quite aware in 1914 what armies were for! The 'ZANG UMB TUUM', in the top left-hand corner, an elliptical quotation from Marinetti's Balkan war epic, where it represents gunfire, is now, in Bohn's reading, no more than a 'metaphor' and 'may represent fireworks in the context of the celebration'.[56] Yet one doubts whether any self-respecting Futurist would have preferred to celebrate anything with fireworks rather than with gunfire! The reference to Austro-Hungary over the Odol sign is glossed as 'probably . . . a newspaper headline',[57] but the crucial issue is what kind of newspaper headline such an allusion would suggest in 1914, given the heated contemporary political climate. From the publication of *Le Futurisme* onwards, the Italian Futurists had adopted a consistently aggressive anti-Austrian stance, particularly in connection with the Interventionist campaigns. The word 'TOT' (which Kozloff, and in his wake Perloff, assumes to be the German word for 'dead' and hence presumably an indication of Italy speaking to the enemy and telling them what state it wishes to help them attain) is correctly identified by Bohn as 'the name of a prominently advertised anti-heartburn pill manufactured in Milan', although there are more local associations to this word than such a surface gloss would suggest.[58]

[54] Manfred Hinz fires off a salvo of damaging charges, in this connection, in the introduction to his *Die Zukunft der Katastrophe: Mythische und rationalistische Geschichtstheorie im italienischen Futurismus* (Berlin, 1985), 1–17.

[55] Bohn, 'Circular Poem-Paintings', p. 267.

[56] Ibid. 266.

[57] Ibid. 268.

[58] Ibid. 267. The reference, which Bohn does not explain, becomes clearer when viewed in the light of Boccioni's *Pittura scultura futuriste*. There, 'Against Landscape and Old-Fashioned Aesthetics' contrasts the attractions of a landscape of advertisements with the passéist cult of Arcadia. 'Have you never experienced the joy of seeing a melancholy line of sentimental cypress trees transformed by a long, violet and insolent inscription: "ACQUA PURGATIVA INFALLIBLE" . . . ? or the green slopes of a gentle hill riddled by numerous pistol-shots

TOT . . . TOT . . . TOT . . .

There are possibilities of "landscape" everywhere' (*Gli scritti*, p. 84). The idea is taken up in the reference to Tot pills in Soffici's poem 'Passeggiata' ('Walk') (*Lacerba* (15 Feb. 1914), 54), which Bohn therefore assumes is being quoted in Carrà's collage; and Marinetti's manifesto 'The Variety Theatre' ('Il teatro di varietà' (1913)) refers to the Giaconda Acqua Purgativa (*TIF* 78). There is also a self-referential dimension to these words, in that the Futurists were fond of associating themselves with purgatives, caffeine, and other metaphorical forms of cleansing and invigoration. Witness the advertisements for the *Almanacco purgativo* on the back pages of *Lacerba* and Marinetti's quotation of the opinion that he is 'la Caffeina dell'Europa!' in *8 anime in una bomba* (*TIF* 802); also the link between the impact of Futurism and both Tot pills and 'les pillules de Pink'. These must have been early associations, for in 1909 the Turin journal *Pasquino* published Manca's cartoon depicting a Futurist parade. In it, the Futurists, dressed up as clowns, are carrying placards bearing the words 'Tot' and 'Pink'. (A reproduction can be found on the back cover of Mario Verdone, *Teatro italiano d'avanguardia: Drammi e sintesi futuriste* (Rome, 1970)).

Helpfully, Bohn identifies seven main sources for Carrà's collage, which, he claims, account for 98 per cent of the printed material in it. They are:

Marinetti's poem 'Correzione di bozze + desideri in velocità' (*Lacerba* (1 Dec. 1913), 269–71),
Boccioni's 'Uomo + vallata + montagna' (*Lacerba* (1 Feb. 1914), 44–5),
Carrà's own '1900–1913. Bilancio' (*Lacerba* (1 Feb. 1914), 38–9),
Gustave Fivé's 'Sports' (*Lacerba* (1 Feb. 1914), 46),
Marinetti's 'Dune' (*Lacerba* (15 Feb. 1914), 51–3),
Soffici's 'Passeggiata' (*Lacerba* (15 Feb. 1914), 54), and
Cangiullo's 'Serata in onore di YVONNE' (*Lacerba* (15 June 1914), 188–92).

Also, apart from these literary *papiers-collés*, 'the piece of sheet music is part of a composition by Pratella (*Lacerba*, Feb. 1, 1914)',[59] namely 'Due pagine di musica futurista' (40–1).

Bohn's main purpose in identifying and enumerating these sources is to counter the alternative (Interventionist) reading of the collage:

> The fact that the *papiers collés* were nearly all snipped out of poems published in *Lacerba* makes it possible to identify their sources. This in turn permits us to reconstruct incomplete phrases and to determine exact meanings by consulting the original text. Thus, for example, the sounds 'traak tatatraak' are taken from Marinetti's 'Correzione', where they represent the backfiring of a car as it slows down. Like 'TRrrrrrrrrrrrr' and 'TRRRRRR' (speeding car noises from the same poem), they have nothing to do with gunfire but are merely the traffic noises of a public square.[60]

However, such a principle of reconstruction seems a questionable approach to apply to *papiers-collés* and it savours of the kind of mechanical geneticism that has long since been outlawed in the study of mainstream artistic collage and montage. In their new aesthetic context, partial quotations—and even fragments of single words—do not necessarily retain their previous connotations and associations; nor do they function as literary pointers to the entire source-work, even within such a contained context as a Futurist appropriation of material from *Lacerba*. Few would wish to 'reconstruct' Max Ernst's or Kurt Schwitters's collages in this way, or even—to take a more political example—John Heartfield's photo-montages.

To review briefly just one or two examples of the difference between source-association and the new meaning in Carrà's collage: the 'HUHUHUHU' from the left-hand side of the work comes from the end of Cangiullo's cabaret-poem 'Serata in onore di YVONNE'. There, it depicts the clamour of the 'victorious public', having, as it were, 'outperformed' the orchestra. In *Festa patriottica*, by contrast, Carrà has twice added the word 'sirene', using the same technique of intercalating single letters of alternating sizes as Cangiullo

[59] Bohn, 'Circular Poem-Paintings', p. 267. [60] Ibid. 267.

employed in 'Serata',[61] and thus clearly making the sound come from a number of festive sirens. Cangiullo's original meaning for the sound therefore no longer applies in the new context. And, on this analogy of noises metamorphosed, one might well wonder what justification we still have to assume that the strip of onomatopoeic lettering running down the right-hand side of the collage (also from Cangiullo's poem) refers back to the noise of the thunderous applause of the *serata* audience, rather than forming part of the jubilant patriotic celebration (in the second version: an Interventionist demonstration) that Carrà's collage claims to be depicting.

To cite one further example to reinforce misgivings about the principle of *exegesis via reconstruction*, the 'echoes' (*echi echi echi*) near the top of the collage and at a number of other points, combined with the city-names in parenthesis (Tokyo, New York, Paris), are culled from the last five lines of Carrà's own earlier free-word poem '1900–1913. Bilancio'. In that context, the words illustrated the way in which, as Pär Bergman puts it, 'the Futurist doctrines reverberate around the globe'.[62] (As far as Bergman is concerned, this is just one more variant on the familiar motif of simultaneity of vision.) To quote this part of Carrà's poem in translation: 'Lyricism of life   our thoughts + actions create concentric waves echoes echoes echoes'. The cities named in the poem as being reached by these concentric waves are Paris, Berlin, Tokyo, New York, Vladivostok, Murzuk, and Wan-fan-ku—the latter mentioned more in a spirit of global totality than for any actual historical accuracy! The dominant image, then, is one of Futurist influence spreading outwards across the world like radio-waves from some mythical Eiffel Tower of the Futurist sensibility and finding echoes all over the world. To begin the list of responding cities with Paris is quite logical; so too is Berlin's number two position, in the light of the role played by Herwarth Walden and the periodical *Der Sturm* in disseminating Futurist ideas in the imperial capital.[63] But it would clearly be ridiculous in a patriotic collage of mid-1914 (even more so if it were to sport the title *Manifestazione interventista*) for Berlin to play the same role as it did in the concluding lines of '1900–1913'. Not surprisingly, therefore, the *echi echi echi*-pattern can only be taken over with modification. Now, inconvenient city-names are removed (although Paris, top left, and Tokyo, mid-left, are still allowed to

[61] For example, the sequence 'PoUrBcBhLeIsCtCrOa' (*Lacerba* (15 June 1914), 192) symbolizes the total identification of orchestra and public at one point during the performance. Carrà appropriates the first five letters of this word-synthesis in such a way that it now either suggests 'pourboire' or could be a random snatch of overheard conversation.

[62] *'Modernolatria' et 'simultaneità': Recherches sur deux tendences dans l'avant-garde littéraire en Italie et en France à la veille de la première guerre mondiale* (Uppsala, 1962), 151.

[63] There may be similar cultural reasons for including Tokyo. The poet Tai Kanbara was in a sense the Herwarth Walden of Japan. Having been introduced to Futurism by Ikuma Arishima and after studying Boccioni's work, he corresponded with Marinetti and dedicated his theoretical contribution to Futurism, *Tai Kanbara's First Manifesto*, to the Italian. See *FF* 495.

figure). In other words, in their new context these *papiers-collés* from *Lacerba* are no longer concerned with the impact of Futurism around the globe; they are more pertinently understood as the echoes of the Milanese patriotic celebration, or Interventionist demonstration, in various sympathetic quarters. By extension, one would now need to question whether the reference to the 'vitt[oria] quotidiana di artiglieri (penetrante gioioso)'—also taken from the final lines of Carrà's earlier *Lacerba* poem—is necessarily, as Bohn interprets it, to be construed innocuously as to the 'atmosphere of the crowd' on a non-political occasion or whether 'joyous daily victories' (the very stuff of militant Futurism, according to the early manifestos) does not acquire very different connotations in this new patriotic, sabre-rattling setting.[64]

To re-contextualize these snippets from various *Lacerba* poems is to ignore the fact that collage is to a considerable degree a deliberate act of de-contextualization in order to force the elements to interact in a fresh aesthetic context. One question which remains, however, is whether Carrà deliberately chose his various *papiers-collés* for this work from such a narrow spectrum of material for specific semantic reasons which have a bearing on the meaning of *Festa patriottica*, whether there is rather merely something of a Futurist in-group reference here, or even whether, on a more practical level, *Lacerba* was simply the most convenient place to search for the sheer range of requisite onomatopoeic and acoustic neologisms needed for the festive word-collage. The latter seems most likely to be the case; the Futurists were not much given to in-jokes of this kind, and the source-material is too variegated and recherché to lend itself to supporting a series of semantic implications for the meaning of the entire collage.

It is also very noticeable that there is no significant measure of thematic correspondence between the various source-poems and the subject of Carrà's collage, no matter which of the two titles it bears. A poem about a theatrical soirée, another about sports, others concerning war and city-life have at most only jubilant crowds of people, intense emotions and activity, noise, and the Futurist ethos as common denominators. And it is noticeable that Carrà by no means extracts the most salient differentiating elements from his sources. In the case of 'Serata in onore di YVONNE', he does admittedly focus on material from the event's ecstatic ending, which at least has something in common with a patriotic festival. But the handling of Fivé's 'Sports' is very different, and more representative of Carrà's practice in most other parts of the collage.

Fivé's 'Sports' is a fairly typical minor piece of free-word racing-car poetry

[64] When Bohn goes on to see these words reflecting at the same time 'the oratory of the speakers celebrating Italy's sovereign independence' ('Circular Poem-Paintings', p. 266)—a sentiment more appropriate to 1911 than to 1914—he has already departed from his guiding principle of reconstruction in seeking to show what 'Carrà is urging his audience to do' (namely 'make patriotism part of their daily life'), which is certainly not what the words meant in their original context.

('quel monstre de vitesse . . . la route est libre . . . CRIC CRAC deuxième vitesse CROC . . . bonne carburation', etc.). With bravado, it naïvely brandishes its technical terminology, automobile onomatopoeia, and superman posture. Yet from the poem Carrà takes little more than its slightly incongruous title—now radiating downwards across the Italian flag—together with various flights of acoustic impressionism ('BREE BREE', 'CRIC CRAC', 'CRAAA') and the—usually telegraphic—word 'STOP', used by Fivé in the road-halt sense. Instead of inserting recognizable extracts from a French Futurist poem in order to conjure up the cosmopolitan nature of the celebration, or even Italy's common cause with France at the time, Carrà merely appropriates the same kind of onomatopoeic material that he elsewhere borrows from Cangiullo and Marinetti. 'Sports' is the only word which obviously betrays its French origins, though out of context it might just as well be English, and its task is ambiguous: it could be signalling the sport-war analogy so dear to the Futurists,[65] or it could point inwards towards the snippets about an aviator breaking a record which, together with the truncated phrase 'Audacia Rom.', is manifestly one of the sources of currently rampant national pride; hence the axis linking the Italian flag via 'Sports' with the aviator-section of the collage.

This aviator element is central to most interpretations of the whole structure and meaning of *Festa patriottica*. Bohn, for instance, argues:

> The center of the spiral contains some highly important information: 'AUDACIA ROMPICOLLISMO | ITALIA | aviatore | battere il record | eliche perforanti'. . . . One suddenly realizes that . . . Carrà's painting has a vertical center. At its apex—far above the celebrating crowd in the square—a daredevil pilot is trying to set a new aviation record of some sort. The words 'eliche perforanti' evoke the loud noise of the airplane's propellers and the piercing spirals of the sirens. They may also refer to acrobatic stunts by the pilot. In any case the airplane is clearly connected with the celebration below and contributes to the general festive atmosphere. Among other things it is now evident that we are viewing the spectacle *through the pilot's eyes*, high up in the air and traveling at a rapid speed. This provides an explanation for the extremely flat perspective of the painting. In fact, not only is the perspective two-dimensional; it is oriented from the center of the spiral facing out. Most of the words are focused on the center: below the center they are upside-down, and on the left and right they are arranged vertically. This means that they can only be read by someone inside the spiral who is directly facing them. Carrà's centralist perspective . . . contributes to the general appearance of disorder. . . . It is more realistic than that of 'Lettre-océan'. . . . The landscape below is distorted because it represents the view from a high-flying, circling airplane.[66]

Such a reading is far more mimetically inspired than Joshua Taylor's reference to *Festa patriottica* as a 'spinning collage' or Linda Landis's suggestion that the

[65] A 1915 front page of *Lo Sport illustrato* (reproduced *FF* 490) shows Marinetti brandishing a shotgun. The caption reads: 'War the World's Only Hygiene and Marvellous Synthetic Sport'.

[66] Bohn, 'Circular Poem-Paintings', p. 270.

structure is 'essentially an image of a rotating airplane propeller'.[67] Whereas, in his letter to Severini, Carrà himself describes this collage as depicting the 'plastic abstraction' (the main stress being on *abstraction*) 'of a civic tumult' and as depending on no more than 'a suggestion of poetic images and colours', Bohn returns the work to a most un-Futurist form of realism—an even greater one than that which was being rejected by the painters and free-word poets at the time.

The most conceivably Futurist perspective that such an approach offers is the possibility that *Festa patriottica* could be an example of aero-painting (*aeropittura*) *avant la lettre*.[68] But this seems rather far-fetched an assumption. There is no more reason why the feats of the aviator mentioned at the collage's centre should actually be taking place in the air above the patriotic celebration—let alone governing the collage's entire structure—than there is for the King and Army, also mentioned in the centre, to be physically present at the occasion. These are all primarily itemized as *reasons* for a sense of national pride, not as features of a realistic scene, be it viewed from the ground or from an aerial vantage-point. In other words, the structure is based on a principle of simultaneity, not pictorial mimesis. As for the impression of two-dimensionality, this is a standard feature of most collage, in contrast to perspectivist painting. It may also be a characteristic of a pilot's perspective, but even in an open cockpit the aviator would hardly receive the civic tumult's noises and written (shouted?) elements all as part of his flying experience.

A far more persuasive approach to the work is offered by Maurizio Calvesi's suggestion that the piece may represent 'in collage-form with its whirling rhythm his [Carrà's] Interventionist state of mind'.[69] With the modification that it would be more fruitful to think of the work as depicting *a* patriotic state of mind—rather than Carrà's own, even though this must have been the collage's source[70]—and using the festival image as an analogy for this

[67] Taylor, *Futurism*, p. 110; Landis, 'Futurists at War', in Anne Coffin Hanson (ed.), *The Futurist Imagination: Word + Image in Italian Futurist Painting, Drawing, Collage and Free-Word Poetry* (New Haven, Connecticut, 1983), 63. Landis's juxtaposition of this collage with Marinetti's 'auto-illustration' *The Propeller* (1915–16?) highlights more differences than similarities.

[68] According to the Manifesto of Aero-Painting, 'the changing perspectives seen from a plane create a new reality, having nothing in common with the traditional perspectives of earthly panoramas' (quoted from Rosa Trillo Clough, *Futurism: The Story of a Modern Art Movement. A New Appraisal* (New York, 1961), 137). Point Two puts the stress on 'unceasing movement', a 'new reality with no fixed point'. But aero-painting really only came to the fore in the post-war period, and in any case it is a nebulous phenomenon. Crispolti brings out the concept's disconcerting elasticity (*FF* 412 f.), though he does distinguish three main tendencies: cosmic projection, aerial fantasies, and aeronautical documentation. Carrà's collage lies, at best, somewhere between the second and third of these kinds of aero-painting.

[69] Maurizio Calvesi, *Il futurismo* (Milan, 1976), 92.

[70] In this connection, Caroline Tisdall and Angelo Bozzolla (*Futurism* (World of Art Library; London, 1977), 183 f.) quote Carrà on 'the excitement of the Interventionist period': 'Patriotic feelings predominated . . . and determined each person's position. For me, love of the fatherland was a moral

atmospheric state, it is possible to see parallels between Futurist *stati d'animo* paintings, including Boccioni's influential triptych with that title, and the principles of such a literary collage as this. Writing in 1912 of his *Stati d'animo* experiments, Boccioni declared: 'we abruptly cross each motif at will with one or more other motifs, of which we never offer the entire development, but simply the initial, central or final notes. . . . we not only have variety, but chaos and conflict between absolutely opposed rhythms, which we nevertheless bring into a new harmony' (*Archivi*, i. 107). This could serve as an apt description of Carrà's collage-method in *Festa patriottica*, though here an added semantic factor is created via the introduction of *papiers-collés* using the printed word. As a result, motifs are now expressed in phrases or verbal fragments (the 'literary'/ free-word element), rather than through a painter's orchestrated shapes alone.

All this is not to deny that there may also be suggestions of a narrower iconic moment in some *parts* of Carrà's collage. Janis and Blesh's description of the collage as 'originating below centre with the *sole bruciaticcio*, the "scorching sun" where dynamism originates in heat, light and movement'[71] presumably derives from the uncontentious assumption that the sun (in fact, referred to top left) is sending a shaft of light down through the centre of the collage: a centre which can be interpreted as an arena or some piazza acting as the focal point for all the patriotic celebrations going on in the bustling city. (With its second, more political title, the collage is likely to suggest Milan's Piazza del Duomo, the scene of one of the major Interventionist demonstrations of September 1914.) The collage's force-lines could similarly be read as streets; the flags might well be thought of as hanging from buildings, as they do in Cangiullo's far more representational word-picture on the same subject;[72] and some of the exploding slogans might be emanating from loudspeakers. But all these possibilities are merely local associations triggered off in the mind by fragments of the collage, not aspects of some rigorously pursued mimetic organizing principle.

Carrà says in his letter to Severini that the impression he wished to create was one more of 'plastic abstraction' than of 'representation'. And for this purpose, the actual words stuck on to *Festa patriottica* are skilfully chosen to function in an equally associative manner. Slogans (patriotic-cum-Interventionist), bits of

entity and could not be considered as an abstraction, but as a spiritual force. Honour, pride, independence and interests constituted the concept of the fatherland which I saw as an "extension . . . and . . . enlarging of the ego", a glorification of individual life' (see 'La mia vita' in Carrà, *Tutti gli scritti*, ed. Massimo Carrà (Milan, 1978), 693). Nevertheless, *Festa patriottica* gains little by being given such a narrowly biographical interpretation.

[71] Janis and Blesh, *Collage: Personalities, Concepts, Techniques*, p. 39. A more local iconographical tradition may be at play here, for in *Le Futurisme* (first published Paris, 1911) Marinetti presents the Futurists as responding to the dictates of 'a new sun, which is certainly not the sun which carressed the placid shoulders of our grandfathers' (*Le Futurisme*, ed. Giovanni Lista (Lausanne, 1980), p. 137). The image also occurs in the manifesto 'Let's Murder the Moonlight!' ('Uccidiamo il Chiaro di Luna!').

[72] See below, pp. 136f.

advertisement (Odol, Tot, *Lacerba*, part of a catalogue), flags and sounds and smells[73] all suggest the vibrant city-context, with the crowds and the Futurist movement celebrating on this occasion, rather than forming part of a consistent scenic depiction. Here, the telegraphic style of words-in-freedom is impressively integrated into a collage which presents patriotism (and the various causes for it) as one of the 'zones of intense experience' that appealed to the Futurist sensibility at the time.

As is clear from Marinetti's letter to Soffici about how Futurist literary collages should be presented, the Italian Futurists were always wary of being accused of plagiarizing from the Cubists, or, almost as bad, having the Cubists poach new ideas from them. In all probability, Apollinaire outlined the idea of collage in general terms to both Severini and to Picasso and Braque at about the same time. And, while Hausmann erroneously believed that Picasso initiated Cubist collage experimentation by 'following the example of the Futurists' with his still lifes of 1912, it is now generally assumed that it was Georges Braque, with his *Le Portugais* (*The Portuguese*) of 1911, who in fact became the first twentieth-century painter to use collage.[74] Nevertheless, as Herta Wescher has demonstrated at some length,[75] the collage technique that is associated with modernism has a substantial prehistory stretching back across a number of centuries, even if the aesthetic underlying earlier forms is radically different. But this is enough to suggest that invention is not really the issue at stake.

In his letter on the subject to Carrieri, one of the few extensive accounts of early collage experiments, Severini recalls the way in which he set out to follow in the Cubists' footsteps:

I was greatly amused creating collages with a vast variety of materials: *Portrait of Paul Fort* with bits of *Vers et Prose*; *Portrait of Arthur Cravan* with small pieces of his review *Maintenant*; a ballerina with forms in relief and many coloured *paillettes*, etc. Amongst my more noted works of this type was the *Portrait of Marinetti* which I did in 1913. In this painting various forms are in relief—a relief obtained by the use of Totin glue and the

[73] While its evocation of an acoustic dimension is one of the memorable achievements of this collage, all Carrà can do when it comes to the olfactory is to scatter the undiscriminating word 'odore' at various points in the picture. In fact, the 'painting of smells', although in theory one of the desiderata of the new Futurist synaesthetic art, remained an elusive goal.

[74] Wolfram, *History of Collage*, p. 16. Also, as Wescher has shown (*Die Collage*, p. 20), the issue is more complicated than such a statement implies, for Braque was already putting *trompe-l'œil* objects into his paintings in 1910 (for example, the nail in *Violin and Palette*). The following point by Susan Marcus is on the whole more pertinent to the development of literary collage: 'in 1911 when Georges Braque stencilled a few letters, numbers and an ampersand on the painting *Le Portugais*, he introduced into Cubism the typographic element that ultimately served as a catalyst for the discovery of collage' ('The Typographic Element in Cubism, 1911–1915: Its Formal and Semantic Implications', *Visible Language*, 6 (1972), 321).

[75] 'Die Vorläufer', *Die Collage*, pp. 7–19.

white of Meudon employed by the gilders. Naturally, several bits of manifestos with titles are also included. The moustache is a real moustache, the starched white shirt is real, and the lapels of the dinner-jacket are of velvet. The result was unforeseen, although pictorially always perfect, because all the diverse elements were naturally arranged according to pictorial criteria and were always outlined and emphasized by lines and zones of colour which humanized them and brought them to life; otherwise they would have been *pieces of dead material*.[76]

Even here, with Italian Futurist collage still practically in its infancy, indications of certain fundamental differences from Cubism can be detected: in the use of 'bits of manifestos with titles' (the foregrounding of a legible typographical element), and, in contrast to the still-life genre, in Severini's suggestion that, if he had not succeeded in imposing on the collage's shapes certain patterns of line and colour that 'brought them to life', they 'would have been *pieces of dead material*'.

The largely static collage-portrait of the contemporary man-of-letters, showing him posing together with various *papiers-collés* from his published work, was well established as a Futurist genre by 1914, although it does not appear to have received any strong impetus from Cubism. Here, the printed word has already become an important ingredient,[77] but usually without acquiring any particular topical or overtly political status, except in the case of occasional references to *Zang tumb tumb* in portraits of Marinetti. The genre is, of course, part of a much longer-standing portraiture convention, dating from the Renaissance, whereby the subject poses surrounded by tokens of his worldly achievements. But, as far as the development of literary collage was concerned, it hardly played an effective role in privileging any significant printed component; also, it invariably had the effect of making the subject appear static and posed.[78] And in this respect at least, it retained affinities with the Cubists' still-life collages.

In addition, the still-life/posed portrait tradition had an even more significant limiting influence on early Futurist collage, apart from its tendency to be non-dynamic. Severini recalls:

we realized later, each of his own accord, that by placing real material in its logical position according to representational logic, one could fall into the error of making others think that the artist had a realistic intention—an intention based on a sort of *trompe-l'œil*—which was open to serious doubts, whilst what was really being sought after with

[76] Carrieri, *Futurism*, p. 118.

[77] Wescher (*Die Collage*, p. 56) notes how 'rich' the written element is in Severini's portrait of Paul Fort, compared with the restrained use of printed material in Cubism.

[78] One notable exception is Severini's dynamic *Ritratto di Marinetti*. Of its typographical element, Wolfram says the 'broken-down text . . . is distributed across the surface in such a manner as if to leap frantically from it' (*History of Collage*, pp. 39 f.). By comparison with Carrà's *Ritratto di Marinetti* (1910–11), Severini's explosive layout makes better use of the typographical element, which is relatively static and arranged in a realistic manner in Carrà's painting.

this technique was unforeseen, poetic results. Thus we began to place the title of a newspaper, for example, not in its normal position, but in another part of the painting. I no longer applied sequins to the dresses of my ballerinas, but where *the painting and not the visible reality required them*. An intensely pictorial function continued to predominate in the creation of *papiers-collés*, but they quickly began to express another value, perhaps at times purely symbolic—a sort of metamorphosis of the material.[79]

The development of Futurist literary collage can, in fact, be viewed as a deliberate, though at first gradual, rejection of the Cubist still-life experiments, or at least an attempt to see what happened if collage were developed in other directions apart from that of representational logic.

'The accent in Cubism', Werner Haftmann notes, 'is on the object. Consequently, the stylistic vocabulary of the first phase of Cubism was almost exclusively developed in still lifes.'[80] As a result, during the years 1912–1914 a whole genre of Cubist still-life paintings using *papiers-collés* taken mainly from newspapers, but sometimes using visiting-cards and other ephemera,[81] had successfully established itself. Such works as Picasso's *Siphon, Glass, Newspaper and Violin* (1912), *Still Life, Violin and Fruits* (1913) and *Still Life with 'Lacerba'* (1914), Braque's *Glass, Carafe and Newspapers* (1914), and Gris's *The Table* (1914) had a strong influence—largely through Severini's mediation—on early Futurist collage. Severini's *Omaggio a mio padre* (*Homage to my Father* (1912)), his already mentioned *Ritratto di Paul Fort* (*Portrait of Paul Fort* (1913)) (fig. 2.3), as well as *Natura morta con Fruttiera* (*Still Life with Fruit Bowl* (undated)), and *Natura morta con 'Lacerba'* (*Still Life with the Journal 'Lacerba'* (1913)), Carrà's *Sintesi circolare di oggetti* (*Circular Synthesis of Objects* (1914)) (fig. 2.4), and *Collage* (1914), and Soffici's *Natura morta* (*Still Life* (1914)) are all in this well-established, essentially Cubist idiom. But, as the frequent recurrence of the phrase 'still life' in the titles suggests, such collages largely lack any sense of Futurist dynamism. One could almost see the genre's perpetuation in Italy as an example of an over-derivative Cubist strain within Futurism. Both French Cubist collages and Italian imitations frequently occur in the pages of *Lacerba*, which also appears to have itself inspired a number of these still-life collages. Thus in 1914, the year in which *Lacerba* began more regularly to include reproductions of paintings and collages in its issues, major French Cubist collages[82] are to be seen alongside a large number of Italian equivalents;[83] and within the

[79] Carrieri, *Futurism*, p. 118.

[80] *Painting in the Twentieth Century*, 1 (London, 1965), 101.

[81] The inclusion of the artist's visiting-card in Carrà's *Sintesi circolare di oggetti* (*Circular Synthesis of Objects*), also known as *Natura morta con sifon* (*Still Life with Siphon* (*Lacerba* (1 July 1914), 201)) is seen by Wolfram (*History of Collage*, p. 42) as a direct imitation of Picasso.

[82] For example, Picasso's *Still Life* (1 May 1914, 131) and Férat's *Still Life*, also known as *Still Life with 'Lacerba'* (14 Feb. 1914, 53).

[83] Soffici is well represented with *Simultaneità di donna carretto strada* (*Simultaneity of Donna Carretto Street* (15 Jan. 1914, 25)), *Cocomero, fruttiera bottiglia* (*Water-melon, Fruit-bowl Bottle* (15 May 1914,

FIG. 2.3. Gino Severini, *Ritratto di Paul Fort*, 1913 (Musée National d'Art Moderne, Paris)

FIG. 2.4. Carlo Carrà, *Sintesi circolare di oggetti*, 1914

space of one or two numbers can be found statically Cubist still lifes (largely by Soffici) and dynamic Futurist collages such as the two we have looked at so far. In fact, Soffici's own personal preference for the still-life genre—he is even reported to have declared that 'the best Italian art . . . is entirely Cubist in character'[84]—may well account for the preponderance of this type of collage in the pages of his journal. In this respect, however, *Lacerba* remains disappointingly unrepresentative of mainstream Futurist collage, which was rapidly developing away from Cubism in two major respects: in its creation of dynamic, explosive patterns[85] and in its accommodation of collage to the goal of pro-war poetry.

149)), *Al buffet della stazione* (*At the Station Buffet* (1 Aug. 1914, 236)), and *Natura morte* (*Still Life* (14 Feb. 1915, 53)). Other examples include: Rosai's *Latrina* (*Latrine* (1 Apr. 1914, 101)), Zannini's *Suonatore ambulante* (*Walking Player* (1 June 1914, 169), and Carrà's *Sintesi circolare di oggetti*.

[84] Quoted from Haftmann, *Painting in the Twentieth Century*, p. 105.

[85] 'To the "landscape" of the Impressionists and the "still life" of the Cubists, Futurist painting has opposed the necessity of a living subject, in direct rapport with the modern artist's state of mind', Fillia wrote in *Oggi e domani* of 18 Aug. 1930 (quoted by Crispolti, *Il mito della macchina*, p. 507).

That a collage using *papiers-collés* taken from contemporary newspapers did not have to conform slavishly to the still-life paradigm associated with Cubism is readily demonstrated by Carrà's *Inseguimento* (*Pursuit*) of 1914, published in the following year in *Guerrapittura* (fig. 2.5). Instead of the stasis of a still life, we have a galloping horse, victoriously chasing the enemy—something akin in intention, if not in achievement, to the dynamism of Severini's series of studies of *danseuses* in swirling motion. As the collage's explicit reference to Joffre indicates, the charging horse is part of the victorious Entente forces' onslaught at the Battle of the Marne.[86] The triangular shapes to the *papiers-collés* take up the basic arrow-shape of the moving horse, although Carrà is less successful in creating the sense of an animal in motion than in some of his other studies of dynamic objects (especially cyclists). The horse appears to be galloping across a stretch of ground made up of an entertainments column from a newspaper, as if to suggest that the pleasures advertised there are passéist and static in comparison with the exhilaration of charging into battle. Nevertheless, although comparatively dynamic and adventurous[87] when set alongside the still-life tradition of the Cubist collage, *Inseguimento* remains unrepresentative of the direction the majority of Futurist collages were taking in one vital respect: its verbal ingredients are still very much subordinate, if not totally sacrificed, to the pictorial aspect. The column advertising entertainments, for example, is hardly meant to be digested line-by-line; even the positioning and size of typeface here are an indication of this fact. On the other hand, most of the leading examples of Futurist collage from *Danza serpentina* onwards, including *Cannoni in azione* (*Guns in Action*) (Severini), the collages of *Guerrapittura* (Carrà), *Conflagrazione* (Buzzi), and those in the periodical *L'Italia futurista* and Marinetti's *Les Mots en liberté futuristes*, *are* meant to be read. Indeed, they so patently occupy a middle-ground between mainstream collage/painting and the *dipinto parolibero* that they have frequently been discussed within both frameworks. Carrà's *Festa patriottica* figures in studies of collage/montage by Janish and Blesh, Wolfram, Jürgens-Kirchhoff, Perloff, and Wescher, while at the same time being treated by Caruso and Martini as an example of free-word poetry. Marinetti's *Le Soir* has also frequently been discussed in both generic contexts,

[86] 'Germany lost the war when she lost this battle', according to B. H. Liddell Hart (*History of the First World War* (London, 1970), 109). But in fact, for all the importance of this 'psychological rather than . . . physical victory' (ibid. 111), the role of the cavalry in it was less decisive than the Futurists appear to have imagined. Foch in fact criticized the French cavalry for failing to be more forceful in its pursuit of the retreating Germans. See C. R. M. F. Cruttwell, *A History of the Great War, 1914–18*² (St Albans, 1982), 34 f.

[87] Jürgens-Kirchhoff refers to 'the recurrence of the same objects repeatedly in the still lifes of the Cubists—painting utensils, musical instruments, newspapers', arguing that, 'the Cubists stuck to the particular problems of their own familiar environment' (*Technik und Tendenz der Montage*, p. 51). In this respect, *Inseguimento* not only departs from the static mode of the still life, but marks a change of thematic emphasis as well.

Fig. 2.5. Carlo Carrà, *Inseguimento*, the *Guerrapittura* version, 1915

and it would be difficult to decide unequivocally whether *Danza serpentina* and *Cannoni in azione* were illustrated free-word poems or pictures with added words. To understand why Futurist collage was moving in this divergent direction from Cubism, it is necessary to appreciate the very different premisses out of which it grew.

Even before the Italian Futurists tried their hands at collage techniques, they had constructed their free-word poetry on a principle of discontinuity which was in due course to lend itself readily to the mixing of media and various forms of collage and assemblage of later years. Writing in the manifesto 'Destruction of Syntax—Wireless Imagination—Words in Freedom' ('Distruzione della sintassi—Immaginazione senza fili—Parole in libertà' (1913)) about the insertion into poetry of onomatopoeic effects, Marinetti calls for a 'swift, brutal, immediate lyricism . . . with no taste of the book about it but, rather, as much as possible of the taste of life' (which already sounds like a recipe for collage). He argues that the proposed introduction of onomatopoeic elements was necessary in order 'to render all the sounds and noises of modern life, even the most cacaphonic' and cites *Bombardamento* and *Battaglia peso + odore* as models of this new use of an extrinsic acoustic dimension which 'abolishes all stylistic bonds, all the brass buckles with which the traditional poets linked images together in their prosody' (*TIF* 66). The idea is further pursued in 'Geometric and Mechanical Splendour and the Numerical Sensibility' in Marinetti's taxonomy of different types of sound, for the aim he specifies, 'to enrich lyricism with brute reality' (*TIF* 90), again seems to be logically pointing in the direction of collage.

Other markedly discontinuous visual effects preparing the way for collage proper include the simultaneous use of different typefaces and of ordinary and bold printing, or the juxtaposition of linear typography with 'designed analogies'. As we shall see in the next chapter, telegraphic lyricism, with its loosening of conventional syntax's hold over the poetic structure, also often creates an effect of juxtaposed (or collaged) autonomous word-clusters. But it is above all Marinetti's concept of the 'wireless imagination', the 'absolute freedom of images or analogies', 'the deep love that assembles distant, seemingly diverse and hostile things' (*TIF* 62), that ultimately prepared the ground for literary collage.

Exploring certain parallels between the radical discontinuities of words-in-freedom and the juxtapositions of collage, Marjorie Perloff has as her starting-point the assumption that the 'Technical Manifesto of Futurist Literature' ('Manifesto tecnico della letteratura futurista' (1912)) and the 'Destruction of Syntax' manifesto, with their emphasis on the bold word-pairings and analogies of the 'wireless imagination', prepare the ground for collage, but only create a superficial similarity between the resultant poetry and visual collage. In a work like *Zang tumb tumb*, she rightly argues, the actual structural principle in operation is 'less that of collage than of catalog'.[88] But there are, among the free-word compositions of the Italian poets influenced by Marinetti's theories, other works which come closer to the collage-technique than the word-strings of Marinetti's battle poems.

On 1 February 1914 *Lacerba* published a poem by Boccioni entitled 'UOMO + VALLATA + MONTAGNA' ('MAN + VALLEY + MOUNTAIN') (fig. 2.6). In it, the description of a descent across a mountainside is so constructed as to fuse man and environment, according to a principle of 'compenetration' which Boccioni had already enunciated in his 'Technical Manifesto of Futurist Sculpture' ('Manifesto tecnico della scultura futurista' (1912)). Futurist poetry, Boccioni reasons there, has already liberated the word from the constraints of archaic syntax, and the Futurist musician Pratella has broken away 'from the chronometric tyranny of rhythm'. Now, sculpture in its turn needs to emancipate itself from a series of aesthetic conventions:

Why should sculpture be the one to lag behind, burdened with laws which no one has the right to impose? Let's turn everything upside-down and proclaim the ABSOLUTE AND COMPLETE ABOLITION OF FINITE LINES AND THE CONTAINED STATUE. LET'S SPLIT OPEN OUR FIGURES AND PLACE THE ENVIRONMENT INSIDE THEM. We declare that the environment must form part of the plastic whole, a world of its own, with its own laws, so that the pavement can jump on to your table, or your head can cross a street, while your lamp twines a web of plaster rays from one house to the next.

Boccioni explains that, in the sculptures created according to this new Futurist principle,

objects will not be juxtaposed with a statue, like explanatory attributes or detached decorative elements, but, following the laws of a new concept of harmony, they will be encapsulated inside the muscular lines of a body. In this way, the cogs of a machine might easily emerge out of the armpits of a mechanic, or the lines of a table could cut a reader's head in two, or a book with its fanned-out pages could intersect the reader's stomach. (*Gli scritti*, p. 27)

This notion of 'compenetration', which Marianne Martin views as 'perhaps the most revolutionary idea contributed by Futurism' to sculpture,[89] is reflected in a number of works of sculpture (*Fusione di una testa e di una finestra* (*Fusion of a Head and a Window*),[90] *Testa + casa + luce* (*Head + House + Light*), *Sviluppo di una bottiglia nello spazio* (*Development of a Bottle in Space*), and *Antigrazioso (la madre)* (*Anti-gracious (the mother)*), as well as in a painting like *La strada entra nella casa*

[88] *The Futurist Moment*, p. 60.

[89] *Futurist Art and Theory*, p. 89.

[90] Significantly, when the work was exhibited in Rotterdam in 1913, its title was given as *Compénétration de tête, fenêtre et lumière* (see Martin, *Futurist Art and Theory*, p. 165). Without doubt, 'compenetration' is a better description of the disjunctive effect than 'fusion'.

Lentamente abbassarsi adagiarsi attraverso la montagna la **testa** toccare il muschio delle vette velarsi nella nebbia le nuvole sulla **fronte** Un ghiacciaio scivolare sotto l' **ascella** (*cuscino cuoio raso cristallo*) una cascata corrermi sotto la **schiena** *scciaccc sclaff sclocchete scloff sclof sccicc sccicc sccicc*) grondare grondare gocciare stillare le gambe sprofondate nel sole della valle Sentire e non toccare Quadrati verdi (*velluto sofficità freschezza elasticità*) rettangoli gialli (*spinosità del grano delle spiche agonizzanti arsura attesa falce-lampo cantare*) rettangoli bruni terrosi (*friabilità ocra gialla terra di siena naturale terra di siena bruciata caffè viola cupo*) Case (*biacca calcina*) dadi tappeto verde Pioppi misantropia meditazione acque lastricati acciaio gorgogliante Alberi confini agli angoli fumi di case al **malleolo** un fiume (*cobalto polvere d'argento*) sotto il **polpaccio** **rotondità** Rumori di zappe rabbiose nei **tacchi** Sentire sentire abbandono penetrazione non toccare Sotto il **ginocchio** un cimitero recinto geometrico di piccoli recinti geometrici piccoli cerchi neri poveri corone numerazione allineazione dimenticanza cancelletto pace modestia Richiami fischi canzoni bau-bau bau-bau beeee... beee... Silenzio sole universale calma orizzontale vette ventose granitiche deserte Il **braccio** sinistro allungarsi sul sistema dorsale delle montagne il **braccio** destro fare ponte su piccola vallata umida fredda buia verde-cupo battaglione di pini macigni accovacciati sparsi lazzaroni immobili Sentire e non toccare Gli armenti pascolare all'ombra viola delle **coscie** e dei **fianchi** Aquila volare cravatta nastro spilla Silenzio del

FIG. 2.6. Extract from Umberto Boccioni, 'UOMO + VALLATA + MONTAGNA', 1914, showing the 'compenetration' of two sizes of type

(*The Street enters the House*). And in 'UOMO + VALLATA + MONTAGNA' (fig. 2.6) the technique finds its poetic equivalent. Boccioni highlights the words for parts of the walker's body so that they stand out from the rest of the text, which describes the setting in free-word formulae. Just as Marinetti often used words in large, bold print to set his acoustic effects off from the descriptive parts of his battle poems, so here different typefaces create a montage-like juxtaposition, a 'compenetration' of man and natural environment.

This technique of intercalation, of 'compenetration', of which Boccioni's poem is but one example among a number,[91] is much closer to the juxtapositions of visual collage and has little to do with the catalogue method of much free-word poetry. And other experiments of the time reveal how close Futurist poetry was to the collage principle. The Venice Palazzo Grassi exhibition showed a poem by Marinetti (*FF* 193), 'Parole in libertà (Bombardamento sola igiene)' (1915?), which takes a column of newspaper print and superimposes on it drawn explosions and shell trajectories, with the result that the printed word

[91] e.g. Armando Mazza's 'compenetrazioni', published in *Firmamento* (Milan, 1920), 65 f., which resembles Boccioni's poem very closely, apart from its more conservative typography. Bergman sees the same principle at work in Cangiullo's 'Addiooo' in *Lacerba* of 15 Nov. 1913 (*'Modernolatria' et 'simultaneità'*, p. 169).

looks like a surface actually being exploded by the bombardment of shapes. Linear and explosive shapes, the juxtaposition of the typographical and the graphic, with the one attacking the other, may not be 'compenetration' in Boccioni's sense, but again it takes free-word poetry substantially towards the domain of collage.

In the manifesto 'The Futurist Cinema' ('La cinematografia futurista' (published in no. 9 of *L'Italia futurista* of 11 September 1916)), Marinetti, Corra, Settimelli, Ginna, Balla, and Chiti outline further synthesizing possibilities which show how liberally the notion of 'literary' words-in-freedom was being construed by this time. Among the proposals is one for:

FILMED WORDS-IN-FREEDOM IN MOVEMENT (synoptic tables of lyrical values—dramas of humanized or animated letters—orthographic dramas—typographical dramas—geometric dramas—numerical sensibility etc.). Painting + sculpture + plastic dynamism + words-in-freedom + composed noises (*intonarumori*) + architecture + synthetic theatre = Futurist cinema.

The principle is one of a fusion of various art-forms, of compenetration, in fact a climate very conducive to the fusion of shapes and words that one finds in Futurist literary collage.

Even without the direct influence of the collage techniques of Cubist painting, in other words, Futurist words-in-freedom can be seen to be moving towards a discontinuity of typographical texture, a collocation of unconnected bold images,[92] and even a transcending of the conventional boundaries between various art-forms. Moreover, the Futurist interest in *simultaneità*, a concept elaborated in counterdistinction to that of the Orphists, was undoubtedly a further factor in stimulating the development of literary collage. As Carmine Chiellino argues, simultaneity is achieved in Futurist painting by the rejection of the classical unity of perspective,[93] which means that a concern with simultaneity is inevitably leading in the same direction as collage. It is one of the more unfortunate ironies of Futurist scholarship that what still remains the most stimulating and wide-ranging investigation into the theory and practice of simultaneity techniques, Pär Bergman's *'Modernolatria' et 'simultaneità'* (1962), should for practical reasons have been confined to the pre-war period. As a result, collage, the Futurist simultaneous art-form *par excellence*, is largely omitted from this study's scope purely on chronological grounds. For, although it is a matter of historical fact that a large number of their Futurist still-life paintings using *papiers-collés* as one of their ingredients had already been

[92] Severini's declaration that the seeds of collage can be found in Apollinaire's poetic practice (quoted in Carrieri, *Futurism*, p. 117) also bridges the gap between painting and literary techniques. See in this connection Francis J. Carmody, *The Evolution of Apollinaire's Poetics, 1901–1914* (Berkeley, 1963), 99.

[93] *Die Futurismusdebatte: Zur Bestimmung des futuristischen Einflusses in Deutschland* (Europäische Hochschulschriften, Ser. 1/252; Frankfurt a. M., 1978), 102.

painted and reproduced in 1913, it was 1914 that saw the real flowering and rapid spread of Futurist literary collage; and it was ultimately the political events of 1914–15 that gave maximum impetus to the exploration of collage's potential by the Italian Futurists. One suspects that, if all this had happened a few years later, the Futurists would have been using photo-montage for political propaganda purposes. (As Calvesi has remarked,[94] there is a strong similarity between the technique of indicating dynamism by means of a series of overlapping forms, a meeting-ground between chrono-photography and Futurist painting, and montage-structure.) But as it is, propaganda paintings and collages and the *dipinto parolibero* were entrusted with much of the propagandizing.

Collage, however, is more than a mere *aesthetic* constructing principle for the Futurists. It is a feature of modern life itself. In the 'Destruction of Syntax' manifesto Marinetti points out that many Futurist reforms, while they may seem innovative within the artistic realm, are really only following the lead given by modern life. The point certainly holds for collage, which they saw already in the world around them and simply wished to transfer to the arts. Thus, without actually using the word, Russolo, in 'The Art of Noises' ('L'arte dei rumori' (1913)), describes the modern city as an acoustic collage.

> Let us cross a great modern capital with our ears more alert than our eyes, and we will get enjoyment from distinguishing the eddying of water, air, and gas in metal pipes, the grumbling of noises that breathe and pulse with indisputable animality, the palpitations of valves, the to-and-fro of pistons, the howl of mechanical saws, the jolting of a tram on its tramlines, the cracking of whips, the flapping of curtains and flags. We enjoy creating mental orchestrations of the crashing-down of metal shutters, slamming doors, the hubbub and shuffling of crowds, the variety of din from stations, railways, iron foundries, spinning mills, printing works, electric power-stations and underground railways . . .[95]

In 'Absolute Motion + Relative Motion = Dynamism', Boccioni refers to the newspaper's dynamism (and the newspaper was to become one of the Futurists' key models of collage structure): 'Today modernity—in advertisements, newspaper sketches and caricature—is using a new kind of fundamental dynamic norm which corresponds more to the truth.'[96] Modern man does not have, then, to turn to Futurist art to find examples of the collage process. In 'the great daily newspaper (synthesis of a day in the life of the world)' (*TIF* 57), he already regularly encounters one of the main paradigms of the new supra-perspectival art-form. The Italian Futurists were not the first to think along these lines. According to Marshall McLuhan: 'The French Symbolists, followed by James

[94] Calvesi, *Il futurismo*, p. 127.

[95] My translation. The passage occurs, in another rendering, in *FF* 561.

[96] *Lacerba* (15 Mar. 1914).

Joyce, saw that there was a new art form of universal scope in the technical layout of the modern newspaper . . . the front page of a newspaper is a superficial chaos which can lead the mind to attend to cosmic harmonies of a very high order.'[97] Such journalistic juxtapositions were, to pitch it no higher than this, perceived as analogues of the new collage structure.

In life as in art, then, collage was in the air. The number of contributory influences on the new literary form was large. But one particular theoretical one deserves special attention. In the early months of 1914 a debate was carried out between Papini and Boccioni in the pages of *Lacerba* which triggered off an incisive and influential discussion of the aesthetic premisses underlying collage's so-called 'return to brute reality', as well as its equivalent in Futurist painting and sculpture. Obviously, the Futurist poets cannot have remained unaware of the importance of the ideas expressed here for the development of a new application of words-in-freedom.[98]

Much of Papini's criticism of the Futurist aesthetic in his initial article 'The Circle is Closing' centres on the implications, as he polemically interprets them, of the collage process:

THE CIRCLE IS CLOSING.
It seems so to me. The circle of the creative spirit. The human addition—voluntary—to natural reality. Art was born as an imitation—but it has advanced as a deformation. Thought arose as interpretation—but it became autonomous in purest metaphysics. There was, on the one hand, the concrete and practical and, on the other, the concept of disinterested freedom. The two worlds met but they were distinct. Today, at the final point of investigation, I seem to see a negation (auto-suppression) of one of these two spheres—the second, the lyrical, the conceptual. There has been a progressive abdication of one in the face of the other. Creation that becomes once again simple action; art which returns to raw nature.

The main supporting evidence for such a disenchanted position that Papini can find comes from the realm of collage-paintings:

In painting: Picasso, who, exasperated by the search for plasticity, takes pieces of wood, cloth and tin plate and sticks them together on to the canvas in order to make a painting.

Severini in a portrait makes use of a real moustache and a real velvet collar. Boccioni adopts wood, iron and glass for his sculpture in order to render values that could not be obtained with the conventional materials.

The list goes on to include Russolo's *intonarumori* machines and various forms of literary iconicity (onomatopoeia, 'designed analogies', and primitive language).

[97] *The Mechanical Bride* (London, 1967), 4.

[98] In 'Mon ami Marinetti', first published in *XXe siècle* (Apr. 1959), and reprinted in *Témoignages*, Severini emphasizes the importance of Papini's article for the collage debate. See, especially, *Témoignages*, pp. 190f.

Whether much of this is mere rhetoric or whether it reflects a genuine anxiety is not always clear, especially given the relations between the *Lacerba* editors and Marinetti at this stage, but Papini goes on, in his strident way, to suggest that this 'evidence' is but a symptom, part of a process whereby the circle leading back to imitation in art is inevitably closing.

At the moment we are at the beginning. Partial substitutions where there is still the part played by the personality in the choice of the raw elements to be mixed with the elaborate elements. But if this method were accepted and pushed to its most rigorous final conclusion, the result would be that the finest of still lifes would be a furnished room. . . .

The circle is closing. Art is returning to reality; thought is once again sinking into action.

On this basis Papini concludes, in a spirit of resignation, that 'the autonomous possibilities of the spirit seem to have been exhausted. The sea of invention completely explored;' 'the springs of the creative personality' have 'dried up'.[99]

As is embarrassingly evident, the value of Papini's polemic hardly lies in the intellectual rigour of its argument. (And in fact, in his eventual rejoinder, 'The Open Circles', he adroitly retreats behind the mask of devil's advocate.) But to its credit, the attack did elicit from Boccioni a vigorous rebuttal of the assumptions underlying Papini's critique: namely that Futurist art was closing the circle by returning to imitation or resorting to *ready-mades*; that the use of collage materials was part of a process moving inexorably towards the stage when life would masquerade as art; that collage was a symptom of atrophied inventiveness. And, in respect of collage, in particular, it led to an explicit statement of the main principles giving rise to the new art-form—and one made in a public context in *Lacerba* at the very time when the techniques of collage-painting were being transferred to the new sub-genre of (political) literary collage.

Boccioni writes in his reply ('The Circle is not Closing'):

as soon as this reality enters and becomes part of the elaborated material of the work of art, the lyrical function that it is called upon to perform, its position, its dimensions, and the contrasts that it provokes transform the objectively anonymous and start it on the path of becoming an elaborated element. . . . you have confused the new elements of reality, which have been introduced into art in order that they may be subjected in time to their inevitable transformation, with a return to raw material.

What Papini fails to appreciate, Boccioni is at pains to point out, is that the circle is in fact still *opening*—to even richer new possibilities; and he is unable to recognize this because he chooses to ignore 'two fundamental principles of the Futurist sensibility':

[99] Giovanni Papini, 'Il cerchio si chiude', *Lacerba* (15 Feb. 1914), 49f.

1. The means of artistic expression handed down to us by culture are obsolete and no longer capable of receiving and interpreting the emotions which come to us from a world that has been completely transformed by Science.

2. The new conditions of life in which we live have created an infinity of completely new, natural elements which, therefore, had never entered previously into the domain of art and for which the Futurists intend to discover new means of expression *at any price*.

There is, thus, in the Futurist system a process of destruction of the old instruments of expression which parallels the process of research to find new ones.[1]

Marinetti's defence of the importation of fragments of 'brute reality' into Futurist poetry and these remarks of Boccioni's about 'new elements of reality . . . introduced into art' find a parallel in Severini's discussion of collage in 'Fantastic Art in the Sacred Realm' ('L'Art fantastique dans le sacré' (1913)), which makes even bolder claims for the new technique. 'The expressive totality of a Futurist work of art', he declares, 'can also require the use of realia as objective elements; for example, pieces of cloth, placards, moustaches, bits of hair, multi-coloured sequins, costume jewellery etc. . . . In our works of art we thus achieve an absolute realism.'[2] But the principal terms in which Boccioni presents collage (although not by name), both in his reply to Papini's attack and also in *Pittura scultura futuriste* and the 'Technical Manifesto of Futurist Sculpture', are of more relevance to subsequent developments in literary collage than the notion of a quest for some elusive form of 'absolute realism'. Thus, for instance, in the 'Technical Manifesto', Boccioni refuses to operate with any simple contrast between art and 'brute reality' or any simplistic notion of art 'returning to reality' *en principe*, but is primarily concerned with a general enriching of the repertoire of materials that an artist has at his disposal:

Destroy the literary and traditional 'dignity' of marble and bronze statues [conclusion no. 4 of the Manifesto exhorts]. Refuse to accept the exclusive nature of a single material in the construction of a sculptural whole. Insist that even twenty different types of materials can be used in a single work of art in order to achieve plastic movement. To mention a few examples: glass, wood, cardboard, iron, cement, hair, leather, cloth, mirrors, electric lights, etc. (*Archivi*, i. 72)

In declaring that an artist may even resort to as many as twenty different materials in a single work, Boccioni is doubtless indulging in manifesto hyperbole to focus attention on the use of materials *as means to aesthetic ends* (here: the achievement of a sense of plastic movement), and not for any primarily referential value.[3]

[1] Umberto Boccioni, 'Il cerchio non si chiude', *Lacerba* (1 Mar. 1914), 67–9. Papini avails himself of the editor's privilege of the last word with 'Cerchi aperti', *Lacerba* (15 Mar. 1914), 83–5.

[2] *Témoignages*, p. 29.

[3] There is a suggestion of this approach in a remark by Severini in his only recently published

Certainly, there are famous enough examples of Futurist collage to demonstrate that the integration of heterogeneous materials from the real world into a painting can have results of a strikingly non-mimetic kind. But for the Futurists the interaction between 'art' and 'reality' is in the majority of cases merely part of a much more fundamental exploration of techniques available and of the relations between various media and genres. There is less interplay between the source and the new context, except in the use of newspaper material, than was the case with Cubism or would occur again later with the Dada *ready-made*.

If one compares the way in which Braque discusses his early collage method with the Futurists' terms of reference, one observes that Cubist collage remains essentially a refinement of the *painting* tradition through the introduction of alien matter, whereas the Futurist concern is more with various intermedia possibilities. Braque reports how he included letters in his paintings of 1911 because 'they were forms which didn't have to be changed in any way. Letters are flat and therefore stand outside space. If they are present in a picture, they make it possible to distinguish the objects in space from those outside it.'[4] By contrast, what began in Futurism as the notion of a repertoire of means at a painter's disposal soon extended beyond the mere inclusion of a number of disjunctive materials to enrich a painting or sculpture and led to the systematic crossing of boundaries between art-forms and genres. From the fusion of words and shapes (literature and painting) we have already considered, the process developed to such extreme syntheses as filmed words-in-freedom, poetry in the sky, typographical ballet, and even architectural 'poetry' in the form of Depero's Book Pavilion of 1927 (see *FF* 419 for a photograph of this alphabetical building). 'The concept of art will have to be enormously widened,' Bruno Corradini and Emilio Settimelli proclaim in 'Weights, Measures and Prices of Artistic Genius' ('Pesi, misure e prezzi del genio artistico' (1914)):

There is no reason why every activity must of necessity be confined to one or other of those ridiculous limitations which we call music, literature, painting, etc. . . . EVERY ARTIST WILL BE ABLE TO INVENT A NEW FORM OF ART, which would be the free expression

'Futurist Ideography' ('Idéographie futuriste') of 1914: 'you will enlarge the means at your disposal' (*Témoignages*, p. 34). But on the whole Severini's theoretical pronouncements about collage—in contrast to his actual practice—are suffused with Cubist ideas.

[4] Quoted from Maly and Dietfried Gerhardus, *Cubism and Futurism: The Evolution of the Self-Contained Picture* (Oxford, 1979), 48. Jürgen Wissmann's influential account of collage in the twentieth century sums up the main difference between Futurist and Cubist techniques in the following manner: 'If Braque's use of a nail as a "reality-signal" had its primary significance as an aid in the localizing of formal interrelationships, in Futurism the emphasis shifted from the realization of painted elements to a presentation of reality itself in the work of art' ('Collage oder die Integration von Realität im Kunstwerk', in Wolfgang Iser (ed.), *Immanente Ästhetik, ästhetische Reflexion: Lyrik als Paradigma der Moderne* (Munich, 1960), 338). See also Jürgens-Kirchhoff (*Technik und Tendenz der Montage*, p. 37) for further documentation of the Cubists' interpretation of elements from the real world in their works.

of the particular idiosyncrasies of his cerebral make-up . . . in which would be found, mixed in accordance with a new measure and scale, the most diverse means of expression—words, colours, notes, indications of shape, of scent, of facts, noises, movements and of physical sensations; I.E. THE CHAOTIC, UNAESTHETIC AND HEEDLESS MIXING OF ALL THE ARTS ALREADY IN EXISTENCE AND OF ALL THOSE WHICH ARE AND WILL BE CREATED BY THE INEXHAUSTIBLE WILL FOR RENEWAL WHICH FUTURISM WILL BE ABLE TO INFUSE INTO MANKIND.[5]

Although most of the movement's manifestos were still devoted to the reform of single art-forms,[6] in reality the Futurists were by 1914 already moving towards the synthesis of various arts—an underrated instance of Futurist practice outstripping Futurist theory. The fact that the term 'collage' does not itself appear to have been in active circulation in Italy during the early years of Futurism[7] fortunately gives one a rare opportunity to see just what features of the genre the Italian Futurists really felt were essential, as they set about describing what they were trying to create.

Over and over again, the idea of a *fusion* of painting and literature occurs in their accounts of collage. Writing about *Danza serpentina* in 'Futurist Ideography', Severini makes it very clear that he sees the breakthrough which the work represents in terms of this new hybrid form: 'With [this] "free-word" drawing . . . I neither attempted to introduce a pictorial, plastic element into literature nor to introduce a literary element into painting. *What I envisaged was the possibility of finding a new form, an autonomous form of literary expression.*' *Danza serpentina* is presented in this contemporary essay as the forerunner, in fact, of a whole 'possibility of creating new ideographic forms which would represent the graphic expression of universal ideas'.[8] And it was as such that Marinetti and Carrà were to welcome the experiment. Severini's caption for *Danza serpentina—recherche d'une expressivité en unissant formes et mots* ('experiment in expressivity blending shapes and words', a title he used for many of the collages

[5] 'Weights, Measures and Prices of Artistic Genius' ('Pesi, misure e prezzi del genio artistico' (Mar. 1914)), published in *Manifesti del futurismo*, iii (Milan, 1919), 114.

[6] As both Flora and Clough point out, there was an inconsistency to the Futurists' on the one hand championing a transcending of conventional distinctions between art-forms, and yet, on the other hand, devising a strategy of issuing separate technical manifestos according to the traditional divisions of sculpture, painting, literature, music, etc. See Francesco Flora, *Dal romanticismo al futurismo* (Milan, 1925), 90 ff., and Clough, *Futurism: The Story of a Modern Art Movement*, p. 173. Something of this tension between the two conceptions of art can be found in Boccioni's individual formulations here and in the 'Technical Manifesto of Futurist Sculpture' ('Manifesto tecnico della scultura futurista'), where the idea of enriching sculpture in the direction of assemblage sits uncomfortably alongside the goal of going beyond conventional sculpture to create new art-forms.

[7] Carrà, for example, cites his *Sintesi circolare di oggetti* as one of his first works executed 'with coloured cards and coloured materials' (*Archivi*, i. 338), when one might have expected the term 'collage' to help characterize his endeavour. He also writes at some length about *Festa patriottica* in his letter to Severini of 11 July 1914 (*Archivi*, i. 341), again without availing himself of the term.

[8] *Témoignages*, pp. 34, 39 (my italics).

republished in *Témoignages*, in 1963—and the Futurists' predilection for hybrid generic concepts—'poema pittorico', 'dipinto parolibero', 'tipomontage',[9] 'poema disegnato', Severini's 'dessin "parolibero"' and 'poème idéographique' (a term taken over from Apollinaire), and 'compenetrazione tipografico'[10]—indicate the extent to which they viewed collage making extensive use of readable *papiers-collés* as an attempt at creating new mixed-media art-forms, rather than simply extending the repertoire of materials available to painting, as the Cubists often did, or available to conventional literature (as was largely the case with the *calligrammes* or the 'designed analogy').

Boccioni may have exaggerated in his suggestion that up to twenty different materials could be used, but there is nevertheless a richness to much Futurist collage. Severini's *Ritratto di Marinetti* soon became notorious for the aesthetic frivolity of its glued-on elements (half a moustache, a piece of felt, bits of manifesto), and his earlier *Geroglifico dinamico di Bal Tabarin* (*Dynamic Hieroglyphic of the Bal Tabarin*) for even resorting to such a provocatively unartistic ingredient as collaged sequins. Never slow to follow his Italian masters, C. R. W. Nevinson produced his *Portrait of a Motorist* (*c.*1914) constructed in the same idiom by the insertion of bits of glass into the paint to indicate the motorist's goggles and with a real button stuck on to the painted coat.[11] But within the vast repertoire of potential collage materials,[12] ranging from the restrained introduction of a small object, for instance the typewriter keys Rosai inserts in his *Dattilografa* (*Typist* (1918)), through to Prampolini's inventive *arte polimaterica* assemblages, such as his *Béguinage* (1914), the category of collaged material which is undeniably the Futurists' special preserve is the verbal collage used for propaganda effect.

Herta Wescher defines one of the main features of Futurist collage as its pronounced sense of topicality:

> As a point of principle, it is worth noting that the *papiers-collés* and the pieces of writing in Futurist paintings introduced a contemporary, realistic element into the works. Whereas for the Cubists the bits of newspaper, labels, and inscriptions aimed at nothing more than the suggestion of mundane reality within the abstract world of forms, in the Futurists' pictures they assumed an active, propagandistic, patriotic function, and it is

[9] Wescher, *Die Collage*, p. 58.

[10] This is the sub-title of Venna Lucio's 'Caffè + Arno' (*L'Italia futurista*, 11 (1917), 33).

[11] For details, see the Arts Council catalogue *Vorticism and its Allies* (London, 1974), 53.

[12] Clearly, if collage represents a play between integration and alterity (see Perloff, *The Futurist Moment*, pp. 47 ff.), then different forms of collage will tend more in one direction or the other. Wissmann's complaint that there has been a tendency to emphasize the 'local value' ('Stellenwert') of the material in the new aesthetic context ('Collage', p. 335) is symptomatic of the influence of Cubism where centripetal forces tend to dominate. (The tendency is also documented in Werner Hofmann, *Grundlagen der modernen Kunst: Eine Einführung in ihre symbolischen Formen* (Stuttgart, 1966), 52). In the case of Futurist propaganda collage, where there is a far more pronounced centrifugal quality leading out to the world of Interventionist politics, a purely formal approach becomes even less satisfactory.

characteristic that they coincided with the period of patriotic fervour. The Futurists often declared that the ideals of revolt and heroics, which they had championed from the very beginning, gained a concrete sense of direction thanks to the war and that concrete means of representation were also appropriate to these new concrete goals.[13]

The point is put succinctly in Linda Landis's account of Futurist propaganda-collage which is simply entitled 'Futurists at War'[14] The fact is that the most important single factor leading to the large amount of *literary* collage among the Italian Futurists was the increasing political radicalization of the movement in the face of the approaching war, and, after war had broken out, during the Interventionist period. With the Futurists, collage gradually assumed more literary proportions because it largely had a more distinctly verbal–rhetorical[15] role to perform. For some time, in fact, political collage became one of the main means of persuasion on behalf of Interventionism for the Futurists.[16] And, with the success of the Interventionist campaign in 1915, one detects a rapid diminution in the output of Futurist literary collages, in contrast to the continuing healthy production of words-in-freedom.

Although in the context of collage Severini's name is invariably—and with justice—associated primarily with *Danza serpentina*, the series of war collages he went on to produce in 1914 and 1915 are far more typical of Futurist developments during this period. His *Cannoni in azione* (1915), exhibited in 1916 at the Paris 'Exposition Futuriste d'Art Plastique de la Guerre', is constructed on a principle of lettered shapes analogous to *Danza serpentina* in some respects, although here the shapes are more representational and less abstractly geometrical.[17] The collage (fig. 2.7) combines a variety of verbal elements, ranging

[13] *Die Collage*, p. 62. The war's significance for the Futurists is well documented in Manfred Hinz, *Die Zukunft der Katastrophe*. Despite the specifics of the Interventionist campaign, he argues that war is seen as 'not for the achievement of specific goals, but for the purpose of destroying goals' (p. 106). The section in which these remarks occur is aptly entitled 'Marinetti's de-politicization of war'.

[14] *The Futurist Imagination*, pp. 60–75.

[15] Marjorie Perloff sees an analogy between the Futurists' concept of the viewer's relationship to a painting and the nature of their collages: 'true to their insistence that the spectator must be placed at the center of the picture, they reconceived collage as a propaganda art, an art that directly bombards the senses' (*The Futurist Moment*, p. 54).

[16] In his introduction to the anthology *Literarische Collagen: Texte, Quellen, Theorie* (Stuttgart, 1981, 18), Volker Hage claims that the politicization of collage only occurred with Berlin Dada, a misconception which comes about partly because in the following anthology his examples are selected from German literature only, whereas the introduction addresses the genre in general. However, this view of political collage's evolution is not untypical.

[17] Writing in 'Symbolism in the Plastic Arts and Literary Symbolism' of the '*qualitative analysis*' necessary to achieve the 'idea-image' of a car's speed, Severini cautions against overemphasizing the abstractness of such a 'realist abstraction' (p. 140) or of so-called 'idéiste realism': 'Speed, like all properties of an object, does not exist in itself for the painter, because we do not want to represent in plastic terms *the speeding car*, but *the speed of the car* in its quality as the dynamism of an idea, emotion, or state of mind. As a consequence, our *realism*, being above all the result of the ideal element and of the

FIG. 2.7. Gino Severini, *Cannoni in azione*, 1915 (Städelsches Kunstinstitut, Frankfurt a. M.)

from epic reports ('the gun sinks down'), descriptive labels such as 'ripping up | rising up | of the earth' and 'gradual curve towards the earth' (twice), describing the parabola of the bombardment, as well as various ritual onomatopoeic effects ('Bboum', 'iiiiii', etc.) and abstract nouns, again in the manner of *Danza serpentina* ('Power', 'immensity', 'freshness'). Of particular importance is the way the word 'FRANCE' is prominently displayed at the bottom of the canvas, identifying the setting as the Western Front—with the French, as so often, being held up by the Milanese Interventionists as an example to still-neutral Italy.

The 1973 Paris exhibition catalogue *Le Futurisme 1909–1916* reproduces this collage with the following passage from Severini's 'Symbolism in the Plastic Arts and Literary Symbolism' printed beneath it:

I have tried to express the idea 'WAR' by a plastic assemblage composed of war's realities: a gun, a factory, a flag, a mobilization-order, an aeroplane, an anchor. According to our conception of 'idéiste' realism, any more or less naturalist description of the

element of *sensory experience*, and although it is often expressed by geometrical shapes, is not an abstraction' (*Témoignages*, p. 47).

battlefield or of the carnage would not be better able to give us the synthesis of the idea 'WAR' than these objects which are the living symbols of war.[18]

Nevertheless, it is clear from the details of this description that the passage is far more likely to be referring, not to *Cannoni in azione*, but to the 1915 collage *La guerra* (*War*), which is part of a whole series of such works that Severini produced at this time, including *Treni dei feriti*, *Il treno dei feriti* (both known in English as *The Hospital Train*), an earlier *La guerra* (1914), *Lancieri italiani a galoppo* (*Italian Lancers at the Gallop*), and *Il treno blindato* (*Armoured Train*). And the distinction between the literary collage *Cannoni in azione* and these other pieces, such as *La guerra*, which often contain small amounts of *papiers-collés*, is not a peripherable one. They are markedly dissimilar kinds of collage, betraying in divergent ways a fundamental dissatisfaction with what Severini critically refers to as 'any more or less naturalist description'.

La guerra (*War* (1915)) (fig. 2.8) evidently reflects the assumption that a collage of symbolic objects, rather than a scenic version of a battlefield, is the best way to communicate the idea 'war'. The result can be viewed either as a simultanist juxtaposition or a collage of emblems, for the flag, the factory producing armaments, the gun, propeller, and anchor (symbolizing the three armed forces, as well as civilian and military war effort), and the mobilization-order are all conjoined under the patriotic slogan 'EFFORT MAXIMUM'. One only has to register the direction in which the mobilization-order is pointing to realize that its agitational message is being directed primarily at the beholder. In contrast to this work, *Cannoni in azione* (more appropriately entitled—in the singular—*Canon en action*, in *Archivi*, ii) is not based on the same principle of juxtaposing emblematic objects to create a quasi-allegorical 'réalisme idéiste'. On the contrary, as its title already indicates, it is much closer (more so than even *Danza serpentina*) to the more or less 'naturalistic' portrayal of a battle-scene. Nevertheless, a certain scepticism towards surface realism is still detectable, above all in the way in which the words supplement the visual dimension and hence create the collage effect.

In his retrospective essay 'Avant-garde Painting: Machinism and Art—Reconstruction of the Universe' ('La Peinture d'avant-garde: Le Machinisme et l'art—reconstruction de l'univers' (1917)), Severini makes the connections between the newly conceived 'réalisme idéiste' and his collage technique very clear, although, as so often,[20] the argument is distorted by a failure to

[18] *Le Futurisme 1909–1916*, catalogue of the 1973 exhibition of the same title at the Musée National d'Art Moderne (Paris, 1973), 126. 'Symbolism in the Plastic Arts and Literary Symbolism' was first published in the *Mercure de France* of 1 Feb. 1916, no doubt timed to coincide with the Exposition Futuriste d'Art Plastique de la Guerre held at the Galerie Boutet de Monvel, Paris, from 15 Jan. until 1 Feb. of that year.

[19] Reproduced in *Archivi*, ii. 333, pl. 75.

[20] Similarly, in his letter to Carrieri, Severini opts for too monolithic a view of the aims of the

FIG. 2.8. Gino Severini, *La guerra*, 1915 (present whereabouts unknown)

distinguish adequately between the painter's own aesthetic position and the differing standpoints of his fellow Futurists.

The precision, the rhythm, the brutality of machines and their movements have without doubt led us to a new realism which we can express without painting locomotives.

All of the efforts of the avant-garde painters tend in the direction of this new realism which, being subordinate to the sensation and the idea, has been defined by me in my earlier article as 'réalisme idéiste', a borrowing of Rémy de Gourmont's concept which seemed to me very precise.

The obsession with penetrating, with conquering by every possible means, the sense of the real, with identifying with life through all the fibres of one's body, is the basis of our experiments and of the aesthetics of all periods of history.

It is necessary to see in these issues, which are of a general nature, the origins of

Futurist collagists: 'I have the impression that deep down in these *papiers-collés* there was more or less on the part of everyone the desire for a *transcendental realism* which later on . . . would coincide with that desire to mount the stream of life to its source which characterized Surrealism' (quoted from Carrieri, *Futurism*, p. 117). However, a rejection of mimesis is not *de facto* 'transcendental'.

precise geometrical constructions, of our placing of various materials on to our canvases, materials such as cloth, sequins, glass, paper; here lie the origins of all those experiments which, unfortunately, were more or less misunderstood or poorly systematized at the time.[21]

One striking feature of nearly all Severini's discussions of his collage-work is the almost exclusive focus on questions of style and technique. Such an emphasis is no doubt largely the result of the exigencies of the painter's one-man campaign on behalf of his so-called 'réalisme idéiste', a goal shared neither by Carrà nor by Marinetti himself. For inasmuch as surface realism is rejected in favour of a conceptualized 'essential' realism, so the First World War becomes 'war', the hideous slaughter on the Western Front is transformed into the emblematic 'gun(s) in action', and the sordid reality of the trenches is presented metonymically, in a way well suited to the collage idiom, with its series of quasi-allegorical signs.

If in its scenic method *Cannoni in azione* is an exception among Severini's war collages, a more typical creation of the year before is his first piece bearing the title *La guerra*.[22] Combining verbal elements once more, and flaunting the noun 'Antihumanisme' with the slogan 'Effort maximum' now radiating outwards vertically from the centre, it is again decked out with symbolic objects: telegraph poles, maps, factory chimneys, scientific equations, and graphs. But the words, instead of being attached to shapes, are this time separate ingredients in the total construct. One item, for instance, is part of the chemical formula for the manufacture of wartime 'gaz asphyxiants', as the collage at one point explains. And this *separation of elements* interacting within the collage (in what Marianne Martin refers to as a 'poster-effect'), rather than the tautological labelling method, is more typical, not only of Severini's work, but of most mainstream Italian Futurist collage.

We encounter it once more in Boccioni's *Carica dei lancieri* (*Cavalry Charge* (late 1914)) which is a juxtaposition of painting and separate newspaper cuttings, although, as has been pointed out,[23] its explicit treatment of the war-theme makes it very much an exception within Boccioni's *œuvre*.

[21] *Témoignages*, p. 57. The article was first published in the *Mercure de France* (1 June 1917).

[22] Galleria Falsetti, Prato. This work is not reproduced in *Archivi*, ii, but can be found as pl. 22 of the 1974 Düsseldorf Städtische Kunsthalle catalogue *Futurismus 1909–1917*: section 11: 'Gemütszustände/Analogien'.

[23] Martin, *Futurist Art and Theory*, p. 200. Noting that *Carica dei lancieri* was first published in *La grande illustrazione* (no. 13, Jan. 1915, p. 4), Giovanni Lista refers to its 'impoverished character', adding that the review paid its contributors and also that it was a special issue 'for the benefit of the refugees in Belgium', this collage being produced '"on commission" . . . at the request of Marinetti or Sibilla Leramo who was in charge of the publication' (introd. to U. Boccioni, *Dynamisme Plastique: Peinture et sculpture futuristes*, ed. Lista, trans. Claude Minot and Lista (Lausanne, 1975), 19, 147). Lista cites Herta Wescher's comment that the 'printed text has, curiously, the same meaning as the narrative scrolls of medieval tapestries, accompanying the composition without entering its structure' ('Collages futuristes', *Art d'aujourd'hui*, 4/34 (1953), 10), as if this were all accounted for by the context.

The choice and positioning of the printed matter in *Carica dei lancieri* clearly indicate that its verbal element is very much part of the collage's semantics. That the German positions are being hard-pressed by the French, and that there is progress being made in the Alsace region, gives a very specific meaning to the depicted cavalry-charge, now revealed as that of the French Army repelling the German enemy. With Paris identified as the source of the information in the inserted newspaper report, albeit one taken from an Italian newspaper, the collage displays what is, from an Italian Interventionist point of view, a patriotic perspective: that of a belligerent anti-German/Austrian nation.

Marianne Martin describes Boccioni's collage as a 'powerful and convincing statement about the dynamics of battle', seeing the painted element as being 'reinforced by the newspaper items employed'.[24] (My point is that the work achieves its true meaning, an Interventionist one, only when we take account of these printed elements.) But this basic structure—of picture (statement) juxtaposed with a 'reinforcing' (newspaper-)report—could be expressed differently, as one of *simultaneity*, for the occurrence that the newspaper-reports are describing is actually coming to life, relived as a present event in the main pictorial part of the collage. As in Marinetti's *Le Soir*, a battle and the description of it are presented simultaneously, although it must be stressed that simultaneity in this context is far from being synonymous with any parity of elements. The painted sections dominate the collage, while the printed word merely functions as a form of subordinate comment, localizing the historical event, just as the code-name 'Joffre' is so frequently expected to do in contemporary Interventionist collages. It is as if traditional passéist modes of description are here again (as in *Le Soir*) assumed to be less effective than the dynamism of a Futurist collage. And yet they are still required; there must be a fusion of words and shapes to give the visual element precise propaganda meaning and to lend a patina of documented historical authenticity to what is above all an agitational, political work of art.

Running down the right-hand side of the collage is a further typographical contribution to the overall effect of simultaneity. References to Austrian/German involvement on the Eastern front and concurrent events in France and Belgium locate the specific Alsace offensive within the wider context of what is happening elsewhere during the war. The only dimension missing—and the perspective necessary to complete the collage's grand conspectus, as it were—is information about what is happening in Italy at this time. This is only implied by the newspaper report *in an Italian newspaper* (in fact, the Interventionist *Corriera della sera*): while the rest of Europe are taking sides and engaging in battle, the Italians, these *papiers-collés* suggest, are ignominiously sitting at home, content merely to read about the momentous events of history in their newspapers. In other words, the collage is implying that Italy should be taking a

[24] *Futurist Art and Theory*, p. 200.

leaf out of France's book and entering the war too, instead of vegetating in its present state of beslippered neutrality. Indeed, one can take Boccioni's essentially Romantic depiction of the dynamic charge of the lancers at the collage's centre as an evocation of the invigorating battle experiences that await the Italians when they do eventually join the French and commit themselves to the anti-German cause.

Since 'Interventionism' itself is a somewhat complex, abstract subject to depict visually, the Futurists sought to give it concrete reality—'idéiste' or otherwise—in a number of specific ways: by portraying the fervour of Interventionist demonstrations (as if these were a prefiguration of the battle experience itself), or by concentrating on France's moments of glorious involvement in the war so far (especially Joffre's decisive Marne victory[25]), in contrast to Italy's neutral stance, or by demonstrating the supposedly fulfilling nature of battle itself as the epitome of Futurist experience. As a result, Futurist Interventionist collages often have a bold, dramatic quality which their largely sloganeering, propagandistic functions might not otherwise necessarily have achieved.

As their verbal collages became more and more geared to political goals, the Italian Futurists found themselves increasingly faced with the practical problem of disseminating them in effective numbers. *Lacerba*, which at its peak purportedly enjoyed a circulation of twenty thousand,[26] unfortunately rarely published more than one collage or picture per issue. Later Futurist journals such as *L'Italia futurista* and *Vela latina*, which were far more open to visually experimental material, never reached the exalted circulation figures of their predecessor; and, of course, they in any case largely came to the fore after the Interventionist campaign was over. Since large-scale exhibiting was also hardly the solution—even the famous Paris exhibition of Futurist 'Art Plastique de la Guerre' of 1916 only reached a limited (and in any case already converted) public—the Italian Futurists resorted to two main tactics. They turned to the publication of individual collages—or sequences of collages[27]—on broadsheets (*volantini*), the manner in which they had, after all, issued many of their original manifestos; and to the gathering together of collage

[25] The French troops' victory at the Battle of the Marne became a frequent symbolic point of reference for the Italian Futurists. It showed France enjoying the kind of military success they felt the Italians should have been having. Moreover, in retrospect, the Marne could be seen as the turning of the tide. But the main perspective at the time lay in the simple fact that the French were engaged in combat, whereas to their shame the Italians were not.

[26] According to Gramsci's letter to Trotsky of 8 Sept. 1922: A. Gramsci, *Scritti politici*, ed. Paolo Spriano (Rome, 1967), 530.

[27] For example, the four-page pamphlet *Parole, consonanti, vocali, numeri, in libertà* (Milan, 1915, discussed below) contained Buzzi's *Bombardamento aereo* (*Aerial Bombardment*), alongside Marinetti's *montagne + vallate + strade × joffre* (*mountains + valleys + roads × joffre*), Govoni's *Il palombaro* (*The Diver*), and Cangiullo's *Le coriste* (*The Choristers*).

propaganda-material in book-form. Because both methods of dissemination had some influence on the actual collage-techniques employed, I should like to consider one or two Futurist broadsheet collages and some of the material from two of the most important collections of the time: Carrà's *Guerrapittura* and Buzzi's *Conflagrazione*.

Buzzi's broadsheet *Bombardamento aereo* (fig. 2.9) was distributed by the Direzione del movimento futurista in Milan and bears the date 11 February 1915. This means that Italy had still not entered the war when it was produced. Although it is attributed to 'PAULO BUZZI, parolibero', the work is much closer to the synthesizing 'free-word-picture' tradition merging words and graphic elements than to free-word 'designed analogies'. It is a mixture of hand-printed words and shapes, with abstract nouns indicating movements, forces, sounds, and emotions during the aerial bombardment. Like so many Italian Futurist works of the period, it is essentially a hymn to war, but with a meaningful glance over its shoulder at the neutrals who do not share this viewpoint. Thus the wedge of 'entusiasmi' on the left is contrasted with the 'freddo paure pacifiste ventraiole' on the right, with an explosion of words radiating from the centre of the picture as gas and petrol bombs land on the enemy placed in the centre. (As so often in Futurist war collages, one gets the impression that the enemy under attack could just as well be the poet's pacifist compatriots as Italy's foreign political foes of the time.) We are informed at the top that the Futurists had prophesied such events in 1910. The enigmatic linking of the date 1915 with the segment marked 'Londra? Parigi? Roma?' would appear to be questioning whether Rome is now prepared to get into (marching) step with the other two cities, and hence whether the same enthusiasm for entering the war is shared by all three of the nations. Although much of the handwritten material retains a basic horizontality, the broadsheet format nevertheless makes it easier than with a hung collage for the recipient to twist the page around and read the various spiralling and vertical pieces of writing. The title may even refer to the broadsheet's most effective method of delivery, as well as its subject-matter. As an unashamedly activist collage, the work has little complexity to its construction. It too 'bombards' the reader with simple ideas and images so that its point can be easily absorbed and the material passed on to other recipients.

A far better-known collage, Marinetti's *montagne + vallate + strade × joffre* (also 1915) (fig. 2.10), consists, by contrast, almost entirely of collaged and printed words,[28] with the addition of a number of large letters to give the rudimentary shape of a setting. The overall effect is very similar to that of some of the sketches in Marinetti's unpublished war-diary,[29] but the structure is more

[28] Useful illustrative material concerning Marinetti's method of composition with such works can be found in *The Futurist Imagination*, pp. 3 f., and *FF* 193.

[29] Cf. Marinetti's parallel collage *Vive la France* (ill. *The Futurist Imagination*, p. 4) where one can also see a use of both drawn letters and glued-on pieces of print.

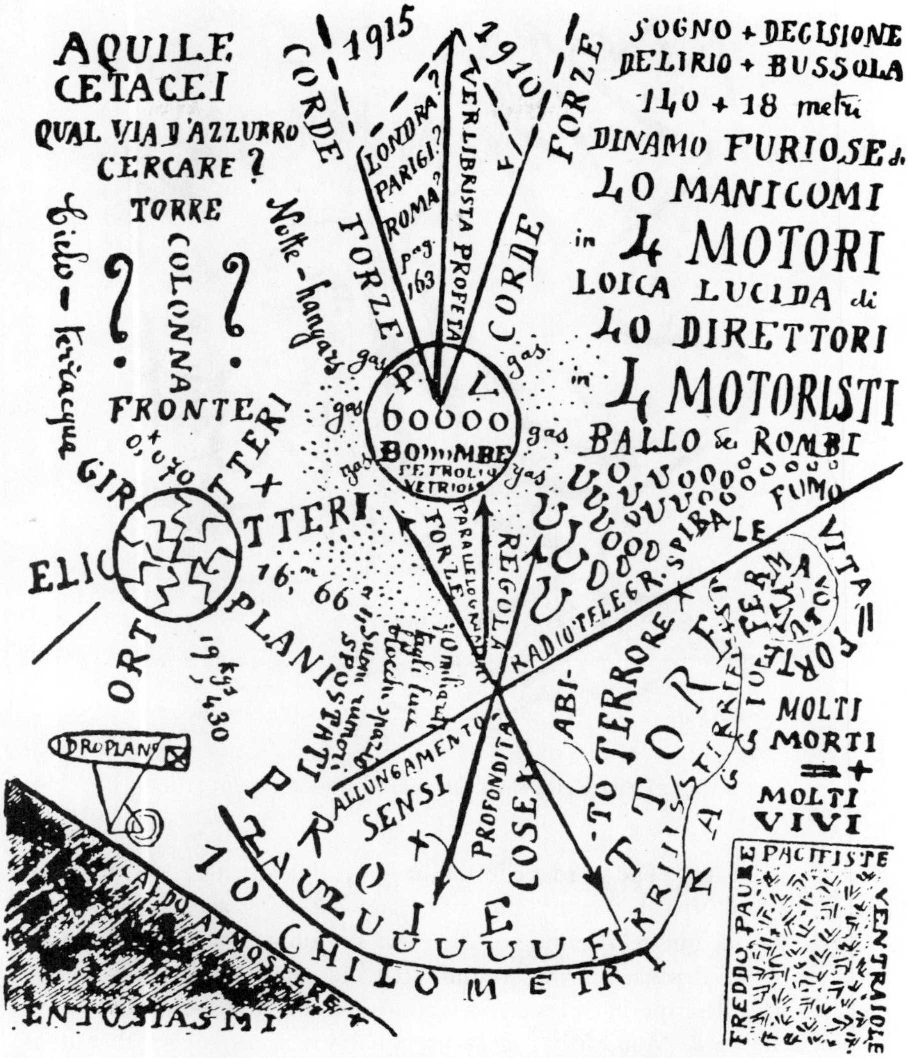

PAOLO BUZZI, *parolibero.* – Bombardamento aereo

Milano, 11 Febbraio 1915.

DIREZIONE DEL MOVIMENTO FUTURISTA: Corso Venezia, 61 - MILANO

FIG. 2.9. Paolo Buzzi, *Bombardamento aereo*, broadsheet, 1915

FIG. 2.10. F. T. Marinetti, *montagne + vallate + strade × joffre*, 1915

verbally associative. The word 'belle' on the left is linked with both 'France'—triggering off 'Vive la France' and 'mort aux boches' near the top M (for 'Marne'?)—and 'guerre'—leading off to a series of onomatopoeic war-sounds directed at the Prussians in the bottom left-hand corner of the collage. The 'leger lourd' description—or stage-directions—at the top would appear to relate to the words 'Mon AMIiiii' and is presumably meant to suggest that this is the voice of the venerable Joffre speaking to one of his heroic compatriots, as is also the 'MaAAAAAa petite'. (The collage plays associatively with the sound-pattern linking the words 'Marne', 'Mon ami', and 'Ma petite' and even, via this last word, the 'tatata' of distant machine-gun fire.) The work is, in fact, as much a *sintesi* scenario as a conventional typographical collage.

One of the features which Marinetti's collage has in common with Severini's *Cannoni in azione* is its recourse to an explanatory verbal element. Not only in the sense that the title is needed, as so often in Futurist collages, to apprise us of the specifics of the historical occasion and to locate details of the setting, but also in the way in which a whole block of print in the bottom left hand corner has to be bracketed and explained as representing the 'dynamic verbalization of the route'. This is what Lázsló Moholy-Nagy has to say about this part of the collage:

> One sees the curves which the motor car followed, the 'dynamic verbalization' of the route of Joffre in the right hand corner. The general's conventional speech to the soldiers is translated into typography. His words are torn to pieces by the accompanying noise of machine gun and cannons. A large number of soldiers is represented by the logarithm column in the upper left, they echo the general's shouting 'Vive la France!' 'Mort aux Boches!'[30]

Not only would many of these details be unrecognizable without the 'verbalization' (clear enough evidence, if Papini ever really needed it, that the circle was a long way from closing), but even the actual shape of the layout here, as opposed to almost everywhere else in the collage, is decidedly undynamic, if not passéist, in its horizontal linearity.

Again, the collage's structure is neither representational nor perspectival, but exploited to create an effect of simultaneity. For, while the plurals in the work's title (mountains, valleys, roads) might conceivably refer to the constituent parts of a single setting at the Front, other aspects of the collage militate against such a traditional interpretation. The block of print identified as the 'dynamic verbalization of the route', for instance, represents a continuum of sounds heard over a period of time as Joffre makes his way along the Front region, just as Joffre's 'Mon ami' and 'Ma petite' are, in Lessing's sense, an epic *Nacheinander* being presented as a *Nebeneinander* by the simultaneity of the collage method. Despite Moholy-Nagy's assumption that we are witnessing Joffre speaking to his troops, whoever is being addressed as 'Ma petite' is unlikely to be a soldier and could hardly reply with 'Mon ami'; so it is reasonable to assume that these phrases are *consecutive* greetings to various people from the only person in the collage actually named: Joffre himself. What we are offered, then, is the simultaneous presentation of the journey to the Front and the unfolding scene of welcome on the Marshall's arrival. The more specific phrase 'après la Marne', in the title of the version published subsequently in *Les Mots en liberté futuristes*, makes it much clearer that this is the visit of a victorious general.

Evidently, this piece was conceived either as an Interventionist collage or, if nothing more, as a tribute to a victorious ally-in-spirit. Joffre stands once again as the exemplary victor halting the German advance, the enemy being depicted here as decimated by gunfire to the point where their numbers are the square-root of what they had been at the outset. The overall effect is not just a military scene of a general visiting his troops; the collage's scope is substantially widened to the national perspective by depicting France's hero being acclaimed by soldiers and civilians alike.

30 *Vision in Motion*, p. 310.

FIG. 2.11. F. T. Marinetti, *Une assemblée tumultueuse. Sensibilité numérique*, 1919

One can gauge the relative politicization of the genre in this instance by comparing the Marne collage with one of the few works in *Les Mots en liberté futuristes* that appears to have been created after the end of hostilities: namely, *Une assemblée tumultueuse* (fig. 2.11). It may well be, as Jane Sharp argues, that the work 'refers to an Italian patriotic celebration at the end of World War I',[31] but there is far less overt political content here than was the case with the Marne collage. Certainly, the date '1918' repeated at the top right-hand corner of the work is the best indication of what may be being celebrated (though, I find it difficult to share Christine Poggi's assumption that some of the shapes near the centre are meant to suggest torpedos; they look more like festive trombones). As the difference between *montagne + vallate + strade × joffre* and *Une assemblée tumultueuse* suggests, a process of abstraction is creeping into Marinetti's collages, to the point where a less propagandistic work like *Une assemblée tumultueuse* looks remarkably similar to Soffici's *Tipografia* (fig. 2.12).

The monochrome publication of Interventionist literary collages on broadsheets and in book form inevitably led to the diminution of the painted element in them, if there had been such an element in the original prior to being litho-

[31] Jane Sharp, 'Sounds, Noises, and Smells: Sensory Experience in Futurist Art', *The Futurist Imagination*, p. 23.

Fig. 2.12. Ardengo Soffici, *Tipografia* from *BÏF§ZF+18: Simultaneità e chimismi lirici*, Florence, 1915

graphed or photographically reproduced. A tendency to work either with rudimentary (outlined) shapes plus printed words or simply with different—sometimes framed—forms of verbal component (Marinetti's collages rely substantially on their typographical component) can be detected in most of the collages of the time. Thus Carrà's *Guerrapittura* work *Cielo di guerra* (*War Sky*) (fig. 2.13)[32] merely augments its central newspaper cutting about France's engagement with the German armies on the Western Front with enough of a suggestion of an aeroplane's wing and tailpiece to indicate a warplane, with part of the *papiers-collés* appearing to drop away towards the terrain below like a cluster of bombs. Otherwise, the collage is mainly supplemented by various onomatopoeic sequences, the 'RRR' and 'VRRRR' of the engine and some sound-patterns of an ambiguous nature, such as the 'zzzzzzz', conceivably representing gunfire heard from a height or the rush of air, though hopefully not the pilot asleep (that code had not yet entered modern language). The figures '300'

[32] *Guerrapittura: Futurismo politico, dinamismo plastico, disegni guerreschi, parole in libertà* (Milan, 1915), 21.

FIG. 2.13. Carlo Carrà, *Cielo di guerra*, 1915

and '365' seem to be used analytically to suggest altitude, although, if this had been a work by Marinetti, it might well have quantified the number of bombs dropped or enemy destroyed. The only other major ingredient is the familiar kind of explanatory labelling used at those points where the capabilities of the iconic alone would have been overextended: the 'Solitude' describing the pilot's cockpit, and the simultaneously presented complex of 'suoni rumori colori odori speranzi' coming up from the soldiers on the ground below. Here we come once more towards an approximation of a verbalized realism which has little to do with Severini's 'réalisme idéiste' and is more akin to some comic-strip version of reality and represents the fusion of free-word poetry with non-verbal collage to create the *dipinto parolibero*.

While some of the larger repeated letters in *Cielo di guerra* have been stencilled on, the rest of the collage is made up exclusively of two forms of printed material: that derived from selected newspaper cuttings and that actually printed on to the page by Carrà himself. By contrast, Carrà's strangely surreal *La notte del 20 Gennaio 1915 sognai questo quadro (Angolo penetrante di Joffre sopra Marna contro 2 cubi germanici)* (*On the Night of 20 January 1915, I Dreamed this Picture (Joffre's Angle of Penetration on the Marne against Two German Cubes)* (1915)) (fig. 2.14), in the same volume, has a far more emphatically sketched quality. The geometrically organized conflict between the dynamic Futurist triangles and force-lines and the stolidly Teutonic cubes, with their menacingly black Iron Cross darkening the sky behind them, is essentially a Cubist-influenced design with added words in the 'free-word' tradition: e.g. the star-shaped *papier-collé* at the centre of the collage and the various onomatopoeic and, again, explanatory labels radiating outwards from it. Linda Landis has interpreted this collage in some detail in *The Futurist Imagination*, and I should like to quote her account at some length here, since it pinpoints a number of details in the work, while at the same time reintroducing the kind of mimetic assumptions from which the Futurists were trying to liberate their art. According to Landis, the collage

> finds its basic structure in aerial maps but, through the use of diagonal lines, wedge shapes, and perspective construction, creates a three-dimensional space. As a map, the collage is divided into halves by the light, vertical strip in the center that denotes the Marne river. The star-shaped newspaper clipping punctuates the importance of the specific battle site with the word 'ici' just as a star or other symbol often is used on conventional maps to designate a particular location. The German position indicated by the Iron Cross in the upper right, thus occupies the area 'east' of the river, while the French army is located to the 'west'.
>
> The engagement of these two armies at the Marne demonstrated the value of reconnaissance aircraft in battle.... In his collage, Carrà alluded to three reconnaissance aircraft, all of which were present at the Battle of the Marne, by depicting their more recognizable features. The two box-shapes ('cubes') identify the German Albatros BII

FIG. 2.14. Carlo Carrà, *La notte del 20 Gennaio 1915 sognai questo quadro (Angolo penetrante di Joffre sopra Marna contro 2 cubi germanici)*, 1915

with its characteristic eight vertical struts defining a two-bay (or two-'cube') wing structure. An unarmed, two-seater biplane, it was used for reconnaissance duties from the beginning of the war until well into 1915. Connected to the Albatros wing section by dashed lines in Carrà's collage is the Iron Cross, the emblem that German pilots had painted on the tails of their airplanes just weeks prior to the Battle of the Marne to prevent being shot down by their own army. . . . Directly beneath the visual abbreviation for the Albatros appears the comma-shaped tail of the well-known German Fokker monoplane which in two of its early forms (the M.5K and the M.5L) saw considerable action during the first weeks of the war, including the Battle of the Marne. The third reference to a war reconnaissance machine in the Carrà work is found on the French side of the Marne in the form of a white, wedge-shaped cutout. The forceful design describes the sharply tapered fuselage of the French Morane monoplane, on which Carrà drew in ink the inverted V-strut and guy wires rising above the airplane's body.[33]

Leaving aside the strange, though not impossible fact that both sides' reconnaissance planes appear to be flying on scouting missions over their own occupied parts of the region, this kind of perspectivism is not what collage, Futurist or any other kind, is really about. There is more abstraction, more simultaneity of images, and less documentation both to the visual dimension of this work and to its title. The reference to the collage as a dream brings out a further possibility to the collage-*stati d'animo* relationship.

In his introduction to *Guerrapittura* Carrà declares that 'today the pictorial masterpiece and the most beautiful poem that has appeared in the world are those that the gun has created'. 'Is true art', he asks, 'the art of authentic innovators, not perhaps always the art which exalts the true Motherland? . . . The war, instead of obstructing, stimulates and encourages the expression and affirmation of artistic genius and increases the courage and power to destroy and create.'[34] As if in direct corroboration of this claim, one of the *Guerrapittura* collages, *Scoppio d'un obice: Espansioni atmosferiche* (*Howitzer Explosion: Atmospheric Expansions*), depicts a series of exploding shapes and words, based on Marinetti's 'zang tumb tuuum' motif, together with the stencilled-on ordnance code 'K4'. This is one of the more radical of Carrà's literary collages, in the sense that no real Italian words appear in it, only onomatopoeic neologisms established two years before by the movement's leader in *Zang tumb tumb*. Yet this abstraction and play with one of Futurism's most famous book-titles contrasts starkly with the callousness of the collage's two-part title. For the juxtaposition of the main caption ('howitzer explosion') with the scientific description of the physical effect ('atmospheric expansions') that such a phenomenon causes displays not one iota of concern for the more tangible human devastation that such a weapon inflicts. There is no irony here, any more than there was in Marinetti's claim that the Futurist was as interested in the 'suffering' of a light bulb as in

[33] 'Futurists at War', *The Futurist Imagination*, pp. 64 f.

[34] *Guerrapittura*, p. 45.

that of human beings; such depersonalization becomes tantamount to a dehumanization of literature.

In accordance with the Futurist obsession with widening the repertoire of available artistic forms, the twelve collages in the first part of *Guerrapittura*[35] are remarkably diverse. They range across a wide spectrum of structures and techniques—from oval designs, reminiscent of some of Picasso's work, and of still lifes by Braque and Gris, mixing patriotic slogans such as 'W Trieste Italiana' with the Italian Army star and various images suggesting disputed territory (*Guerra nell'Adriatico* (*War in the Adriatic*)), to a fusion of exploding word-fragments and dynamic shapes (*Scoppio d'un obice*) and a mixture of drawing and collaged photo-fragments (*Ufficiale francese che osserva le mosse del nemico* (*French official observing enemy movements*)). The volume even includes completely abstract collage-free drawings with a collage-structure, nevertheless, such as *Ritmi d'un cannone trainato al galoppo* (*Rhythms of a Gun pulled along at the Gallop*) and *Volontario garibaldino, ventre a terra (sparando)* (*Garibaldini Volunteer, lying flat (shooting)*). Yet one only has to compare the rich colours and sheer riot of effects in Carrà's *Festa patriottica* or Severini's *Nord-Sud* (*North–South*) with the monochrome restrictions on the material in *Guerrapittura* to appreciate the extent to which some Futurists were willing to make sacrifices and contain their experiments in order to achieve a wider circulation for their war propaganda.

One of the untitled collages in Buzzi's *Conflagrazione* (fig. 2.15), after *Guerrapittura* the most important collection of Futurist war-collages, bears the bold words 'I furibondi combattimenti nei Carpazi' and includes in *papiers-collés* form an actual map of the region, fragmented into three parts and interwoven with various other word-balloons to illustrate the scene of the offensive. The collage consists, apart from the map, of what appear to be two newspaper headlines, the one already quoted (which could be an invented caption), together with 'Contrattacchi austro-tedeschi sul Latorcza', with the addition of a series of containing outlines and shaded areas. Unlike most Futurist collages about the war, however, this work remains distinctly static. Such drama and dynamism as it does possess lie primarily in its verbal elements, not in the actual visual collage structure, whose main role is one of documentation.

A more accomplished, though less typical, *Conflagrazione* collage (fig. 2.16), which can for convenience be referred to as *Mulhouse* (after the town-name at the bottom of the work) creates a subtle sense of movement and confusion through the interpenetration of the words 'NEVE' and 'TEMPESTE' (the only other verbal element being the 'Dai Monti Tatarka' at the top and the town-name at the bottom). *Mulhouse* gives the same conspectus of a battle-situation as one has on a map, and it is significant that something as artificial as a map, rather than a

[35] The twelve collages are followed by a miscellany of theoretical statements, a section of free-word poetry, and a 'Futurist Political Programme'.

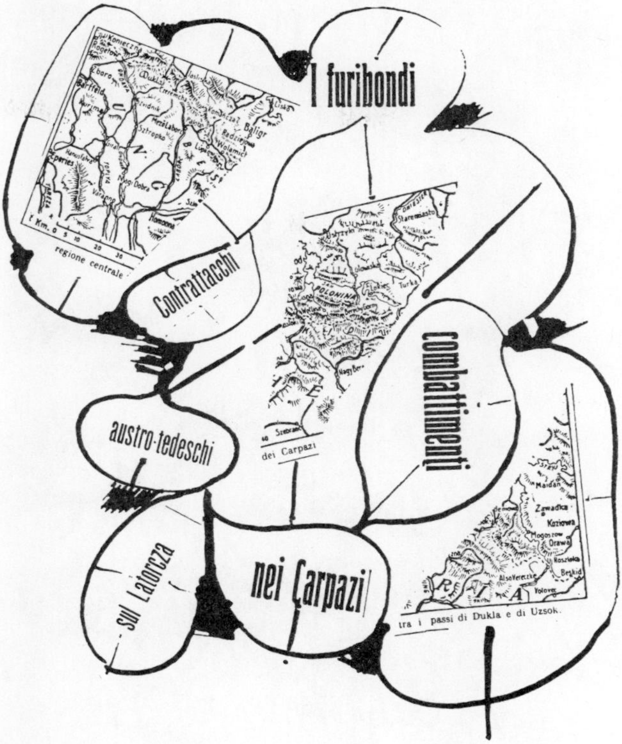

FIG. 2.15. From Paolo Buzzi, *Conflagrazione*, 1915–18

pictorial representation of the scene, is chosen. It is as if Marinetti's technique of creating a free-word *carta sincrona* had been transferred from shaped poetry to literary collage.

Other works in *Conflagrazione* range from boldly laid-out slogans in the form of grafitti, or reminiscent of the poster quality of some of Severini's work (e.g. 'Russia rivoluzione guerra × libertà') and a series of descriptive phrases based in the form of a question mark, asking about Italy's future ('rottura sul fronte italiano L'Angua del Destino morde L'Italia') to more conventionally literary free-word poems such as Buzzi had been writing for a number of years.

With the exception of the magnificent collages in *Les Mots en liberté futuristes*, there is an unevenness to most of the Futurist collections of collages and books of liberated typography in the Interventionist cause—as if the desire to put a political statement quickly into circulation at times conflicted with the longer-term aesthetic need to allow a balanced collection of good collages to materialize. As a consequence, some of the best Futurist collages exist alongside political ephemera.

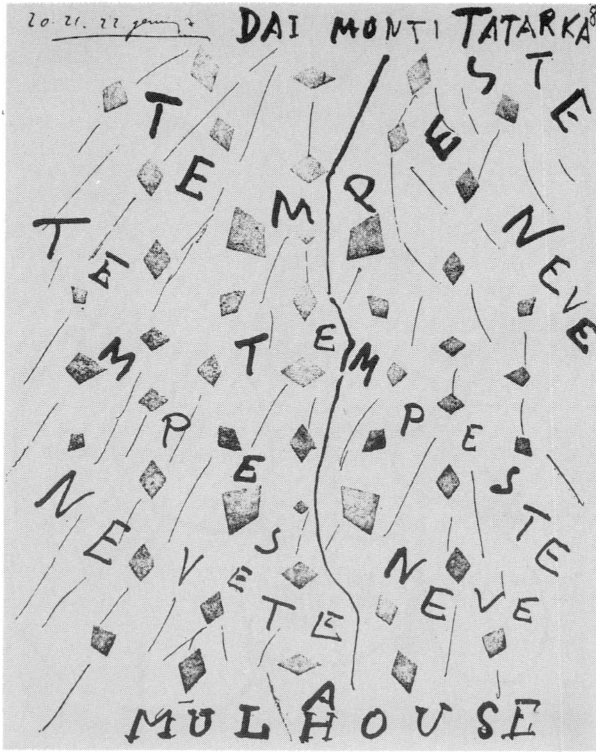

FIG. 2.16. From Paolo Buzzi, *Conflagrazione*, 1915–18

One work, which remained unpublished at the time and yet represents one of the most impressive 'fusions of words and shapes' the period produced, is Cangiullo's *Milano-Dimostrazione* (1915) (see fig. 1.2). Cangiullo, whose name is primarily associated with the animated letters and figures of *Caffèconcerto: Alfabeta a sorpresa*, had always displayed a penchant for the whimsically mimetic in his work and never experimented with the kind of simultaneous, a-perspectival collage that Marinetti, Carrà, and Severini were interested in. In this respect, *Milano-Dimostrazione* shows Cangiullo as remaining true to form. But it nevertheless reveals how effective (even if un-Futurist) the much lampooned pictorial style can be. It is a depiction, using a spirited combination of densely packed longhand writing, representing the demonstrating crowds—presumably at either an Interventionist or a Fascist rally—and the background drawing of the Piazza del Duomo in front of the Galleria Victor Emanuel. Patriotic slogans such as 'long live war', 'down with Austria and Germany and Turkey', 'Long live the King', and 'Mussolini' (for 1915, a very early reference to Futurist sympathies for nascent Fascism) gradually fade in the distance into a general blur of single letters and shapes. The Italian national flag joins together the crowd and the surrounding buildings; indeed the collage nowhere operates with any simple contrast between the writing that represents the crowd (and their shouts) and the drawn part of the work, for even the patterns on the edge of the buildings turn out to be made up of repeated numbers and letters. The fact that this piece figured in our consideration of Futurist iconic effects in poetry indicates just how much of a grey area such techniques represent. It is no more a collage, in the narrower sense of the word, than *Danza serpentina* is; in fact, it is constructed according to the same process of fusing words and graphic elements that Severini is supposed to have discovered. But here the pictorial retains the upper hand, and, in this respect, the work is something of a maverick among Futurist synthesizing experiments.

If 1914 was the *annus mirabilis* for Italian Futurist collage, 1915 was already the last year to see the genre being produced in any great quantity. Although dated 1919, and interpretable as a celebration of Italian victory, *Les Mots en liberté futuristes* was largely a retrospective collection of pieces that had individually been produced a number of years before. As we have seen, the anachronism of the Interventionist cause to some extent accounts for the waning of a particularly propagandistic mode of collage, although other forms of collage and assemblage continued to be produced during the Second Futurist phase, and the work of Prampolini, Rosai, and Sironi matches the achievements of the earlier Futurist collagists.

The end of the Interventionist campaign can, however, only be one factor in this process. After all, the next logical step would have been the development of patriotic pro-war collages which could have carried on being produced right up until 1918. But they did not materialize, with the exception of some of the works by Severini we looked at earlier in this chapter. In all probability, other reasons are also at play: the fact that Italy's entry into the war meant that military service curtailed the Futurists' creative activities, and in some cases their lives; paper shortages; the possibility that the Futurists, during the Interventionist phase, had already to a considerable extent exhausted the possibilities of interplay between shapes and readable word-elements, at least as far as political rhetoric was concerned. (It is really with Berlin Dada political collage, and even more effectively, with satirical photo-montage, that a comparable form comes to the fore again.) The main form of collage developed in Italy after this period by Prampolini, Rosai, and Sironi is non-literary and is largely in the assemblage idiom. Moreover, it lacks the dynamic geometry that one associates with the collages of the First Futurists.

Symptomatically, when Prampolini writes about the use of various materials, including newspaper fragments, and other elements borrowed from reality, he stresses that they have been totally liberated from their original meaning and function: 'We have reached the stage where we can replace the conventional components of a picture with real elements which we grasp in their authentic materiality. In doing so, we naturally exclude their normal function and in fact give the fragments of objects new life by arranging them according to their plastic–architectonic efficacy.'[36] It might be countered that Cubist collage had in a sense reached that stage some years before, but, if one looks at the actual quality of these new assemblages, it is clear that Second Futurist collage is gradually moving in the direction of Dada and Surrealist juxtapositions and not looking back to Cubism. Yet, whereas Schwitters keeps the literary, verbal element alive in his collage-work, it all but disappears in Second Futurism.

No doubt due, on the one hand, to the close connections between French Cubism and Italian Futurism, which did much to foster the spread of interest in collage, and, on the other, to the relative cultural isolation, chauvinism, and different preoccupations of the Russian Futurists, collage did not take root in Russia with the same literary component as one finds in Italy. To be sure, there was a collage tradition in Russia, using superimposed numbers, letters, and words somewhat in the Cubist manner. Works such as Goncharova's *The Cyclist*, Popova's *Italian Still Life* (which includes a fragmentary allusion to *Lacerba*), Malevich's *An Englishman in Moscow*, Rosanova's *Geography*, and Udaltsova's *At the Piano* all give evidence of a Cubist-orientated interest in the interaction between the word (or the number) and the main graphic component. But there is no comparable fusion of the poetic and the visual, no equivalent to the *dipinto parolibero*.

When Khlebnikov argued that there was 'an analogy between his formal pursuits in the sphere of poetic language and the system of modern French painting', what he had in mind was not the use of a poetic element within collage, but the analogy between the French Cubists' fragmentation of objects and his own procedures with the word. It is in this respect, above all, that the analogy implied by the term Cubo-Futurism is most meaningful.[37] A linguistic dissecting procedure comparable in method to the techniques of the Futurist painters predominates, but not any specific attraction to collage on the part of the Russian poets.

[36] *Noi*, ed. E. Prampolini and B. Sanminiatelli (Milan, 1917), no. 1.

[37] Quoted from N. Khardziev, 'Mayakovskiy i zhivopis'', in *Mayakovskiy: Materialy i issledovaniya*, ed. V. Pertsov and N. Serebryanny (Moscow, 1940), 381. In Krystyna Pomorska's assessment, 'Russian Futurism transmitted the principles of Cubism into poetry', by which she means the 'refusal to make "images of reality"' and 'the domination of the material itself', but not specifically any collage technique (*Russian Formalist Theory and its Poetic Ambiance*(The Hague, 1968), 38). Similarly, Vahan D. Barooshian argues that 'the notion of the self-contained word in particular was derived from Cubism' (*Russian Cubo-Futurism, 1910–1930: A Study of Avantgardism* (The Hague, 1974), 111). On the relationship of Russian Futurism to Cubism, see M. Grygar, 'Kubizm i poeziya russkogo i cheshskogo avangarda', in Jan van der Eng and Mojmir Grygar (eds.), *Structures of Texts and Semiotics of Culture* (The Hague, 1973), 59–101.

One exception to this general pattern is the collage-work of Yuri Annenkov. Annenkov had visited Paris in 1912 and 1913, had seen the Futurist exhibition at the Galerie Bernheim-Jeune, and had himself participated in the Salon des Indépendents. Annenkov, who began experimenting with collage techniques using *papiers-collés* in 1915, was associated with the St Petersburg Cubo-Futurists, although he does not seem to have performed any substantial mediating role between Paris and Russia.

Of the few collages by him that have survived, his *Rolls Royce* and *Exlibris*[38] use fragments of readable print, though for non-political purposes. Thus, the collaged fragment of an advertisement used in *Rolls Royce* is simply meant to be recognized as a token of the commercialized metropolis; and the Cyrillic element in the *Exlibris* collage functions simply as the sign for *book*. Similarly, the use of *papiers-collés* in Annenkov's *Petersburg* is much more in the Cubist painting tradition, which also had a strong hold in Russia, than that of Italian Futurist literary collage, with its combination of dynamism and readable, agitational content.

A nowadays much better-known literary experiment using the collage method of combining words and shapes with Russian Futurism, though again devoid of any political function, comes with Vasily Kamensky's so-called 'ferro-concrete' poetry. Gerald Janecek sees similarities between Kamensky's 'Konstantinopol'' ('Constantinople') and Marinetti's words-in-freedom,[39] although a closer analogy, albeit hardly one likely to have been an influence, is to be found in Severini's collages. Certainly, Kamensky's use here of geometric outline shapes filled with words is generally reminiscent of the *carta sincrona*, although it is *Danza serpentina* and a later Futurist work like Depero's 'Tramvai' (see fig. 4.1) that come closer to Kamensky's collage-like visual effect.

Kamensky's 'ferro-concrete' poems were, like so many Italian Futurist examples of the newly discovered *dipinto parolibero*, such a hybrid form that they were treated both as visual art and as poetry. They were first shown at the 'No. 4' Moscow exhibition of art (April 1914), but were also reproduced in the literary pages of the *Pervyy zhurnal russkikh futuristov* (*First Journal of Russian Futurists*). If Susan Compton is correct in her suggestion that 'Kamensky [chose] the word "ferro-concrete" to describe the arrangement of words on the page in direct contrast to the Italian futurists' "words-in-freedom"',[40] then this was no doubt partly because Marinetti's visit to Russia had left the majority of Futurists there with an unduly mimetic impression of the Italian's sound-poems and

[38] For illustrations, see Wescher, *Die Collage*, p. 104.

[39] *The Look of Russian Literature: Avant-Garde Visual Experiments, 1900–1930* (Princeton, New Jersey, 1984), 133.

[40] *The World Backwards: Russian Futurist Books, 1912–1916* (London, 1978), 83. Susan Compton also suggests here that 'there may be . . . a deliberately futurist pun in the fact that Kamensky's own name contains the ordinary word for "a stone" (*kamen'*), a traditional material of which ferro-concrete could be seen as a modern, man-made equivalent' (p. 112).

'designed analogies'. (It is clear from Livshits's *Polutoraglazyy strelets* (*One and a Half-Eyed Archer*) that Marinetti was regarded there, as he was in England, as the purveyor of crudely onomatopoeic and pictorial devices,[41] hence it is understandable for a Russian visual poet to want to dissociate himself from such an apparently limited experimenter.) But the distinction runs deeper than this. If the stress in much Italian typography and collage is on liberation, explosiveness, and dynamism, Kamensky's effects are, by contrast, contained and static.

The words of a Kamensky poem have, as it were, already been prepoured into moulds—the lines containing them representing the equivalent of the iron rods of a ferro-concrete construction (except that they are on the outside, instead of reinforcing the structure from within). The result is that the poems are clearly divided into a number of semantic, interpretive segments, and, as the overall analogy with building would suggest, the general visual impression is one of rigid stasis, by comparison with free-word layouts. In some cases, as for example in the poem dedicated to David Burliuk in the *Pervyy zhurnal russkikh futuristov*, this segmentation has a semantic, but largely non-iconic function. In others, such as 'Konstantinopol'' ('to be pasted on walls and fences'), a series of distinctly iconic impressions are created.

In 'Konstantinopol'' (fig. 2.17),[42] Kamensky's visit to the Turkish capital is reflected by means of a series of imitative signs, including > and < to denote the ebb and flow of the crowds on the Bosphorus waterfront, architecturally shaped columns and words in mosque-shapes, including a triangle for the Hagia Sofia, and even, as Kamensky explains in his detailed analysis of the poem to Shemshurin,[43] a verbal structure in the shape of a fez. According to the poet, the work's language is a deliberate fusion (within a collage framework) of Russian words, snatches of Turkish, misunderstood foreign sounds, bird-cries and shouts, as well as various portmanteau constructions and neologisms. As Vladimir Markov points out, such effects are 'essentially impressionist',[44] a point which could equally well be made of many of Marinetti's collage-works. But there is a further complication, since it is Constantinople *recalled*, rather than the city as experienced at the time: one is reminded of Boccioni's view that a Futurist picture, and that includes collages, 'must be the synthesis of what one remembers and what one sees'.[45] The occasional use of foreign and invented words here, in fact, is beginning to shift the work away from simple referentiality in the direction of transrational language (*zaum'*) and a focus on what the

41 See Kruchenykh's gibes: *Manifesty*, p. 71.

42 For an excellent detailed reading of the poem, see Janecek, *The Look of Russian Literature*, pp. 124–33.

43 *Strelets*, 1 (1915). The main points of this relatively inaccessible analysis are reproduced in Vladimir Markov, *Russian Futurism: A History* (Berkeley, 1968), 197 f.

44 *Russian Futurism*, p. 198.

45 Preface to the Sackville Gallery catalogue (*Archivi*, 1, p. 109).

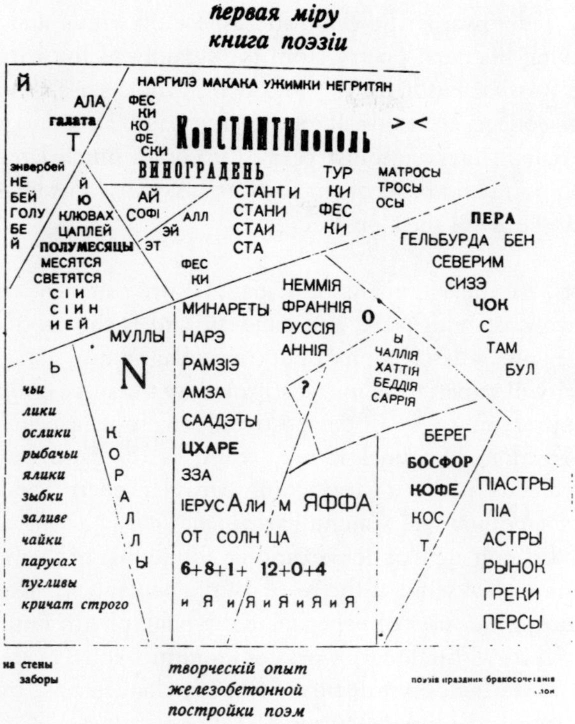

FIG. 2.17. Vasily Kamensky, 'Konstantinopol'', 1915

Russian Futurists liked to call the 'word as such',[46] individual words, here often in shrinking columns, whose importance comes through the act of dissecting isolation, rather than the kind of shape-labelling function we encountered in the work of Severini and Buzzi.

However, the 'ferro-concrete' poems marked no more than a brief episode in the history of the Russian Futurists' liberated typography; and, even within Kamensky's general output, they are but a phase.[47] Moreover, they do not appear to have had an influence on other poets' visual work, in the way that there was a great deal of cross-fertilization among the Italian Futurists and the Cubists. (The possible exception, as we shall see when we consider his work later, is the Ukrainian Panfuturist Mykhaylo Semenko, whose layouts do at times resemble Kamensky's.) Russian Futurist poetry, even in the new highly

[46] On this feature of Kamensky's work, see Pomorska, *Russian Formalist Theory and its Poetic Ambiance*, p. 109.

[47] The chapter on Kamensky in Barooshian's *Russian Cubo-Futurism* (pp. 98–107) gives a more general account of his entire poetic development than Janecek does.

variegated forms that Janecek has documented and analysed so impressively, was largely directed towards the linguistic aspect of words and their root associations. Having liberated poetry from the tyranny of linearity and tired typography, the Russian Futurists did not on the whole seek to fuse art and painting in new collage creations. If one really wants an analogue for the collages of the Italian Interventionist period, it comes much later: with the painted Agitprop trains and boats of the early 1920s, with Mayakovsky's Rosta posters, and with the art of the Constructivists.

In some respects, the development of Italian Futurist literary collage runs parallel to the work of both poets and painters, and is the product of both groups. Christine Poggi's description of Futurist collage's aim as the 'recreation of the simultaneity of experience, untransformed by the imposition of logical transitions or arbitrary unity' (i.e. of perspective),[48] nicely brings out the parallel with words-in-freedom. At times it may seem, as the Futurists no doubt intended, difficult to make a distinction between poetry and painting. Severini's *Danza serpentina* and Cangiullo's *Milano-Dimostrazione* are cases in point, especially difficult areas of demarcation since neither of them is a collage in the strict, etymological sense of the word. But in general, whereas 'designed analogies' and most other uses of experimental typography in Futurist poetry remain limited effects within longer works of literature, collages are by definition self-contained aesthetic entities. Both forms are characterized by a mistrust of passéist language and mimetic forms of representation. Yet, contrary to Severini's fears at the outset, they can still be read.

If Jeremy Adler is right in his claim that 'representational examples' of concrete poetry, such as Reinhard Döhl's frequently cited *Apfel* (*Apple*), are 'untypical' of the general tradition and that 'the greater part of concrete poetry and most visual poetry after *Un coup de dés* . . . is *non*-representational',[49] then it is Apollinaire's *calligrammes* and Marinetti's 'designed analogies' which are historical throwbacks rather than the Futurists' collages. The short-lived genre of Futurist literary collage, even if it was much more in the mainstream of subsequent experimental developments in modernism, never achieved the fame, even notoriety, that words-in-freedom with their 'designed analogies' enjoyed at the time. It has only been in the 1970s and 1980s, in the wake of Caruso's and Martini's two-volume anthology of various forms of visual poetry and a more liberal displaying of Futurist collage at recent exhibitions, that the genre has been recognized for what Marinetti rightly held it to be: an 'absolutely original fusion of plastic dynamism and free-word poetry' (*Archivi*, i. 342).

48 'Marinetti's *Parole in Libertà* and the Futurist Collage Aesthetic', *The Futurist Imagination*, p. 4.

49 'Technopaigneia, carmina figurata and Bilder-Reime: Seventeenth-century figured poetry in historical perspective', *Yearbook of Comparative Criticism*, 4 (1982), 147.

3.
Telegraphic Lyricism and the Cult of the Minimal

SURVEYING more than a decade of Italian Futurist poetry in 1924, Enrico Prampolini came to the conclusion that 'lyricism was tending towards telegraphy'.[1] Three years earlier Iwan Goll had drawn attention to the same tendency, formulated this time programmatically in respect of contemporary German poetry:

Our aim is to compress the most profound of experiences into telegrams . . . a Spartan cablegram of the soul that can be read in the lift. All the pathos of congealed grammar, all the logic of unnatural syntax must disappear from poetry. We must become simple! . . . Shorthand is a matter of training. . . . Modern telegraphic man only reads titles, suggestive nouns. . . . Simplicity entails . . . the jettisoning of grammar, which slows us down, and the ultimate reduction to the bare essentials.[2]

In his often stinging account of the pre-war period of Russian Futurism, Kornei Chukovsky records a similar sense of contemporary impatience with the wordiness of conventional language:

We no longer have the time in the chaos of the cities to explain ourselves at length and with verbal abundance, to lavish ten words where two or three are sufficient. Words are becoming compressed, abbreviated, coagulated. The result will be lucid, expressive words. Anyone who knows what America has done with the English language in the past two decades [i.e. before 1922] will understand what I'm talking about: what a Yankee says in a minute would take a Russian five times longer to say. We are suffering from a terrible waste of verbal energy and we have a need for economy: *we don't have the time*, that is the universal watchword of the present; it will transform our lazy language, which is sluggish and provincial . . . into a rapid, telegraphic language.[3]

The sentiment was echoed in the same year (1922) by the Polish Futurist Bruno Jasieński in his prologue to *Pieśń o głodzie*:

agencje reutera, havasa, paty,
długie, kilometrowe papierowe zwitki.
komunikaty.

[1] 'Atmosfera scenica futurista 1924', *Arte viva*, 2 (1958), 33 f.

[2] *Die Neue Rundschau*, 2 (1921), 1082–4.

[3] Kornei I. Chukovsky, *Futuristy. Igor' Severyanin.—Al. Kruchenykh.—Vas. Kamenskyy.—Vl. Mayakovskyy* (Peterburg, 1922), 22.

miasto słyszy wszystko.
wie za kogo wychodzi księżniczka hiszpańska
i co spiskują w gdańsku niemcy wciąż niesforni,
o budowie w himalajach nowego wiaduktu.
radiodepesze z kalifornii
i stan pogody w timbuktu.
o wszystkim pisze miasto w swoich 40-szpaltowych poematach:
strajki w elektrowniach. przejechania. samobójstwa.
oto jest prawdziwa gigantyczna poezja. . . .
jak śmieszne są wobec niej wszystkie poezje.
poeci, jesteście niepotrzebni!

agencies: reuters, havas, pat,
kilometre-long rolls of tickertape.
messages.
the city hears it all.
it knows whom the spanish princess is going to marry
and about the latest german conspiracy in danzig,
what new viaduct is being built in the himalayas.
radiotelegrams from california
and weather reports from timbuktu.
all this the city writes in its 40-columned poems:
the strikes at the electrical plant. travels. suicides.
this is true gigantic poetry. . . .
how ridiculous is all other poetry in the face of it.
poets, you are superfluous![4]

Mayakovsky begins his poem 'Khorosho' ('Good')[5] with another variation on this sentiment, this time emphasizing, as the Futurists often did, that the new telegraphic style is not so much a personal idiom as *the* style of the modern technological age, and one that is destined to replace the anachronistic language of previous literature:

Время—
вещь
необычайно длинная,—
были времена—
прошли былинные.
Ни былин,
ни эпосов,
ни эпопей.
Телеграммой
лети,
строфа!

[4] Draft prologue to *Pieśń o głodzie* (Cracow, 1922), quoted from Zbigniew Folejewski, *Futurism and its Place in the Development of Modern Poetry: A Comparative Study and Anthology* (Ottawa, 1980), 234 f.

[5] 'Khorosho: Oktyabraskaya poema', *Sochineniya*, 3 (Moscow, 1970), 257.

Time—
    is a thing
        unusually long—
The times of the sages—
            they've been and gone.
No sagas,
    no epics,
        no legends.
Fly,
  verse,
    like a telegram!

Back nearer the beginning of this period of excitement about the possibilities of a new form of 'telegraphic' language, one finds Apollinaire confessing (as early as 1914): 'I have done my best to simplify poetic syntax and have succeeded in certain instances, especially in the poem *Les Fenêtres*. . . . When it comes to speed and economy, the telegraphic style offers us resources which ellipsis will endow with a marvellously lyrical power and quality.'[6] Across the Channel, one finds Wyndham Lewis fired with a similar enthusiasm about the new elliptical style of his *Enemy of the Stars* (1913):

> It was my object to eliminate anything less essential than a noun or a verb. Prepositions, pronouns, articles—the small fry—as far as might be, I would abolish. Of course I was unable to do this, but for the purposes of the novel, I produced a somewhat jagged prose.[7]

Louis MacNeice uses the term 'telegraphese' for this kind of writing (which he finds much in evidence in Auden's first volume of *Poems*), a style with 'the less important words such as articles and conjunctions, even demonstrative and relative pronouns . . . often omitted'. Yet, in MacNeice's opinion, Auden was aiming 'at an economy difficult to attain in English, which is an uninflected language'.[8]

At our present distance, it is virtually impossible to pinpoint at just what stage in the evolution of modernism the idea of a radically elliptical, even minimal style began to seize hold of the literary imagination. As Nils Åke Nilsson points out, 'the discussion of the parts of speech is, in fact, very old. Many grammarians and philosophers . . . have participated in it. [But] for a long time it was a genetic problem, concentrating on nouns versus verbs. Which

[6] Letter of 1 July 1913 to Madeleine Pagès (*Œuvres complètes*, ed. Michel Décaudin; vol. iv, ed. André Balland and Jacques Lecat (Paris, 1966), 478). Apollinaire elsewhere uses the phrase 'abbréviations sténogrammatiques' (*Lettres de Guillaume Apollinaire à sa marraine, 1915–1918* (Paris, 1948), 14).

[7] *The Letters of Wyndham Lewis*, ed. W. K. Rose (Norfolk, Conn., 1963), 45. Giovanni Cianci analyses the indebtedness of Lewis's *Enemy of the Stars* to the telegraphic lyricism of Italian Futurism in 'Un futurismo in panni neoclassici: Sul primo Wyndham Lewis vorticista', in *Wyndham Lewis. Letteratura/Pittura*, ed. Cianci (Palermo, 1982), 25–66.

[8] *Modern Poetry: A Personal Essay* (Oxford, 1938), 146.

part of speech was the oldest.' It was only 'later that interest shifted to the importance of the parts of speech in the sentence from a grammatical point of view',[9] and eventually, as we shall see, from there to the question of which parts of speech should be eliminated and which foregrounded.

It is clear that, unlike the time-honoured emphasis on the value of ellipsis in manuals of rhetoric or an individual writer's cultivation of verbal economy and the *mot juste* in more recent times, the idea that extreme simplification, even an artificially systematic reduction of language, would furnish the avant-garde idiom *par excellence* is an early twentieth-century phenomenon. Apollinaire's almost casual reference to 'the telegraphic style' suggests that by 1914 he was scarcely advocating something new, but simply taking up an idea already in circulation. Indeed, he even advanced the claim, in the same year, that a pre-existing French 'telegraphic style' had had a substantial impact on the free-word experiments of the Italian Futurists.[10] But Theo Hermans has, with a good deal more justification, interpreted Apollinaire's telegraphic style, not as part of a French influence on Futurism, but as something largely derived from Marinetti:

> Apollinaire's . . . 'telegraphic style' . . . relies on juxtaposition, rapid semantic shifts, and a degree of discontinuity combined with a simplification of the syntactic apparatus. . . . It seems . . . obvious . . . that in spite of Apollinaire's serious reservations regarding the Futurist innovations . . . whatever limited sympathy he has for them follows his recognition of the 'wireless style' enunciated in Marinetti's manifestoes.[11]

The French poet's presentation of this 'telegraphic' or 'wireless' poetry in terms of speed and economy is not just reminiscent of various formulations in Marinetti's theoretical pronouncements of the previous year or so; it was undoubtedly influenced by these.

Yet, although Marinetti was highly active in promoting the idea of a new 'telegraphic lyricism',[12] the history of the subject is substantially complicated by two factors. First, not all of its advocates understood 'telegraphic' writing in the sense in which Marinetti meant the term. (Indeed, we shall discover that the style has almost as many connotations as it has practitioners.) And, second, the story of the growth in popularity of 'telegraphic' poetry—and to a lesser extent, prose and drama—is by no means exclusively confined to a specifically Futurist

[9] *The Russian Imaginists* (Stockholm Slavic Studies, 5; Stockholm, 1970), 75. Nilsson's account of 'Verbless Poetry' (pp. 74—91) reveals that, while there is a sporadic anticipation of systematic grammatical pruning in earlier individual works, such as Afanasy Fet's pre-Symbolist verbless poem 'Shepot, robkoe dykhan′e' ('Whispers, Timid Breathing'), the main radicalism comes in the early part of the twentieth century, with the Futurists and the Imaginists.

[10] 'Nos amis les futuristes', *Les Soirées de Paris* (15 Feb. 1914), 78 f.

[11] *The Structure of Modernist Poetry* (London, 1982), 70.

[12] Marinetti defined the new 'lirismo telegrafico' as a 'swift, brutal and immediate lyricism' ('Destruction of Syntax', *TIF* 66).

influence radiating out across the Continent from Italy. Long before Marinetti played his decisive part in promoting it, the cult of the 'telegraphic' was being prepared and rehearsed across Europe. Its mushrooming popularity owed much to two things: the discoveries of Guglielmo Marconi, and contemporary mythification of the Eiffel Tower.

Even granting Futurism an instrumental role in disseminating and popularizing telegraphic writing among the Orphists, the Expressionists, and the Vorticists, in the case of such a widespread modernist phenomenon one has to allow for the eventuality of polygenesis. Indeed, the possibility that the attractions of such a style had occurred almost simultaneously to various avant-garde writers in different countries is a strong one. The German Expressionist Franz Richard Behrens, for instance, apparently hit upon the idea of a *Telegrammstil* while on guard-duty one night during the First World War, yet we possess no evidence of any direct influence from the Italian Futurists having turned his thoughts in this direction.[13] And even if the Ukrainian Panfuturist Mykhaylo Semenko, who *was* familiar with a number of Marinetti's writings,[14] wrote his 'Kablepoema za okean' ('Cablepoem beyond the Ocean') at a time when it was at least conceivable that he was conversant with some examples of Italian 'telegraphic lyricism', there is no certainty that he was influenced in his own work by this Futurist phenomenon. And there were many other proponents of such a style at the time who, while in many cases being indebted to Futurism, may well have discovered the technique for themselves and only found confirmation of its modernist appropriateness in the movement's manifestos, and who, in any case, took the idiom in markedly new directions. One thinks of Mayakovsky, who in 1914 declared that 'the nervous life of the cities requires quick, economical, abrupt words',[15] Blaise Cendrars's experiments with a form he called the 'télégramme-poème',[16] what the 'paroxyste' Nicolas Beauduin figuratively referred to as his 'wireless' poems (his 'poèmes sans fil'), the Polish Futurist Stanisław Młodożeniec's telegraphic poems (e.g. 'XX wiek' ('The Twentieth Century'), which will be considered later) or Iwan Goll's repeated call for German poetry to accommodate to the idiom of the 'Telegraph Agency'.[17]

[13] According to Gerhard Rühm, the entry for 26 Aug. 1915 in Behrens's unpublished field diary reads: 'Discovered on night-watch my single-word poetry. Concentrated brevity. Telegram. To write poetry is to abbreviate' (Franz Richard Behrens, *Blutblüte: Die gesammelten Gedichte: Werkausgabe*, ed. Gerhard Rühm, i (Munich, 1979), 371). Other entries in the field diary (e.g. 'Lichtenstein und Marinetti' and 'A.1 œuvre Futuristik!', i. 42, 10) indicate that Behrens was at least familiar with some aspects of Italian Futurism. Further details are given in Kurt Möser, *Literatur und die 'Große Abstraktion': Kunsttheorien, Poetik und 'abstrakte Dichtung' im 'STURM' 1910–1930* (Erlangen, 1983), 288.

[14] On Semenko and Marinetti, see Myroslava M. Mudrak, *The New Generation and Artistic Modernism in the Ukraine* (Studies in the Fine Arts: The Avant-Garde, No. 50; Ann Arbor, 1986), 167.

[15] The remark, originally in 1914 in the Moscow newspaper *Nov'*, is quoted from Vladimir Markov, *Russian Futurism: A History* (Berkeley, 1968), 182.

[16] See, for example, his 'Dernière heure', *Œuvres complètes*, ed. Michel Décaudin (Paris, 1966), i, 68.

[*See p. 148 for n. 17*]

In his prose-work 'La Guerre électrique' ('The Electric War') of 1911, Marinetti singles out the invention of wireless telegraphy as one of the great milestones in civilization's progress towards the Futurist electric millennium of the twenty-first century. And it also figures in the 'Destruction of Syntax' manifesto ('Distruzione della sintassi—Immaginazione senza fili—Parole in libertà' (1913)) as one of the great modifiers of the modern sensibility.[18] In a similar vein, Kamensky's poem 'Ot ieroglifov do A' ('From Hieroglyph to A') spreads the gospel of 'THE RADIO TELEGRAPH', ranking it alongside aviation, X-rays, and a number of other on the whole predictable manifestations of modernity.[19] Like the radio signals themselves, the message that the new world of telegraphy had arrived was in the air at this time, and the antennae of the modernist poets were among the first to receive the signals.

The year 1909 had seen the award of the Nobel Prize for Physics to Marconi on account of his various achievements in the realm of wireless telegraphy. This was a popular award: his discoveries clearly excited the public imagination and could be seen to have a tangible influence on the quality of everyday modern life. The establishment of cross-Channel wireless contact between England and France in 1899 had seemed to many, caught up in the fervour of centurial mysticism, to presage boundless possibilities of communication for the new age. Marconi's subsequent success with transatlantic signals (in 1901) reinforced this new feeling of power by demonstrating that even over greater stretches the curvature of the earth's surface was no impediment to long-distance communication. Thus Europe was seen, as in Jasieński's poem, to be able to communicate—at least potentially—with *anywhere* on the globe. News could be transmitted and received virtually instantaneously thousands of miles away. The telegraphic dissemination of the new word of science became the technological equivalent of the spreading of the Gospel, and writers treated the subject with an appropriate religiosity, often using the imagery of the biblical word.

The exciting possibilities of the telegraph were summed up in Beauduin's essay 'The Poetry of the Age' ('La Poésie de l'époque' (1914)): 'nowadays ideas go around the world with the speed of lightning. Mankind lives in each one of us, becomes more integrated in us from day to day . . . from hour to hour, from second to second, in the uninterrupted flow of events transmitted by cablegrams and by wireless stations.'[20]

During the 1920s hopes originally focused on wireless telegraphy were transferred to radio in general, and the search for a telegraphic idiom was

[17] 'Our language must be abrupt: abrupt, narrow, stoney . . . above all unambiguous, for the telegraph agency has no time to transmit empty phrases: electricity is expensive' (*Die Neue Rundschau*, 2 (1921), 1082).

[18] *Le Futurisme*, ed. Giovanni Lista (Lausanne, 1980), 141. See also *TIF* 57.

[19] *Dokhlaya luna* (*Croaked Moon*)[2] (Moscow, 1914), 74.

[20] *Mercure de France*, 107 (Jan. 1914), 282.

bound up with both of these new means of communication. One of the most rhapsodic and detailed Futurist hymns in praise of the new wireless deity is to be found in Velemir Khlebnikov's 'Radio budushchego' ('The Radio of the Future'), written in 1921 while the poet was working, albeit as a humble night-watchman, for the Russian Telegraph Agency (ROSTA).[21] The essay begins by evoking the way in which the radio of the future will change people's lives. The argument is strongly dependent on natural imagery, as if to suggest that the new discovery is a logical extension of man's previous attributes (a variant on the Futurist concept of man 'multiplied' by his new technological discoveries):

> The Radio of the future—the main tree of consciousness—will open up a knowledge of countless tasks and will unite all mankind.
>
> Around the Radio's central station, this iron castle where clouds of wires stream out like strands of hair, there will surely be posted a skull and crossbones, with the familiar warning: 'Danger!' For the slightest pause in the workings of the Radio would bring about a spiritual fainting in the entire country, a temporary loss of consciousness. . . .
>
> Around the clock, from this point on the globe, flocks of news items from the life of the spirit scatter like the spring flight of birds.
>
> In this stream of lightning-birds, the spirit will prevail over force, good advice over intimidation.

The radio's task, according to Khlebnikov, will be not only to unite people, both socially and politically, but also to bring together mankind's disparate activities into a new quasi-religious unity: 'the works of the artist of the pen and the artist of the brush, the discoveries of the artists of thought (Mechnikov, Einstein) suddenly transporting mankind to new shores.' After this comes the paradoxical part of Khlebnikov's vision, part progress and part regression. He

[21] The acronym ROSTA (Rossiyskoe Telegrafnoe Agentstvo) was, in Khlebnikov's eyes, itself a 'telegraphic' word: 'mankind . . . rejects long-sounding words. . . . the Russian language has taken a bold leap by moving over to short, artificial words such as ROSTA.' In fact, according to Gary Kern (who quotes the above remark in *Velimir Khlebnikov, Snake Train: Poetry and Prose*, ed. Kern (Ann Arbor, 1976), 263), a variant of this piece is entitled 'ROSTA of the Future'. When Chukovsky talks about the transformation of Russian 'into a rapid, telegraphic language', the examples he gives are markedly dissimilar to what this would mean in the Italian context, yet his thinking is very reminiscent of Khlebnikov's reaction to the acronym ROSTA. His examples of 'the Americanization of the Russian language' (*Futuristy*, p. 23) are all confined to the single word: either portmanteau-formations or examples of the new Soviet obsession with acronyms. What is more, they come from everyday language, not the literary discourse of Futurism. 'Such an Americanization of language is normal and necessary,' Chukovsky claims, 'in fact, when we call an automobile an auto and the metropolitan system the metro, when . . . in place of The Society for the Commerce in Pharmaceutical Products in Southern Russia we say Yuratat' or 'K. D. in place of the Constitutional Democratic Party', this is 'using the same methods of compressing and accelerating language that . . . the Ego-Futurists . . . introduced into poetry' (p. 23). Futurist poetry's development of language in the direction of the 'compressed and laconic' (p. 25) is for Chukovsky a matter of single-word compression, not the making more skeletal of the traditional sentence. On the contrary, the sentence-structure in which such neologisms occur, and in which the ROSTA of the future is discussed, remains relatively conventional.

foresees that some embryonic form of television would be the logical next step from the radio as a purely acoustic medium, yet, as a poet, he at the same time remains anxious to focus the power of the visual on the transmission of the printed word, not non-verbal images:

The peaks of waves in the scientific sea will sweep through the country to the Radio's sub-stations, so that on the same day they can become letters on the dark screens of enormous books which rise up on the village squares, stand higher than the houses and slowly turn their own pages. . . . These books of the street are the Radio's libraries! They frame the village centres with their gigantic dimensions, they perform the tasks of all mankind, with the one daily spiritual wave which sweeps over the country inundating with a rain of scientific and artistic news—the Radio has accomplished this task with the aid of lightning. On the enormous shady books in the villages, the Radio has imprinted the story of a favourite author, an article on the fractional powers of space, an account of aeroplane flights and the news about neighbouring countries. Everyone reads what he likes. The book, one and the same for the whole country, stands in every village, constantly surrounded by readers, its type neatly set, a silent library in the village-centres.

And now a piece of momentous scientific news appears in black letters on the books: The chemist X . . . has discovered methods for obtaining meat and bread from wide-spread types of clay.

The crowd becomes excited and thinks: 'What next?'

Earthquake, fire and accident will be printed around the clock on the books of the Radio. . . . The whole country will be covered with its stations.

Khlebnikov's prophetic piece does admittedly, in the section 'The Radio and Exhibitions', grudgingly allow for the transmission of visual images. Here, he conjures up a vision of the latest Moscow art exhibition being transmitted 'by impulses of light and . . . repeated in thousands of mirrors through all the Radio's networks'. But, through a focus on the voice, the radio becomes a quasi-mythical, god-like agency.

The iron mouth of the auto-voice has transformed a surge of lightning, picked up and relayed it, into loud colloquial speech, into song and the human word.

The entire village has gathered to listen.

The lips of the iron trumpet carry the news of the day, the affairs of state, weather-reports, reports on the stormy life in the capitals.

There is something quaintly rustic about this vision of Russian villages listening attentively for news from the cities (in Italian Futurism, the perspective tends to be much more metropolitan). And one suspects from this strange amalgam of written and read language, that universal literacy is not yet being predicted for the foreseeable future.

It seems that some giant is reading the gigantic book of the day. But this is an iron reader, this is the iron mouth of the auto-voice. Sternly and distinctly it reports the news of the morning, received from the signal tower of the Radio's central station.

With its overtones of paternalism, centralization, and control, Khlebnikov's radio of the future constantly teeters on the brink of assuming the role of Big Brother. The claim that 'radio has accomplished the task which the Church as such could not accomplish', and the way in which it assumes the dimensions of an all-controlling *Weltgeist*, indicate the messianic, and politically dangerous, hopes being invested in the new discovery.

Khlebnikov's Radio of the Future is, to all intents and purposes, both symbol and organ of the new Soviet Union.

> Proud skyscrapers plunging into the clouds, a game of chess between two people located at opposite ends of the globe, a lively conversation between a man in America and a man in Europe. . . . Now the libraries have darkened, and . . . the whole country receives the communion of the word bórne in silence and isolation, the communion of bubbling fountains.

The radio will bind society together (just as Khlebnikov hoped that his new transrational language would one day). 'Contemporary doctors will treat distant patients by suggestion'; 'the programme of national education will be put into the hands of the Radio'. And, more sinisterly, radio will be used to increase the people's contentment and productivity in this brave new world.

> In the middle of winter, the honeyed smell of lime, mixed with the smell of snow, will be the Radio's real gift to the country.
>
> It is known that certain notes, like A and B, increase muscular ability, sometimes as much as sixty-four times by compressing the muscle for a given time-interval. In days of intensified labour, summer harvesting, and the construction of huge buildings, the Radio will transmit these sounds throughout the entire country, increasing its strength many times over.

Khlebnikov's vision concludes with the promise that 'the Radio will forge the unbroken links of the world-soul and fuse all mankind together'. Darker associations creep in with the assertion that all this 'will give the Radio even greater power over the consciousness of the country'.[22] There is no hint of scepticism as to what 'consciousness' may lie behind the apparatus, forcefully harmonizing the nation, or even the world. Technology is given wholehearted approval.

[22] *SP* iv. 290–5. The idea began life in Khlebnikov's series of Futurist 'Predlozhenie' ('Proposals') concerning the new medium (*Vzyal: Baraban futuristov* (1915)): 'To use the radio to transmit readings from the Principal University to the village schools. Any school at the foot of a green hill may receive the news of learning, and the teacher will be the ear-trumpet of the attentive village. The language of lightning as the conductor of scientific truth' (quoted from Gary Kern's translation in *The Ardis Anthology of Russian Futurism*, ed. Ellendea Proffer and Carl R. Proffer (Ann Arbor, 1980), 13). The sections 'Skybooks' and 'Means of Communication: Sparkwriting' in Khlebnikov's 'The Steppe of the Future' embroider on the same idea. In the former section, crowds of people are described forming 'the creative commune, by means of shadeprint on shadebooks, transmitted by the latest news, casting the necessary shadeletters from the beaming eye of the spotlight' (*The Ardis Anthology of Russian Futurism*, p. 133).

One encounters the same rapturous excitement about the possibilities of radio in Balla's 'Dialogue between a Passéist and a Futurist' of 1928:

Radio.—Who can tell what would happen if an inexhaustible source of indestructible and formidable energy, which is contained in a small volume, could be possessed, known, dominated and applied by Man? Would human audacity still have limits imposed on its conception of the impossible and the improbable? Not only to dominate distances, but also space with the simplicity, certainty and almost the same speed as that of thought. To multiply life by reducing the effort of animal life to its lowest and to speed up a hundredfold the creative work of genius. . . . The network of communications, simplified and multiplied, would lead the intensity of life into an atmosphere which is for the moment inconceivable.[23]

If one goes back to the pre-Revolutionary, even to the pre-war era in Europe, one finds little but unreserved praise for the new discoveries. Nicolas Beauduin once more gives voice to the exuberant spirit of the early period, this time in poetry:

Le téléphone vient m'apporter les nouvelles
Des quatres points des continents.
Je vibre avec l'Asie, l'Europe et l'Amérique . . . .
Toute la vie mondiale et frénêtique
Exalte en moi ses ouragans et remous . . . .
Je suis l'homme nouveau, l'Homme Cosmogonique
En qui s'intègre l'univers.[24]

The telephone has just brought me news
From the four corners of the earth.
I vibrate with Asia, Europe and America . . . .
All the frenetic life of the globe
Exalts in me its storms and its eddies . . . .
I am the new man, Cosmogonic Man
In whom the universe is integrated.

Translated from scientific fact into poetic myth, Marconi's discoveries soon acquired special emphases. The letters TSF (sometimes the actual phrase 'télégraphie sans fil' or 'telegrafia senza fili') figure prominently in the poetry of Apollinaire, Marinetti, and various contemporaries. And for good reason. For it was not such issues as the scientific challenges of the earth's curvature or atmospheric interference that concerned the literary imagination, but the simple, quasi-magical realization that communication could now be made *without wires*.

Delaunay reflects the aura of mystery to the process with his suggestion that,

[23] 'Dialogo di un.passatista con un futurista', *L'Impero* (1 Mar. 1928), 3; this is inevitably the Futurist speaking. See also Folgore's poem 'Radio' in *Ponti sull'oceano* (Milan, 1914), 66.

[24] *La Vie des lettres*, 3 (Oct. 1913), 406.

as a transmitter, the Eiffel Tower possesses a new kind of poetry which 'communicates mysteriously with the entire world'.[25] (*Du monde entier* was the title of one of Blaise Cendrars's books at the time.) In Apollinaire's 'Lettre-Océan', in Marinetti's *Zang tumb tumb*, in the poetry of Beauduin, Cendrars, Depero, Govoni, Semenko, Młodożeniec, the Berlin *Sturmkreis*, and Goll, one finds wireless telegraphy developing into one of the fetish-objects of the period. It is not realistically, and certainly not very scientifically, presented; rather it is cloaked in the aura of mystery which Delaunay attributes to it. In *Pittura scultura futuriste* (*Gli scritti*, p. 201), Boccioni refers to radio-waves ('le onde Hertziane') and their magic powers, when it comes to making distances on the globe disappear. Jannelli coins a telegraphic epithet with just such a connotation when describing an impetuously signalling love as an 'amore marconigrafico';[26] and Govóni speaks, in a strange amalgam of Crepusculism and modernism, of bell-towers as 'stazioni di telegrafia senza fili | delle anime'.[27] The idea is taken up in Fillia's *Lussuria radioelettrica* (1925), Gerbino's *Telegrafo e telefono dell'anima* (1926) with its section of telephone-call poems and poems presented as 'telegrammi', Emilio Buccafusca's *Realradioculla* (1934), Farfa's *Marconia* (1937), Buzzi's *Poema di radio-onde* (1940), Pino Masnata's *La radia* (1941), and in Depero's *Liriche radiofoniche* (1934), especially 'La voce dell'antenna' which ends with the lines:

> Le antenne metalliche delle radio-stazioni scattano da questi tappeti vasti ed intricati, come alti pistilli meccanici, di luce, fuggenti dalla terra.
> Luci ed aste che si flettono al vento:
> su su su sottili
> su su su d'acciaio
> su su su intrecciate.
> Dalle loro punte estreme partono le acustiche membrane delle onde di Hertz:
> tìnn tìnn tìnn tìnn
> peee peee peee peee
> espandersi — circoli — onde — timbri . . .[28]

[25] *Du cubisme à l'art abstrait* (Paris, 1957), 111. Pär Bergman assembles a wealth of illustrations relating to this contemporary telegraphic *topos* (*'Modernolatria' et 'simultaneità': Recherches sur deux tendences dans l'avant-garde littéraire en Italie et en France à la veille de la première guerre mondiale* (Uppsala, 1962), 263 ff.), but the subject is approached from a predominantly thematic point of view. Background information about the symbolic status of the Eiffel Tower can be found in Roger Shattuck, *The Banquet Years: The Origins of the Avantgarde in France* (London, 1969); Virginia Spate, *Orphism: The Evolution of Non-figurative Painting in Paris, 1910–1914* (Oxford, 1979); Marjorie Perloff, 'Deus ex machina: Some Futurist Legacies', *The Futurist Moment: Avant-Garde, Avant Guerre, and the Language of Rupture* (Chicago, 1986), 195–213, and *Literatur im Industriezeitalter*, ed. Peter-Paul Schneider *et al.*, 1 (Marbach, 1987), 283–9.

[26] 'Messina', *Lacerba* (15 Feb. 1914), 58. The 'Marconigram', also referred to in Cangiullo's poem 'Caffè' (*Lacerba* (1 Mar. 1914), 78), brings together the currents of modernism, Italian patriotism, and the fetichism of a new cult word. Armando Mazza may well have chosen to refer to an aerial as an 'antenna marconica' ('transatlantico', *Firmamento* (Milan, 1920), 35) for similar patriotic reasons.

[27] *Poesie elettriche* (Milan, 1911), 204.

[28] Depero, *Liriche radiofoniche* (Milan, 1934), 23.

The metal aerials of radio-stations spring up from these vast and intricate carpets, like high, mechanical light pistils, fleeing from the earth.
Lights and bars bending in the wind:
up up up thin
up up up made of steel
up up up interwoven.
From their extreme ends the sound membranes of Hertz waves start:
tinn tinn tinn tinn
peee peee peee peee
expand—circles—waves—timbres . . .

The same cult of the radiophonic and telegraphic is to be found among the Russian Futurists: in David Burliuk's American manifesto *Radio-Style* of 1926, in the radio-poems published by Edvard Strikha in *Nova generatsiya* and *Avangard* during the late 1920s, as well as his radio- and cable-exchanges with Semenko in the same journal. A 'telegraphic' collection like *Zustrich na perekhresti* (*Meeting at the Crossroads* (Kiev, 1927)), containing poems by G. Shkurupiy, M. Bazhan, and Mykhaylo Semenko, indicates that the telegraphic was not simply a product of western European avant-gardes coming directly under the influence of Paris and Italian Futurism.

The works just cited display a thematic concern with the new means of communication, but the telegraphic not only influences the subject-matter of much of the poetry of the time; it also leaves its mark on lyric poetry's figurative language. Seeking to match the Northern Lights with some equally powerful contemporary image, Cendrars comes up with

Tu resplendis avec toute la magnificence
de l'aurore de ta télégraphie sans fil.[29]

The letters 'TSF' soon became a shorthand, itself 'telegraphic' signal that could establish the modernist credentials of a certain style of poetry. The newfound sense of power at being able to communicate *without wires*, together with a feeling of elation at the cosmopolitan, even global, possibilities of wireless telegraphy and the kind of *telegraphic idiom* that such communications demanded, rapidly became leitmotifs of a certain Futurist lyricism, the poetry of the 'telegraphic voice'. Repeatedly,[30] poets had recourse to this imagery of an almost

[29] 'Tour', *Du monde entier: Au Cœur du Monde: Anthologie nègre* (Paris, 1963), 81. Since we shall later be concerned with influences on *Sturmkreis Telegrammstil*, it is worth noting that this poem was published in German translation in *Der Sturm* as early as 11 Nov. 1913 (125 ff.).

[30] Cf. Philippe Soupault's lines 'La Tour Eiffel lance ses rayons aux Îles Sandwich | Gutenberg 24–19' ('Rose des Vents', *Au sans pareil* (Paris, 1920), 45). Paul Ginestier, *Le Poète et la machine* (Paris, 1954), in particular the chapter 'La Magic noire du monde moderne' (pp. 137–63), is a mine of information on this subject. Among the poems using the 'TSF' motif cited there are Drieu la Rochelle's 'Fond de Cantine' (*Nouvelle revue française* (1920)), Pierre Mac-Orlan's 'Villes' (*Nouvelle revue française* (1929)), and Emmanuel Aergeter's 'TSF' (in *Le Voilier aux diamants* (Paris, 1935)). According to Gerald Janecek

superhuman ability to transcend the obstacle of distance, as if wearing Seven League Boots,[31] and of the awareness of being an integral part of what was becoming, in Marinetti's words, 'a centralized network of acoustics from all over the world'.[32] Or, in the more abrasive tones of Cendrars's poem 'Crépitements':

> Les arcencielesques dissonances de la Tour dans sa télégraphie sans fil
> Midi
> Minuit
> On se dit merde de tous les coins de l'univers.[33]
>
> The rainbow dissonances of the Tower with its wireless telegraphy
> Midday
> Midnight
> People are exchanging the greeting 'shit' from all the corners of the universe.

A contemporary poem by Soffici, entitled 'Studio', describes the artist's studio—and, by extension, modern man himself—as becoming a 'radiotelefantastic cabin open to all messages',[34] the sense of wonder here being transmitted by the portmanteau neologism: 'radiotelefantastica'.

The period's obsessive fascination with the telegraphic only makes sense within the symbolic context of the much-fêted Eiffel Tower. For, as Pär Bergman, Virginia Spate, and Marjorie Perloff have demonstrated, it had, as a transmitter, become a powerful image of modernity; its 'wireless station was felt to communicate with the entire universe'.[35] In work after work—by Apollinaire, de Torre, Huidobro, Däubler, Goll (see figs. 3.1 and 3.2)[36]—the

(*The Look of Russian Literature: Avant-Garde Visual Experiments, 1900–1930* (Princeton, New Jersey, 1984), 124), Vasily Kamensky also wrote a poem entitled 'Radio-Telegraph', though this is now lost.

[31] Seven League Boots are alluded to in Marinetti's *Le Monoplane du pape*, when the telegraphist transmits words 'qui ont des bonds d'un kilomètre' (Paris, 1912, 273).

[32] 'Manifesto dell'aeropoesia' (1931); here quoted from Rosa Trillo Clough, *Futurism: The Story of a Modern Art Movement. A New Appraisal* (New York, 1961), 179.

[33] *Du monde entier*, p. 66.

[34] Cited in Raffaele Carrieri, *Futurism* (Milan, 1963), 96.

[35] Spate, *Orphism*, p. 9.

[36] Apart from fig. 3.1, the poetic word-picture of the Eiffel Tower in Apollinaire's *Calligrammes: Poèmes de la paix et de la guerre* (Paris, 1925), 78, and fig. 3.2, the cover of Iwan Goll's *Der Eiffelturm: Gesammelte Dichtungen* (Berlin, 1924), examples include: Vicente Huidobro's 'Tour Eiffel', *Nord–Sud*, 6–7 (Aug.–Sept. 1917), 24; Guillermo de Torre's 'Torreiffel', *Cosmopolis*, 9, 34 (Oct. 1921), 200–3 (also published in a French version in *La Vie des lettres* (July 1921)); Apollinaire's 'Zône', which addresses the Tower in person; Goll's 'Paris brennt', in which the Famous 'Monsieur Eiffel' invites all the poets of Europe to a dinner at which he entertains them with a cosmic 'son et lumière' display ('Symphoniekonzert der Wolken | Akustik des Kosmos') and sends them each off with three shares in his Company (*Der Eiffelturm*, p. 23); and Goll's 'An den Eiffelturm' (*Technische Zeit*, ed. H. Küpper (Essen, 1929), 39). In the 1920s the Russian poet Ivan Alexandrovich Aksenov, who had already included two poems on the subject in his *Neuvazhitelnye osnovaniya* (*Invalid Fountains* (Moscow, 1916)), planned a volume called *Eiffelaya*, to be illustrated by Lyubov Popova (details in

FIG. 3.1. Eiffel Tower 'auto-illustration' from Guillaume Apollinaire, *Calligrammes. Poèmes de la paix et de la guerre*, Paris, 1925

Tower/transmitter figures as the symbol of the new sensibility, the ultimate personification of the powers of the 'TSF', or of Paris, the European modern city *par excellence*. The most well-known contexts in which it is celebrated as such are Delaunay's influential cycle of paintings and *Du cubisme à l'art abstrait*. The Tower also figures in Severini's picture *Metropolitana di Parigi* (see *Archivi*, ii. 331, pl. 70), but this is, as far as I can ascertain, a rare allusion to the pride of Futurist Milan's most obvious rival-city. The cult of the Eiffel Tower is, as might be expected, far more intensively practised in the French capital. Cendrars, who gave a lecture on the Eiffel Tower at the time,[37] also dedicated a poem to the structure which contains the following fairly typical sentiments:

Russian Avant-Garde Art: The George Costakis Collection, ed. Angelica Zander Rudenstine (London, 1981), 412); however, the work failed to materialize. In prose, one of the more sympathetic non-French evocations of the Tower comes from the pen of Theodor Däubler, a leading mediator between Italian Futurism and German avant-garde circles: 'it is more removed from space than any other building. It is a scaffold and a skeleton of the Future. It is seemingly functionless. Serving no purpose. . . . The Eiffel Tower has a soul: it is an artist. It is also Delaunay's father' (*Der neue Standpunkt* (Dresden, 1916), 181 f.).

[37] *Œuvres complètes*, iv. 198–9.

FIG. 3.2. Cover of Iwan Goll, *Der Eiffelturm*, Berlin, 1924. Designed by G. Salter after a drawing by Robert Delaunay

Tu es tout	You are everything
Tour	Tower
Dieu antique	Ancient deity
Bête moderne	Modern beast
Spectre solaire	Solar spectre
Sujet de mon poème	Subject of my poem
Tour	Tower
Tour du monde	Tower of the world
Tour en mouvement.[38]	Tower in movement.

For all the triteness of such versified eulogies, their stance is instructive; it is that of the enthusiastic observer—a perspective which is nevertheless still too much on the outside to be of direct relevance to the birth of the new telegraphic style.

Of more significance for the creation of an avant-garde telegraphic idiom are the various works in which the poet so comes to identify with the Eiffel Tower that he either locates his perspective up in the mast itself, as Apollinaire

[38] Ibid. i. 55.

had done in 'Lettre-Océan',[39] or at least demands that Telegraphic Man should try to assume such an Olympian vantage-point.

In his 'Vision de la Cité Future', Beauduin proclaims:

> Notre place est là-haut sur la Tour, en vigie,
> C'est là que nous devons veiller et nous tenir.
> Pour ranimer le feu divin des énergies
> Et guetter l'Homme Avenir.[40]
>
> Our place is up there on the Tower, on the lookout.
> That is where we should stand and keep watch.
> To revive the divine fire of energy
> And lie in wait for the Man of the Future.

The Eiffel Tower combines associations of being at the centre of a vast communications network, of enjoying a new kind of Promethean (in some cases, Zarathustran) overview over the world, and of having constructed a modern deity appropriate to 'l'Homme Avenir'. Marinetti's Mafarka, one feels, ought to have been the creative spirit behind Paris's Tower.

In one respect, it made more sense for contemporary *writers* to discover such a profound affinity with the Tower than for a painter to do so, for the poet, like the mast itself, is a sender and receiver of verbal messages. A painter's perspective on to such a subject will tend, by contrast, to be that of the onlooker, though a concept of 'simultanisme' such as Delaunay's can allow for many other perspectival possibilities. But if the painter is generally an observer, a writer like Richard Meyer, in his poem 'Paris' (appropriately dedicated to Apollinaire in thanks for 'Zône'), can adopt an even more convincingly empathic relationship to the radio-mast:

> Ich glühe tief und weiß:
> *ich bin der Eiffelturm*! . . .
> Ich empfange dankbar drahtlose Depeschen
> aus Kanada,
> Timbuktu,
> Nauen
> und Adrianopel.
> Die liebe Mitwelt sorgt ganz mütterlich für Sensationen:
> Nicht wahr? Paris liegt gar nicht an der Seine![41]
>
> I glow strong and know:
> *I am the Eiffel Tower*! . . .

[39] For a detailed analysis, see Willard Bohn, 'Circular Poem-Paintings by Apollinaire and Carrà', *Comparative Literature*, 31 (1979), 246–71, rev. and repr. in *The Aesthetics of Visual Poetry, 1914–1928* (Cambridge, 1986), 9–24.

[40] *La Vie des lettres*, 1 (Mar. 1913), 266.

[41] *Der Sturm*, 11 (Aug. 1913), 86 (my italics). That one of Meyer's telegraphed messages comes from Adrianopolis leaves little doubt about at least one of the influences behind this telegraphic poem.

I receive with thanks wireless telegrams
From Canada,
Timbuktu
Fishing boats
And Adrianopolis.
The dear world around me makes
Maternally sure I receive sensations:
Isn't that so? Paris does not
Lie on the Seine at all!

There soon came into being, however, one further—and from our point of view, far more important—mode of identification with the Eiffel Tower: with the Tower not as sender and receiver of messages from all parts of the globe (as Meyer and Goll have it), but as *transmitter in a certain kind of modern idiom*. The title 'le Sans Fil' accorded to Cendrars by the Futurists[42] reflects this sense of identification between poet and mast. The volume of Goll's poems published using the title *Der Eiffelturm* epitomizes the same sense of affinity to the transmitter, a kinship echoed in Richard J. Pioli's recent translations of Marinetti with the section-heading 'An Eiffel Tower of Words'.[43] The Czech Futurist Stanislav Neumann's lines 'We, the telegraph wires, ever hungry | for the food of electric impulses'[44] move in the same direction. (By contrast, Shershenevich's 'I am an important telegram that the world must deliver to the future'[45] is a much rarer and less productive variant on the poet/telegraph–transmitter motif.)

The point was that, if poets felt such a strong affinity with the Eiffel Tower, why, they reasoned, should they not begin to speak like it: in some form of telegraphic language? If, as Cendrars says of the Tower's transmitter, 'it told the time of day to all the ships on the high sea, why not also to the poets?'[46] Hence, the 'language of the TSF', the 'oceanic messages' much-vaunted in Delaunay's *Du cubisme à l'art abstrait*[47] were soon to offer a model for various new forms of poetic diction.

A voice can be heard, for example, speaking in the new idiom, in Depero's 'La voce dell'antenna' ('The voice of the aerial').

[42] The title was accorded Cendrars by the journal *Le Futurisme* on 9 Jan. 1924.

[43] *Stung by Salt and War: Creative Texts of the Italian Avant-Gardist F. T. Marinetti* (Reading Plus, 2; New York, 1987), 17–35.

[44] Neumann, 'Songs of Wires', quoted from Folejewski, *Futurism and its Place in the Development of Modern Poetry*, p. 249. Bruno Jasieński's poem 'Morse' (published in *But w butonierce* (*A Boot in a Buttonhole*) (Warsaw, 1921), 53) compares man with a telegraph-station. An Italian poem of related interest, which seeks to take the motif away from the overtly French associations of the Eiffel Tower, is Luciano Folgore's 'Le antenne', *Il canto dei Motori* (Milan, 1912), 111–13.

[45] Quoted by Markov (*Russian Futurism*, p. 377).

[46] *Blaise Cendrars vous parle*, ed. Michel Manoll (Paris, 1952), 220.

[47] *Du cubisme à l'art abstrait*, p. 62.

Orecchie della terra, sradicate e sospese nelle altitudini per udire, ascoltare i richiami, le voci dei pericoli, dei naufraghi, dei disperati.

Timpani d'aria, attenti alle più misteriose vibrazioni.

Tinn tinn tinn tinn

peee peee peee peee

S.O.S. 35 nord 4.7 ovest.

I contatti battono i denti della febbre. Le sillabe escono ad una ad una come quelle di un moribondo

. . . . gol-fo a-gi-ta-tissimo e-qui-paggio dispe-rato ten-ta salva-re na-ve sbandata, capi-ta-no scom-par-so peee peee S.O.S. in-cen-dio a bor-do ciurma e pas-seg-geri im-paz-ziti Peee-peee-peee S. O. S. 35 nord 4.7 ovest . . . .[48]

Ears of the earth uprooted and suspended in the altitudes to hear, listen to the calls, the voices of danger, of those who are shipwrecked and desperate.

Air ear-drums, attentive to the most uncanny vibrations.

Tinn tinn tinn tinn

peee peee peee peee

S.O.S. 35 north 4.7 west

The contacts chatter feverish teeth. The syllables come out one by one like those of a dying man

. . . . ex-tre-me-ly rough gu-lf de-spe-ra-te crew try sa-ve li-sting ship, cap-tain vani-shed peee peee S.O.S. fi-re on bo-ard crew and pas-sen-gers go-one mad Peee-peee-peee S. O. S. 35 north 4.7 west . . . .

It is at this historical juncture, where the cult of the telegraphic ceases to be just a literary theme (and often a rather fatuously treated one, at that), and is on the point of evolving into the elaboration of certain forms of telegraphic writing, that Marinetti and the Italian Futurists enter the story.

Like so many avant-garde innovators, Marinetti liked to play down similarities between his own ideas and those of writers close in spirit to him (the classic example being his failure to mention one of his principal mentors, Mario Morasso, even once in his writings). In fact, Futurist promotional strategy usually demanded that an artist be seen to be breaking radically with tradition, rather than merely building on earlier developments. (Even Carducci's early thoughts on the cultural significance of telegraphic communication are ignored.) So, despite the fact that 'telegraphic' writing was generally 'in the air' in European avant-garde circles at this time, it could hardly have suited Marinetti's purposes to present his appropriation of the idea in too derivative a context.

[48] *Liriche radiofoniche*, p. 24.

Marinetti had lived for some time in Paris, playing a conspicuous part on the French literary scene. Even after he took up residence in Milan, he returned to Paris so regularly that he could not have failed to be aware of the developments just considered. Nevertheless, various contretemps with the Orphists and other Cubists, Apollinaire's own ambivalence towards Futurism, and, above all, the unfortunate fact that the Eiffel Tower was a French achievement, were among factors hardly making it politic for him overtly to acknowledge any indebtedness. But, by associating the cult of the telegraphic exclusively with Italian Futurist words-in-freedom, Marinetti was conveniently forgetting not only the developments in Paris, but those in other metropolises as well. When he took his proposals about a new 'telegraphic lyricism' to Berlin, for example, he was to be informed in no uncertain terms that all this was nothing new. Various writers, including Arno Holz, Georg Kaiser, and Carl Sternheim, had in their various ways experimented with some form of *Telegrammstil*; and its central place within the experiments of the *Sturmkreis* certainly owed as much to Holz as it did to Marinetti. For all that, Marinetti cannot be dismissed as no more than unoriginal in this matter. His personal contribution to the evolution of 'telegraphic writing'—and it was a substantial one—was to take up the widespread, but somewhat nebulous idea and to give it a more sharply formulated, specifically *linguistic*, and, in places, even a more idiosyncratic interpretation. The key to this new direction lies in Marinetti's battle-reporting experiences.

In September 1911 the poet left Europe in order to cover the Libyan War as special correspondent for the Paris newspaper *L'Intransigeant*. Carrieri refers paradoxically to 'the long lyrical telegrams' sent back from the theatre of combat. 'Some of these telegrams', he remarks, 'were authentic poems impregnated with all the smells and sounds of a desert battle,'[49] the implication being that here were the immediate forerunners of the later telegraphic style of 'Battaglia peso + odore' ('Battle weight + smell') and *Zang tumb tumb*. Actually, despite the fact that one can observe certain thematic links between Marinetti's war poetry and his telegraphed reports to *L'Intransigeant*, there is little in these newspaper accounts, written in predominantly conventional syntax, as one might expect them to be, given the context, to indicate that Marinetti had yet gone very far along the path that would lead to his later radical 'telegraphic lyricism'.

Carrieri quotes the following extract from an on-the-spot report relating to the Battle of Tripoli. Of the series of accounts published in *L'Intransigeant* between 25 and 31 December 1911, it is one of Marinetti's more flowery communiqués.

It is the hour of the sunset-director of the orchestra who, with an enveloping gesture, unites all the scattered flutes of the bird's in the trees and the mournful harps of the

[49] Carrieri, *Futurism*, p. 79.

insects, and the creaking of the branches, and the stridulation of the stones. And it is he who at a certain moment silences the mess-tin-timpani and the discordant guns in order to permit the full voice of the open-armed, upright stars to soar above the united orchestral instruments against the background of the sky. And then there suddenly appears the prima donna of the occasion, wearing a very low-necked gown; the desert, in fact, displays its immense breasts with their liquefying curves covered with pinkish rouge and studded with the cascading gems of the prodigious night.[50]

In this deep purple essay in *journalisme romancé*, an extensive set of analogies between the battle-zone and an orchestral performance, of the kind also central to Marinetti's later poem 'Bombardamento', are well in evidence. But there is as yet little to justify calling such an absurdly overwritten piece of war-reporting 'telegraphic', even though it may have been literally telegraphed back from the front.

In retrospect, Marinetti was fond of presenting his new brand of telegraphic writing as if it had somehow grown immediately and 'intuitively' out of the conditions he worked under as a war correspondent. In both 'Destruction of Syntax' and the preface to *I nuovi poeti futuristi* he tries to pretend that his Futurist 'need for laconicism', the compulsion to present things 'telegraphically', arose as a direct result of the 'economic rapidity that the telegraph imposes on reporters and war correspondents for their journalistic reports'.[51] But after reading some of his 'star-studded' effusions for *L'Intransigeant*, which must have had a very indulgent editor,[52] one may doubt whether he could have been any more expatiative, even if no reporting or telegraphing restrictions had been in operation. We are still a long way from Marinetti's ultimate telegraphic economy, whether in the poetry of reduced syntax or the drawings, executed on telegram-forms, with no verbal message on them at all,[53] that he was to send during the First World War. In fact, Marinetti's telegraphic style was much slower to develop than legend permitted him to admit. It did not automatically grow as straightforwardly out of war-reporting conditions as he was afterwards so eager to claim.

[50] Carrieri, *Futurism*, p. 79.

[51] See *TIF* 61, and *F. T. Marinetti presenta I nuovi poeti futuristi* (Rome, 1925), 10.

[52] Strictly speaking, *L'Intransigeant* was not a newspaper but, in Noëmi Blumenkranz-Onimus's words (*La Poésie futuriste italienne: Essai d'analyse esthétique* (Paris, 1984), 15), a 'feuilleton poétique'. This may help account for Marinetti's freedom to write so self-indulgently, but it makes his claim to be a war reporter subject to normal journalistic restrictions seem something of an opportunistic fiction.

[53] Linda Landis describes Telegrams 41 and 69 (in the Marinetti Archive at Yale) as consisting of 'long arrows (cannon and anti-aircraft fire), tight zigzag lines (columns of black smoke trailing fallen aircraft), X shapes (targets and landing fields) and the mathematical equations which sum up the battle experience' ('Futurists at War', in Anne Coffin Hanson (ed.), *The Futurist Imagination: Word + Image in Italian Futurist Painting, Drawing, Collage and Free-Word Poetry* (New Haven, Connecticut, 1983), 73 f.). One of the telegrams is reproduced on p. 73 of this catalogue.

Both the theory and the practice of Marinetti's telegraphic lyricism actually emerged during the summer of 1912, i.e. well after his Libyan Campaign experiences (which were reported in French, still his first language at this time) and some months before his departure to observe the Bulgarian siege of Adrianopolis (now Edirne) during the First Balkan War. (*Zang tumb tumb*, the major work which eventually grew piecemeal out of these experiences, bears the subtitle 'Adrianopolis, October 1912'.) This means that the 'Technical Manifesto of Futurist Literature' ('Manifesto tecnico della letteratura futurista'), of 11 May, and the Supplement containing Marinetti's first genuinely telegraphic poem ('Battaglia peso + odore'), published 11 August, both precede the experiences at Adrianopolis. Therefore, if the telegraphic style which they launch can in any way be thought of as having evolved out of his war-reporting experiences, then this can only be in the sense that Marinetti developed his new immediate telegraphic idiom *out of dissatisfaction with the method of the* Intransigeant *reports*—not, as he was later fond of implying, because the new style had automatically resulted from the severe word-rationing imposed on war correspondents. Still, by the time of his Siege of Adrianopolis mission in late 1912, Marinetti had at his disposal a new 'laconic' style which could be used to report the sensations of battle in a striking way. Obviously, there is a time-lag: both *Zang tumb tumb* and the earlier 'Battaglia peso + odore' represent a substantially retrospective condensation of previous experiences, they hardly constitute the new mode of poetic 'reportage' that Roman Jakobson took them to be.[54]

By and large, the theory of telegraphic writing pre-dates its appearance. Many of Marinetti's language-reforming proposals had already been publicized in the 'Technical Manifesto of Futurist Literature' before the Supplement came up with an example in the shape of 'Battaglia peso + odore'. Since these manifesto proposals were in themselves as influential as the first poetic works in the new idiom, it may help to look at Marinetti's theoretical position first. The main proposals are to be found in the 'Technical Manifesto of Futurist Literature', in 'Destruction of Syntax' and in 'Geometric and Mechanical Splendour and the Numerical Sensibility' ('Lo splendore geometrico e meccanico e la sensibilità numerica' (1914)).

It becomes evident from Marinetti's various writings on the subject that the framework within which the new abbreviated style is conceived is not just that of war, but modernity in general, of which large-scale warfare is but one

[54] According to Jakobson, Marinetti's proposals represent 'a reform in the sphere of reportage, not poetic language' ('Noveyshaya russkaya poeziya (Nabrosok pervyy) Viktor Khlebnikov', in *Texte der russischen Formalisten*, ii. *Texte zur Theorie des Verses und der poetischen Sprache*, ed. Wolf-Dieter Stempel (Munich, 1972), 25, 31). Underlying this reaction is a contrast between 'emotive and poetic language' which sees the former as a communicative form, whereas the poetic word is to a certain degree 'objectless'. See Peter Steiner, *Russian Formalism: A Metapoetics* (Ithaca, 1984), 10.

manifestation. In his colourful catalogue of the 'elements of the new Futurist sensibility' (in 'Destruction of Syntax'), Marinetti lists a number of results of the modern appreciation of speed, including the 'new touristic sensibility' and a concomitant 'sense of the world' as a global village (*TIF* 59 f.). In other words, it is not just the impact of the telegraph or war reporting which underlies his linguistic-cum-stylistic reforms, but the whole dynamism of a world newly interpreted, though the telegraph remains an apt symbol for a writer to choose for this new sensibility. The arts, Marinetti suggests, are already following life in this respect, now they simply need to pursue it even further. Thus, in modern painting he sees the artist's 'habit of visual foreshortening and synthesis' as being 'caused by the speed of trains and cars that look down on cities and the countryside' (*TIF* 60).[55] Already, most modern forms of communication display, according to Marinetti, a general 'dread of slowness, pettiness, analysis, and detailed explanations'. The Futurist 'love of speed, abbreviation, and the summary' leads to the attitude: 'Quick, give me the whole thing *in two words*.' It is this that poetry has to respond to stylistically.

In Marinetti's inventory, only one further item follows the excerpt just quoted, and that is a symptomatic reference to the Futurist 'love of depth and essence in every exercise of the spirit' (*TIF* 60). The juxtaposition of brevity with the 'essential'—i.e. that which relates to the essence—is crucial, for it points to one of the most fruitfully misguided myths of the period: the idea that brevity is more than just a virtuous economy reflecting the pace of modern life, that it is in some way the key to greater truths and thus renders most passéist writing not only anachronistic but also superficial.

The evolutionary, teleological context within which Marinetti and many of the other Futurists saw their radical proposals for a new language is most clearly summed up at the beginning of the discussion of 'Free expressive orthography'. Here, Marinetti speaks of '*the historical necessity*' of such liberated spelling being 'demonstrated by the successive revolutions that have continuously freed the lyrical powers of the human race from shackles and rules' (*TIF* 69). We are thus allegedly dealing, not with one individual writer's (or even one movement's) attempt at discovering a style adequate to the times, but with a phenomenon interpreted as part of a process of liberation from the hold of the past which is an inevitable feature of the modern world. This goes some way towards explaining the imagery of social emancipation used: for example, the reference in the 'Technical Manifesto' to the 'pressing need to liberate words, to drag

[55] Ferdinand Léger makes a similar point in 'Les Réalisations picturales actuelles' (1914): 'When one crosses a landscape in a car or a train, [it] loses in descriptive value, but gains in synthetic value. . . . Modern man registers a hundred times more impressions than an eighteenth-century artist. . . . All this has influenced the condensation, variety and fragmentation of forms of modern painting' (*Functions of Painting* (London, 1973), 12). Marinetti probably knew this lecture, although a common Unanimist source for the idea cannot be ruled out.

them out of their prison in the Latin period' (*TIF* 40).[56] Looking at the ethos underlying such works as *Mafarka le futuriste* and *Zang tumb tumb*, one might well conclude that Marinetti felt more compassion for incarcerated words than for most of his non-Futurist fellow human beings.

To clinch his argument that brevity has become a modern necessity, Marinetti resorts, in addition to a specious historical teleology, to anecdotal underpinning:

> Now suppose that a friend of yours gifted with this faculty of expressing himself telegraphically finds himself in a zone of intense experience (revolution, war, shipwreck, earthquake, and so on) and starts right away to tell you his impressions. Do you know what this lyrical, excited friend of yours will instinctively do?

Note that the use of 'right away' and 'instinctively' already contrasts sharply with what Marinetti, with all his paraphernalia of manifestos and extensive theorizing, is attempting to engineer artificially.

> He will begin by brutally destroying the syntax of his speech. He wastes no time in building sentences. Punctuation and the right adjectives will mean nothing to him. He will despise subtleties and nuances of language. Breathlessly he will assault your nerves with visual, auditory, olfactory sensations, just as they come to him. The rush of steam-emotion will burst the sentence's steampipe, the valves of punctuation, and the adjectival clamp. Fistfuls of essential words in no conventional order. Sole preoccupation of the narrator, to render every vibration of his being. (*TIF* 61)

On this analogy, poetry should, if it is also to devote itself to 'zones of intense experience', entail 'the uninterrupted sequence of new images', even the 'foreshortening of the image to the single essential word' (*TIF* 41).

This foregrounding of 'single, essential' words operates on a redundancy principle whereby some words, belonging to specific grammatical categories, are deemed necessary and others are slotted out as superfluous. The result is a form of telegraphic diction. Yet, as Christine Poggi points out, Marinetti's actual free-word telegrams 'violate all the rules of proper telegram writing. These rules, listed on the back of the form under the heading "*Avvertenze*" . . . advise the writer to make the message brief and clear. In order to avoid confusion, the writer is urged to use approved abbreviations.'[57] In the particular instance Poggi is discussing, the violation of the conventional rules for telegrams lies in Marinetti's use of diagrammatical pictures instead of any words. But even more mainstream, verbal forms of Futurist free-word writing deviate

[56] See, in this connection, Manfred Hinz, *Die Zukunft der Katastrophe: Mythische und rationalistische Geschichtstheorie im italienischen Futurismus* (Berlin, 1985).

[57] 'Marinetti's *Parole in Libertà* and the Futurist Collage Aesthetic', in *The Futurist Imagination*, p. 7.

from the principles of real telegram-formulation in a number of significant ways:

1. While in real telegrams the reduction of utterances to the minimally decodable can lead, incidentally, to a series of sentences with only a skeletal grammar retained, the Italian Futurists were first and foremost concerned with the destruction of syntax as *a means to* the dynamic.
2. In Italian Futurism, specific grammatical forms (adjectives, adverbs, articles, particles) were judged to be axiomatically superfluous. Such a categorical, linguistic approach to the problem of conciseness is neither that of the Italian Post Office's *Avvertenze* nor that of general stylistics; in both contexts, what is essential and what is redundant remains a matter of the specific content and context of a given utterance, not of categorically applied linguistic principles.
3. Real telegrams are designed to be functionally palimpsestic, i.e. they represent reduced versions of conventional, longer utterances which can in due course be reconstructed without either difficulty or ambiguity. (Even Cary Grant must have known what the telegram 'How old Cary Grant?' really meant!) Yet this unequivocal decodability is seldom a feature of modern telegraphic poetry.

In various theoretical writings of the pre-war period, Marinetti set out a series of proposals concerning a set of highly specific linguistic reforms which would facilitate the 'essential brevity' that the Futurist sensibility supposedly demanded as its medium. Yet, although he presents these reforms, largely for purposes of impact, as an out-and-out '*destruction* of syntax' and couches most of his other suggestions in equally anarchic formulations,[58] what Marinetti is advocating is really no more than a series of substantial, but nevertheless selective modifications to existing syntax. The *OED* defines 'syntax' as 'the arrangement of words (in their proper forms) by which their connection and relation in a sentence are shown'. The Futurists cannot be thought of as going as far as to abolish syntax, understood in this sense. (The Russian Futurists' idea that they were 'shaking syntax loose' (*Sadok sudey* (*Manifesty*, p. 52)) is a more accurate description of what is going on in both countries.) In fact, the bulk of the theory merely seeks to legislate anew about what 'in appropriate forms' and 'arrangements' means. To this end, Marinetti puts forward a series of proposals concerning *which words*, invariably interpreted restrictively as *which parts of*

[58] e.g.: 'I speak of destroying the canals of syntax' (*TIF* 62); 'Destroy the I in literature. . . . We must drive man from literature' (*TIF* 44). Such ideas are presented in much the same iconoclastic tone as characterizes the Founding Manifesto's declarations about the need to destroy museums and other bastions of passéism. Cf. Johannes Becher's image of 'Alogical bombs destroying traditional academic sentence-structure' (letter to the *Frankfurter Zeitung*, quoted in Walter Höllerer (ed.), *Theorie der modernen Lyrik: Dokumente zur Poetik 1* (Hamburg, 1965), 284).

speech, should disappear from Futurist free-word arrangements. Moreover, his attention appears to have been confined to literary discourse. Although there are apocryphal reports that Marinetti sometimes spoke in a form of 'italiano telegrafico',[59] it is doubtful whether he ever intended the new idiolect to become universally or even widely used. Unlike Goll, who conceived of his *Telegrammstil* as the idiom of the fourth millennium, Marinetti apparently harboured no illusions about words-in-freedom becoming the *lingua franca* of even a distant future. In 'Destruction of Syntax' he declares that 'philosophy, the exact sciences, politics, journalism, education, business, however much they seek synthetic forms of expression, will still need to use syntax and punctuation'. He adds, tellingly: 'I am obliged, for that matter, to use them myself in order to make myself clear to you' (*TIF* 57). Only in one manifesto, 'The Variety Theatre' ('Il teatro di Varietà' (1913)), and then only in the final paragraph, does Marinetti employ telegraphic writing in a non-literary context. Having extolled the attractions of the variety theatre, of clowns, acrobats, and Futurist performance, he transcends his own prose manifesto style by launching into a series of verbal pyrotechnics that are clearly meant to communicate the spirit of an evening in the variety theatre.

Because, and don't forget it, we Futurists are YOUNG ARTILLERYMEN ON A TOOT, as we proclaimed in our manifesto 'Let's Murder the Moonshine', fire + fire + light against moonshine and against old firmaments war every night   great cities to blaze with electric signs   Immense black face (30 meters high + 150 meters height of the building = 180 meters) open close open close a golden eye 3 meters high SMOKE SMOKE MANOLI SMOKE MANOLI CIGARETTES   woman in a blouse (50 meters + 120 meters of building = 170 meters) stretch relax a violet rosy lilac blue bust   froth of electric light in a champagne glass (30 meters) sizzle evaporate in a mouthful of darkness   electric signs dim die under a dark stiff hand come to life again continue stretch out in the night the human day's activity courage + folly never to die or cease or sleep   electric signs = formation and disaggregation of mineral and vegetable centre of the earth   circulation of blood in the ferrous faces of Futurist houses increases, empurples (joy anger more more still stronger) as soon as the negative pessimist sentimental nostalgic shadows besiege the city   brilliant revival of streets that channel a smoky swarm of workers by day   two horses (30 meters tall) rolling golden balls with their hoofs GIOCONDA PURGATIVE WATERS   crisscross of *trrrr trrrrr*   Elevated *trrrr trrrrr* overhead trrrombone whissstle ambulance sirens and firetrucks   transformation of the streets into splendid corridors to guide push logic necessity the crowd toward trepidation + laughter + music-hall uproar   FOLIES-BERGÈRE EMPIRE CRÈME-ÉCLIPSE tubes of mercury red red red blue violet   huge letter-eels of gold purple diamond fire Futurist defiance to the weepy night   the stars' defeat warmth enthusiasm faith conviction will power penetration of an electric sign into the house across the street *yellow slaps* for that gouty, dozy bibliophile in slippers 3 mirrors watch him   the sign plunges to 3 redgold abysses open close open close 3-thousand meters

[59] G. A. Borgese, *La vita e il libro* (Turin, 1911), 127.

deep horror quick go out out hat stick steps taximeter push shove *zuu zuoeu* here we are dazzle of the promenade solemnity of the panther-cocottes in their comic-opera tropics fat warm smell of music-hall gaiety = tireless ventilation of the world's Futurist brain.[60]

One of the few other instances in Italian Futurism of a telegraphic style being used outside a directly literary context is in the case of the diary that Boccioni wrote while at the front in 1915: his 'Diario di guerra (agosto-novembre 1915)' (*Gli scritti*, pp. 313–23). For example, the entry for 13 October includes the sequence 'Ora 1 pom. *Malcesine* 55 Km cartuccie galletta confusione, scarico 120 b. visto per la prima volta, = fierezza, + orgoglio, + guerra' (*Gli scritti*, p. 314) ('1.00 p.m. *Malcesine* 55 km cartridges biscuit confusion, unloading 120 b. seen for the first time, = boldness, + pride, + war'). This time the idiom occurs in a context which suggests that it was unlikely to have been intended for public consumption. Whereas Marinetti's 'Variety Theatre' manifesto is still using the style for rhetorical effect—and there are those (including most recently Marjorie Perloff) who would argue that his manifestos are his most successful aesthetic creations—Boccioni appears to be using the telegraphic as a form of personal shorthand, such as diarists have used for centuries.

Even if Marinetti is wrong to assume that syntax can be meaningfully destroyed, his more cautiously formulated view that 'only the unsyntactical poet who unlinks his words can penetrate the essence of matter' (*TIF* 46) is a revealing one. A comparison with the way the English Imagists set about cultivating brevity, with much of Apollinaire's and Auden's 'telegraphese', or even Hemingway's later 'cable-ese', would prove that there is no inherent reason why the quest for such a new 'essential brevity' need necessarily go hand in hand with the wholesale rejection of certain grammatical categories.

Ironically, in their very formulations Marinetti's proposals highlight the fact that syntax has not been totally *destroyed*.[61] His principal demands—'use infinitives', 'abolish the adjective', sometimes even 'abolish the verb' (*TIF* 41 f.)—all

[60] 'Il teatro di Varietà' (*TIF* 77–9). This final free-word paragraph appears in the version published in *I manifesti del Futurismo* (Florence, 1914), but not in the one in *Lacerba* (1 Oct. 1913), 211, which means that the English translation (first published in *The Daily Mail* of 21 Nov. 1913) boasts a more experimental conclusion than the one in *Lacerba*, which ends after the word 'moon' (*luna*). There can be few instances of a paper like the *Daily Mail* outstripping the radicalism of *Lacerba*.

[61] Franco Gisuti remarks in this connection: 'Marinetti gives up several grammatical categories, i.e. adverb, adjective, verb conjugation, and an auxiliary system like punctuation, and in so doing undermines the very structure of language. But also, he rejects the principle of a linguistic operational order by denying to both syntagmatical linkage and paradigmatical selection all of their *langue* value, so that the analogical relations between apparently heterogeneous elements of actual reality and between them and the semiotic units and operations that are used to cognize and represent them can no longer be appropriately rendered. . . . In destroying all structural ties between elements he prevents himself from getting to the analogical common ground to which disparate elements can be led back to acquire cohesion' (*Italian Futurism as a Sociolinguistic Phenomenon* (Ann Arbor, 1981), 11 f.).

rest on a tacit recognition that the various parts of speech will still be recognizable as such, otherwise the poor Futurist iconoclast would not be able to identify which elements to prune and which to retain. Looked at in linguistic terms, the Futurist style of telegraphic lyricism could be thought of as a form of idiolect based on the language-reducing principle of the lipogram,[62] now removing whole word categories instead of individual letters, in the way that Otto Nebel and Georges Perec have done. In this respect, it is a parasitic form, generated from—and only comprehensible to someone with competence in—traditional (grammatical) language. Many of the successful effects of Futurist poetry depend, as a result, on an interaction in the reader's mind between the pattern of expectation established by conventional language and what Ian Dunlop has called 'the shock of the new'.[63] This focus on what can best be thought of as an 'alternative syntax' is a clear indication that Marinetti's reforms, no matter how rigorously put into practice, cannot *per se* lead, as he seems to have believed they would, to an inevitable revitalization of poetic diction and automatically result in good individual Futurist poems. They can no more do this than the existence of a conventional linguistic repertoire can of itself be a guarantee of good or bad poetry. To expect this is to confuse the medium with its potential—which is one of the fundamental fallacies of much Futurist aesthetic theory. All that Marinetti's proposals could establish was a new (telegraphic) *repertoire*. Ellipsis, or any other form of telegraphic style, cannot, *pace* Apollinaire, on its own create powerful poetry. That would be the equivalent of asking of language as a whole (or any of its idiolects) what could really only be expected of the individual practitioner. There is a difference between what reforms can automatically bring about and what they may, at most, facilitate. And the blurring of these two issues not only represents one of the blindspots of Futurist theory, but was surely one of the main underlying causes of much Futurist bad writing.

In support of his proposals about what should be abolished and what retained in the new idiom, Marinetti also advanced a number of idiosyncratic arguments concerning the effect of different parts of speech on a poetic utterance (arguments that have generally been reiterated and accepted uncritically by the majority of commentators on Futurist literature). We are told that

> except for needed contrast or a change of rhythm, the different moods and tenses of the verb should be abolished, because they make the verb into a stage-coach's loose wheel adapting itself to rough country roads, but unable to turn swiftly on a smooth road. *The infinitive verb*, on the other hand, *is the very movement of the new lyricism*, having the fluency of a train's wheel or an aeroplane's propeller.
>
> The different moods and tenses of the verb express a prudent egotism, a high and

[62] See Ciro Vitielo, 'La nozione di liposegno, tra teoria e prassi nel sistema parolibero di Marinetti', in A. Longati *et al.* (eds.), *F. T. Marinetti: Futurista* (Naples, 1977), 135–62.

[63] Ian Dunlop, *The Shock of the New* (London, 1972).

low force, and tiredness, of desire and delusion, of pauses, in other words, in the trajectory of hope and will. The infinitive verb expresses optimism itself. . . . when I say *to run*, what is that verb's subject? Everyone and everything: that is, the universal irradiation of life that runs and of which we are a conscious particle. . . . The infinitive is the passion of the *I* that abandons itself to the *all*, the heroic disinterest of continuity of the joy and effort of acting. Infinitive verb = the divinity of action. (*TIF* 87 f.)[64]

Elsewhere, Marinetti would have us believe that 'every noun' is 'a vehicle or belt set in motion by the verb in the infinitive' and that the infinitive 'constitutes the very speed of the new style' (*TIF* 65).

It is, of course, possible to drive a passéist coach and horses through the premisses of many of these pronouncements when they are set out in such absolutist terms. One only has to consider what would happen to Marinetti's claim concerning the infinitive, if, say, he had chosen as his illustrations the verb *to languish* or *to vegetate* instead of *to run*. An equally Futurist work like Alberto Viviani's nihilistic free-word poem 'Stati d'animo di un suicida' (discussed later in this chapter) shows that the linguistic reforms mooted by no means necessarily result in the kind of dynamic lyricism promised by the early theory.

It might be objected that the reforms so far mentioned have not been presented within any specifically 'telegraphic' framework. Indeed, we hear more of 'steam-pipes', 'clamps', and 'wheels' than of the 'TSF'. But the full title of the manifesto in which many of these reforms are presented is 'Destruction of Syntax—*Wireless Imagination*—Words-in-Freedom' ('Distruzione della sintassi—*Immaginazione senza fili*—Parole in libertà' (my emphasis)). In fact, the most explicitly 'telegraphic' section of the manifesto is the one in which Marinetti, obviously availing himself of the conveniently current 'TSF' motif, refers, not primarily to the creation of a new pared-down syntax, but to what he calls 'l'immaginazione senza fili' ('the wireless imagination').[65] To talk, as Hermans does, for example,[66] of Marinetti's concept of a '"wireless" style' in this connection can be somewhat misleading, for it perpetuates the manifesto's own lack of clear distinction between the general syntax of telegraphic lyricism

[64] Gisuti argues that 'Marinetti's verb tends to melt into the pounding nominalized seriality rather than to activate it. The verbal function is too vitally linked to its aspects and its bearers (the nouns) to stand a chance of significance once those ties are cut. We can say that, far from achieving its claimed dynamic function, Marinetti's verb solidifies into an object-like form, its energy choked by a nominal layer. . . . The transcending impulse that Marinetti struggles to release does not take shape due to his incapability for mediation. . . . his style is largely a nominal one, as we see the "renewed" verbal function come to a stop in its actual noun-like style' (*Italian Futurism as a Sociolinguistic Phenomenon*, p. 114).

[65] Although Marinetti claims in 'Destruction of Syntax' (*TIF* 63) that the 'Technical Manifesto of Futurist Literature' 'invented . . . the wireless imagination', it only did so by advocating that 'every noun should have its double; that is, the noun should be followed, with no conjunction, by the noun to which it is related by analogy' (*TIF* 41). There is, admittedly, a passing reference to the 'wireless imagination' (p. 46), but the idea is not elaborated in any detail until 'Destruction of Syntax'.

[66] *The Structure of Modernist Poetry*, p. 70.

and what Marinetti has to say, more *pro domo*, about the role of bold analogies in his conception of poetry. Whereas the cult of the telegraphic was a widespread phenomenon, the 'wireless imagination' is the preserve of Marinetti and a very small number of his imitators.

The Italian title of the section 'Immaginazione senza fili' has been variously rendered as 'imagination without strings', 'imagination without sequence', 'untrammelled fantasy', and 'wireless imagination'.[67] This last designation comes closest to the spirit of Marinetti's phrase, which is intended to suggest the letters SF (as in TSF) and to capitalize on the modernist associations the epithet 'wireless' enjoys—though, as the other translations imply, there are also additional connotations to the phrase. Pär Bergman manages to capture the main association nicely, however, with his abbreviation 'ISF', coined in deliberate analogy to 'TSF'.[68]

The telegraphic *style* is, in Marinetti's eyes, almost a by-product of the 'wireless imagination', which means 'the absolute freedom of images or analogies, expressed with *unhampered words* and with no connecting wires of syntax and with no punctuation' (*TIF* 63). Just as wireless telegraphy had proved itself able to transcend the limitations of geography and magically bring together Paris and Adrianopolis, Berlin and Timbuktu, so the 'wireless imagination' of the Futurist poet was at liberty to juxtapose the most incongruous of ideas and images.

> Up to now writers have been restricted to immediate analogies. For instance, they have compared an animal with a man or with another animal, which is almost the same as a kind of photography. Others, more advanced, might compare the same trembling fox terrier to a little Morse Code machine. I, on the other hand, compare it with gurgling water. In this there is an *ever vaster gradation of analogies*, there are ever deeper and more solid affinities, however remote.

Doubtless, it remains a matter of subjective judgement whether one finds the analogy with the Morse Code machine or that with gurgling water a 'vaster', more 'remote' one. But the general underlying point that the wireless imagination creates bold analogies need not stand or fall on account of the eccentric illustrations chosen for it.

'Analogy', according to Marinetti, 'is nothing more than the deep love that assembles distant, seemingly diverse and hostile things. An orchestral style . . .

[67] These translations are taken, respectively, from: *Futurist Manifestos*, ed. Umbro Apollonio (London, 1973), 99; Carrieri, *Futurism*, p. 79; Caroline Tisdall and Angelo Bozzolla, *Futurism* (World of Art Library; London, 1977), 95; and Hugh Shankland, 'Futurism in literature and the theatre', in *Futurismo 1909–1919: Exhibition of Italian Futurism* (Newcastle, 1972), 81. The English translation of the manifesto—in *Poetry and Drama*, 1 (3 Sept. 1913), 262–4 and 291–305—is entitled 'Wireless Imagination and Words at Liberty'.

[68] *'Modernolatria' et 'simultaneità'*, p. 191.

can embrace the life of matter only by means of the most extensive analogies' (*TIF* 42).[69]

This notion of a poetry of the 'wireless imagination' in fact combines two principles, the one largely thematic, the other linguistic. First, there is the idea of the imagination's freedom to create bold analogies, what he has in mind when he goes on to declare 'With words-in-freedom we will have: CONDENSED METAPHORS. TELEGRAPHIC IMAGES ... COMPRESSED ANALOGIES' (*TIF* 63). Second, there is the actual *style* of words-in-freedom, which will render these compressed analogies (hence the reference to their being 'expressed with unhampered words and with no connecting wires of syntax'). Since analogy is only one feature of Marinetti's poetry, one which occurs sporadically only, it follows that the impact of the wireless imagination will be a matter of local effects, rather than a totally new idiom, in the way that telegraphic writing was for many of Marinetti's contemporaries. In any case, the relationship between the 'wireless imagination' and the telegraphic style is not as clear-cut as Marinetti implies. The *Zang tumb tumb* lines, *treno-pascecane immagliarlo spingerlo nel ferryboat–balena* ('the train–shark netting it pushing it in the ferryboat–whale') (*TIF* 564 f.) may contain two bold juxtapositions, but it is the verbal element which is the really innovative part of this sentence. The double-noun functions rather as traditional subject, but the unconjugated Italian verbs (reflected in my English translation by the use of gerunds) break the confines of traditional syntax—though such an effect could just as well have been achieved by a single subject.

In the second, stylistic sense, 'wireless' means little more than the ideas already set out in connection with the 'destruction of syntax' brought about by words-in-freedom.[70] In the first sense, it defines a much more eccentric notion of quasi-Surrealist juxtaposition which had only been hinted at in the previous pronouncements about words-in-freedom and which bears little resemblance to what came to be thought of in the majority of European avant-garde circles as the 'telegraphic' style.

As will be apparent from the following discussions of 'Battaglia peso + odore' and 'Bombardamento', Giovanni Lista's view that *the wireless imagination* was 'never realized' and remained 'a secondary hypothesis in Marinetti's mani-

[69] Writing of Marinetti's figurative language in *La Vie charnelle*, Jean Royère remarks: 'il ne file pas la métaphore, il la câble' (*La Phalange*, 3–4 (1908–9), 265).

[70] Just how ambiguous Marinetti's exposition could be is indicated by Tisdall's and Bozzolla's assumption that he reserved the term 'wireless imagination' for 'the stage at which the first term of each analogy could be omitted to leave no more than an uninterrupted sequence of second terms' (*Futurism*, p. 95). In fact, he nowhere goes this far, but merely forecasts: 'some day we will achieve a yet more essential art, when we dare to suppress all the first terms of our analogies and render no more than an uninterrupted sequence of second terms' (*TIF* 46). There is no suggestion that only this later stage is construed as representing the 'wireless imagination' at work.

festos' is open to question.[71] Certainly, this harnessing of a theory of bold analogies to a general notion of modish telegraphic lyricism was very much restricted to Marinetti's early Futurist phase and to the work of a limited number of his emulators, whereas telegraphic writing in some form was more widespread, even within Futurism.[72] The fact that the 'Destruction of Syntax' manifesto was issued as a broadsheet with its title abbreviated to *L'immaginazione senza fili e le parole in libertà*[73] may have helped give a measure of emphasis to the link between the leaps of the imagination and an abbreviated idiom; and it certainly seems to have led among Marinetti's contemporaries to a degree of confusion between the 'wireless imagination' and 'telegraphic lyricism', a confusion which has persisted to this day.

There are various features to Marinetti's theoretical promulgation of telegraphic lyricism which delimit the idea in a new way, if not actually give it somewhat esoteric overtones: first, there is the association of an abbreviated style with specific linguistic reforms; second, the assumption that the new laconicism is related to the use of bold analogies; and, third, the association of the telegraphic idiom with the Futurist quest for 'essentiality'. This final issue is something that will be considered later in the context of Futurist primitivism, although one or two observations may help at this stage.

Within the Futurist aesthetic, the phrase 'essential brevity'[74] means more than just a *necessary* stylistic economy; it signifies the brevity that communicates the essential. Describing what he calls 'the plunge of the essential word into the water of sensibility', Marinetti sees this plunge as being 'minus the concentric circles that the word normally produces' (*TIF* 64). What this means is that, ostensibly freed from the restrictions imposed by conventional grammar, words-in-freedom will penetrate to the depths, the essence of material reality. Hence, Marinetti can claim that 'only the unsyntactical poet who unlinks his words is able to penetrate to the essence of matter' (*TIF* 46). Brevity is by this token the key to the essential, and the essential for Marinetti means pertaining to 'the essence of matter' (*TIF* 64).

In the light of this, one needs to distinguish carefully between Marinetti's

[71] *Marinetti* (Paris, 1976), 111.

[72] Bergman (*'Modernolatria' et 'simultaneità'*, p. 320) notes that Cendrars tried to obtain an 'effect of violent contrast, by putting side-by-side, "using no wires", ideas, feelings and images which surprise the reader' and that this technique must have been primarily influenced by Delaunay. Yet a closer analogy can be found in Marinetti's theory and practice of the 'wireless imagination'. Cendrars's indebtedness to Italian Futurism still awaits detailed investigation, although there is enough material in Perloff's *The Futurist Moment* to suggest that such an investigation would yield significant results.

[73] Milan, May 1913. French and German translations were also issued on broadsheets using this shorter title.

[74] See Point 9 of 'Geometric and Mechanical Splendour and the Numerical Sensibility' (*TIF* 91).

quite specific notion of 'essentiality' and the more general Futurist—Expressionist and Imagist—search for an 'essential' language in the sense of the requisite style for the new technological age or adequate to the modernist spirit. Frequently, the reception and widespread impact of Marinetti's ideas during the Futurist period were based on simplifications and misconceptions, albeit sometimes creative ones. The catch-phrases 'essential brevity' and 'telegraphic lyricism' both meant something very different for Marinetti than they usually did during the period in general, either in Italy or abroad. For Marinetti, the telegraphic was not some kind of all-purpose style, but the expression of a particular mode of Futurist sensibility.

With 'Battaglia peso + odore' (1912) and *Zang tumb tumb* (1914), Marinetti was able to present his new innovative form of telegraphic writing, using it in both cases to describe battle-scenes.

In order to highlight the differences between the reduced syntax of his elliptical words-in-freedom and conventional passéist language, Marinetti followed his 'Destruction of Syntax', when reissued in *Noi futuristi* (Milan, 1917), with an illustration of the way in which a free-word poet would describe a battle-scene, juxtaposed with the way in which it would read in a conventional account.

ESEMPIO DI PAROLE IN LIBERTÀ

Avanguardie: 20 metri battaglioni-formiche cavalleria-ragni strade-guadi generale-isolotto staffette-cavallette sabbie-rivoluzione obici-tribuni nuvole-graticole fucili-martiri shrapnels-aureole moltiplicazione addizione divisione obici-sottrazione granata-cancellatura grondare colare frana blocchi valanga.

TRADUZIONE DI QUESTE PAROLE IN LIBERTÀ IN FORMA TRADIZIONALE

20 metri di distanza dividevano l'avanguardia italiana dall'avanguardia turca. Sull'immensità delle sabbie, i battaglioni s'allungavano come interminabili cortei di formiche. Uno squadrone di cavalleria procedeva sulla duna come una lunga fila di ragni. Il deserto sembrava il vasto estuario di un fiume, che si potesse guadare soltanto per due o tre minuscole strade. S'ergeva su un poggio il generale italiano, simile a un isolotto, e si staccavano d'intorno a lui, a quando a quando, delle staffette balzanti come cavallette.

Il vento e la fucileria rivoluzionavano le sabbie. Gli obici scoppiettavano in cielo gesticolando colle loro lunghe braccia d'oro e con grandi urli, come tribuni dall'alto d'invisibili balconi, sull'insurrezione della battaglia. Più in alto cresceva col sole l'ardore bruciante delle nuvole, simili a graticole.

Nell'accanirsi del combattimento, i fucili scottanti sembravano soffrire come martiri, fra le mani dei soldati. Ma gli shrapnels ornavan loro continuamente la testa di aureole gloriose. V'era dovunque una moltiplicazione di volontà e d'istinti, un'addizione e una divisione continua di sforzi e d'angosce. Gli obici, di tanto in tanto, sottraevano dalla battaglia intere compagnie. Ecco una granata cancellare tutto uno squadrone. Il sangue grondava dovunque. Si videro allora colare a poco a poco giù per le dune due battaglioni nemici in fuga. Cominciava il franare degli obici della Sicilia; cadevano come blocchi ad uno ad uno, sempre più presto, finché la valanga diede il segnale della vittoria italiana.

The 'EXAMPLE OF WORDS-IN-FREEDOM' given here is recognizable as the seventh paragraph of 'Battaglia peso + odore' of five years before, whereas the 'TRANSLATION OF THESE WORDS-IN-FREEDOM INTO TRADITIONAL FORM' is no more than an overwritten new piece, evidently concocted, no doubt with some relish, in order to satirize the self-indulgent longueurs of passéist literature[75]—though the 'traditional form' is not as far removed from the purple writing of his *Intransigeant* reports as Marinetti would no doubt have wished us to believe. To his credit, however, Marinetti does quote a passage from his earlier *Intransigeant* style in the 'Technical Manifesto of Futurist Literature' and describes it as being 'still masked and weighed down with traditional syntax' (*TIF* 43: the passage, entitled 'Battaglia di Tripoli', is there quoted in French).

An examination of the Libyan campaign reports reveals, however, that the problems are not simply centred on the author's being still hampered by traditional syntax. After all, he can still write his manifestos in traditional syntax without indulging himself in this kind of writing.

The free-word paragraph translates:

Vanguards: 20 metres battalions-ants cavalry-spiders roads-fords general-small island dispatchriders-grasshoppers sands-uprising howitzers-tribunes clouds-gratings guns-martyrs schrapnels-halos multiplication addition division howitzers-subtraction grenades-deletion streaming trickling landslide blockages avalanche.

This mercifully brief extract from an excruciatingly bombastic and repetitively formulaic early exercise indicates the extent to which telegraphic lyricism at this initial stage was synonymous in Marinetti's eyes with the compressed analogies of the 'wireless imagination'. The paragraph from 'Battaglia peso + odore' is dominated by nominal juxtapositions, with verbs in the infinitive (again I have used English gerunds to reflect this) only taking over from nominal asyndeton in the final description of slaughter. At this time in Marinetti's development, the telegraphic seems far removed from any essentiality, in the poet's own sense. Instead of focusing on the dynamic reality of the battle-scene, the reduced syntax becomes a means of indulging in impressionistic pairings, with the hyphen masking a truncated simile.[76] The

[75] Quoted from *TIF*, p. lxxxix. The second passage begins: 'A distance of 20 metres separated the Italian advanced guard from that of the Turks. On the great wastes of the sands, the battalions pushed forward like an interminable procession of ants. A cavalry squadron spread out across the dunes like a long spider's web. The desert looked like the vast estuary of a river . . .'; the piece continues for some time in the same passéist vein, replete with obvious similes and over-descriptive sequences.

[76] Khlebnikov's 'Khudozhniki mira' ('Artists of the World') also contrasts two modes of expression, in a way rather reminiscent of Marinetti's juxtaposition: 'Instead of saying: "The hordes of Huns and Goths, having joined together and gathered around Attila, full of enthusiasm for battle, moved onward together, but having been met and defeated by Aetius, the protector of Rome, they dispersed into many bands, halted and settled peacefully on their land, after spreading out into the steppes, filling up their emptiness"—should one not say: "Sha + so Huns and Goths, ve Atila, cha po, so do but bo + zo Aetius, kho Rome, so mo ve + ka so, lo sha steppes + cha"' (*SP* v. 220).

As Raymond Cooke points out (*Velimir Khlebnikov: A Critical Study* (Cambridge, 1987), 81 f.), what is

style is 'orchestral' in the sense that it orchestrates a set of analogies created in the mind of the observer–poet, even though they are less bold than the ones cited in Marinetti's theoretical account of the 'wireless imagination'.[77]

One of the things which Marinetti's satirical juxtaposition of words-in-freedom and traditional writing highlights—especially given the description of the longer passage as a 'translation' of the free-word piece into traditional language—is the nature of the telegraphic idiom as a form of modernist discourse.

Michel Décaudin has explored various drafts of Apollinaire's poem 'Zône' to reveal how the final elliptical phrase 'Soleil levant cou tranché' had grown, as a telegraphic reduction, out of two earlier, more conventional lines:

1. Le soleil est là avec sa tête coupée
2. Le soleil est là c'est un cou tranché.[78]

In this case, we know from the evidence of the drafts that the process of making telegraphic was a stage in the poem's genesis. So here, one could think of the earlier phrases eventually becoming distilled into some form of telegraphic *translation* ('Soleil levant cou tranché'). When MacNeice tries to explain Auden's 'telegraphese' with the words 'instead . . . of saying "We can speak of trouble" or "One can speak of trouble" he will start off with the verb "Can speak of trouble"',[79] we are dealing with a different situation. 'Can speak of trouble' cannot, on the grounds of any genetic evidence, be thought of as a 'translation' of preconceived (conventional) material. MacNeice is simply comparing how another kind of poet would have expressed something with the way Auden does.

In general, it is important to recognize that telegraphic writing is not simply created out of normal utterances by some process of reduction in the way real telegrams usually are. The *New Republic* for 9 September 1957 did, it is true, publish a poem by Robert Graves entitled 'Wordsworth by Cable' which reduced 'The Solitary Reaper' to the sequence 'SOLITARY HIGHLAND LASS REAPING BINDING GRAIN STOP MELANCHOLY SONG OVERFLOWS PROFOUND VALE'. But such a

at stake here is not the economy or telegraphic superiority of the second version, in Italian Futurist terms, but what Khlebnikov takes to be its comprehensibility, since *zaum'* is being presented as the prototype of a universal language.

[77] Alfred Döblin voices outspoken criticisms of this passage in his 'Open Letter to F. T. Marinetti' (*Aufsätze zur Literatur* (Olten and Freiburg, 1963), 13 f.): 'You have associations, and these act as connections, but you are unable to express these connections in any way. . . . You do violence to your ideas, you leave them incomprehensible, in order not to infringe your linguistic principle. . . . What does your adjective-free, adverbless "general-island" have to say to me? . . . Everything, almost everything remains hanging in an indefinite void—perhaps in your head each thing is in the right place, but it doesn't come out that way, it is not yet made poetic, it has not become a linguistic reality . . . it is your theory that condemns you to such abstraction.'

[78] Michel Décaudin, *Le Dossier d''Alcools'* (Paris, 1965), 81.

[79] *Modern Poetry: A Personal Essay*, p. 170.

jeu d'esprit is an exception rather than the typical pattern of telegraphic creation. Telegraphic writing and the writing of telegrams are essentially different activities. Even when Italian Futurists used the noun *rarefazioni* in the subtitles of their volumes, the process of creation was not one of rarefying—or progressively pruning—conventional sentences.

A telegram is under normal circumstances an abbreviated version of a precisely decodable statement. 'WORK FINISHED + STOP + COMING TUESDAY + STOP + MEET AT STATION + STOP + SEVEN' would be convertible back into conventional language—something like 'I have finished the work and will be coming to see you next Tuesday, please meet me at the station at seven o'clock'. Perhaps because of the Futurists' fondness for sprinkling the word 'STOP' around their telegraphic poems,[80] commentators have often treated such telegraphic language as if it worked on a comparably crude principle of removing superfluous elements from preformulated material. Thus, in his elucidation of Marinetti's 'Technical Manifesto of Futurist Literature', Armin Arnold writes: 'In German we would say "Tomorrow the soldier will give his life for the Fatherland". The Futurist would prefer: "Tomorrow soldier gives life Fatherland".'[81] Such straightforward mechanical reconstructions from the telegraphic back to normal idiom are misleading. Even a comparison of the free-word and traditional passages in *Noi futuristi* shows that there is no conceivable way in which someone trying to decode the first passage as one 'translates' a telegram into an unequivocal message would be able to arrive at a reconstruction of the second passage, for the simple reason that the second passage contains more details and says more than the first one does. The 'wireless' analogy 'battalions–ants', for instance, may still be central to the passéist sentence 'On the wastes of sands the battalions stretched out like endless processions of ants'. But if only the equation 'battalions–ants' is at the reader's disposal, it is impossible to deduce which quality of ant-behaviour (organized activity, frenzied eating, sheer proliferation, or being on the march) may be intended as the key feature of the juxtaposition; or even, thinking of fox terriers and Morse Code

[80] Very near the beginning of *Zang tumb tumb* come the following lines, as if to alert the reader to the provenance of the new idiom:

> tutto cio fuori di ma *anche in*
> *me* totalità simultaneità sintesi assoluta =
> superiorità della mia poesia su tutte le
> altre STOP (*TIF* 564)
>
> all that outside me *as well as in*
> *me* absolute totality simultaneity synthesis =
> the superiority of my poetry over all the
> others STOP

'STOP' is liberally sprinkled throughout the rest of the volume, usually at the end of a sequence of ideas (e.g. *TIF* 573, 579, 583, 597, 600).

[81] 'Zur Linguistik des Expressionismus: Von Marinetti zu August Stramm', *Die Literatur des Expressionismus: Sprachliche und thematische Quellen* (Stuttgart, 1966), 22.

machines, whether the analogy is so remote that it would be more appropriate to look for radical dissimilarities instead of common denominators. If the telegraphic style is intended to prune redundant elements, the adverbial phrase 'on the vast wastes of sand' is presumably deemed unworthy of retention because of its purely natural descriptiveness and the risk that it will distract attention from the essentials of the battle-experience. But the overall result of this procedure is that we are offered two very different kinds of statement and ultimately two very different battle-scenes in these passages. Possibly, this is to present the pairing 'battalions–ants' too negatively. Instead of it standing, in telegram fashion, for one single, unequivocal longer utterance, the phrase might be better described as triggering off various interpretive possibilities. In fact, one of the main achievements of telegraphic lyricism was not the creation of a series of brief, unambiguously decodable finite utterances in a compressed, asyntactical form, but to present richer associative plays on words. Christa Baumgarth's linking of telegraphic writing with a more longstanding associative style[82] rightly gets to the heart of the openness and the subjective potential of the innovation.

The way in which Marinetti describes the economic virtues of telegraphic writing in 'Destruction of Syntax' and the preface to *I nuovi poeti futuristi* suggests that he himself sometimes erroneously assumed the style to be decodable in the manner of a telegram. Telegraphic communication between a free-word poet and his Futurist audience, Marinetti suggested in the latter preface, is rather like an exchange between two old friends. A few words, a gesture, a glance are sufficient to make clear exactly what either of them has in mind.[83] The analogy implies that there is no room for doubt about what they wish to communicate; as with real telegrams, the implication is that we are dealing with a finite, 'closed' message (in Eco's sense). But if it is already questionable whether such a model applies to the 'wireless' analogies of 'Battaglia peso + odore', by the time we get to *Zang tumb tumb* it certainly no longer holds.

The battle description from the 'Bombardamento' section of *Zang tumb tumb* (fig. 3.3) shows how Marinetti's conception of telegraphic writing has changed since the first illustrative poem. Gone is the obsession with the bold juxtapositions of the 'wireless imagination'. The style remains predominantly nominal, but with a strong injection of acoustic neologisms. The layout, with suggestions of longer and shorter pauses, now becomes an integral part of the telegraphic rhythm. Verbs are no longer kept axiomatically in the infinitive; and the adverbial element in the passage assumes fresh importance (a concession to the fact that certain parts of speech are not to be abolished as redundant *per se*, but will be required—or not—depending on the exigencies of the task in hand). Although there is no evidence either way, it is tempting to speculate that the

[82] *Geschichte des Futurismus* (Reinbek, 1966), 89.

[83] *I nuovi poeti futuristi*, p. 10.

cing **ciak** *[PRESTO]* **ciaciaciaciaciaak**
su giù là là intorno in alto attenzione
sulla testa **ciaack** bello Vampe
vampe
vampe *vampe*
vampe *vampe*
vampe ribalta de forti die-
vampe
vampe
tro quel fumo Sciukri Pascià comunica te-
lefonicamente con 27 forti in turco in te-
desco allò **Ibrahim** **Rudolf** **allô** **allô**

ching chiak *[QUICKLY]* **chiachiachiachiachiaak**
up down there there around in above watch out
above your head **chiaack** beautiful Flames
flames
flames *flames*
flames *flames*
flames footlights of the positions be-
flames
flames
hind that smoke Shukri Pasha communicates by te-
lephone with 27 positions in Turkish in Ger-
man hellò **Ibraham** **Rudolf** **hellô** **hellô**

FIG. 3.3. From F. T. Marinetti, *Zang tumb tumb*, Milan, 1914

Marinetti of *Zang tumb tumb* may have responded fruitfully to objections such as Alfred Döblin's concerning the limitations of his earlier 'Battaglia peso + odore'.

The ambiguities in this passage (*TIF* 694 f.) can hardly be accounted for by the 'old friends' or the telegram–palimpsest paradigms. One cannot be sure, for example, whether the sounds at the beginning still represent the 'shouting of officers slamming like brass plates', mentioned just previously in the poem,[84] or whether this signifies some other noise (rattling equipment, as the men march faster, gunfire, or the sound of supplies being brought up). The adverbs of place ('up down there there') could stand for gunfire going off all around the soldiers, or other possible threats, such as snipers, booby-traps, or barbed-wire. And the

[84] Whereas a 'Battaglia'-inspired version of this would no doubt have entailed the 'wireless' juxtaposition 'officers–brass plates', Marinetti has now retreated to a conventionally constructed simile.

call 'attenzione' is not elucidated by any explicit reference to what has to be noticed. Similarly, the collective phrase 'footlights of the positions behind' has a non-specific, generalizing quality which is difficult either to visualize or to supplement with extrinsic information of the kind real telegram communication relies on. (Which positions, how many, where are they, what does their being seen in terms of the footlights image add up to as an account of the battle situation?) In this poem, the elliptical style is skilfully accommodated to the uncertainties and chaos of battle, whereas the juxtapositions of 'Battaglia peso + odore' have a certain Olympian, even stage-managed quality, as well as resembling the abstraction of a baroque conceit. In 'Bombardamento' the reader knows little more than the combatants, and the piece's new telegraphic style neatly combines such partial vision with a sense of dramatic pace and a powerfully acoustic rhythm. To argue that the literally interpreted telegram-model is inappropriate to our reading of such a passage scarcely constitutes a criticism of such a work (at most it pinpoints one of the limitations to Marinetti's theoretical position). What is thereby suggested is that a different model of reading is needed. And one is available.

One of the more pragmatic ideas put forward in recent years of speculative reader-response criticism has been the notion of the 'gaps', the elements of indeterminacy, which all texts possess to a greater or lesser extent, but which tend to be more pronounced in modernist forms of writing. Wolfgang Iser has, in a variety of largely fiction-orientated contexts, explored 'the extent to which the "unwritten" part of a text stimulates the reader's creative participation'.[85] These 'inevitable omissions . . . the gaps [*Leerstellen*] left by the text itself' mean that 'each individual reader will fill in the gaps in his own way', will engage, that is to say, in acts of consistency-building, ideation, or supplementation of information in some other respect. In the case of telegraphic writing, one could think of the gaps as being linguistically accentuated and systematized. 'With "traditional" texts', Iser points out, 'this process was more or less unconscious, but modern texts frequently exploit it quite deliberately.'[86]

There can be no doubt that Apollinaire's 'telegraphic style', the various Futurist forms of 'telegraphic lyricism', and Expressionist *Telegrammstil* represent highly conscious forms of gap-creation. Even if Marinetti's theoretical pronouncements are sometimes inadequate to account for what is going on, his practice gives evidence of a less procrustean attitude to the possibilities of the telegraphic. The following consideration of the 'Bombardamento' section of *Zang tumb tumb* shows Marinetti becoming much more inventive in his use of a telegraphic idiom than in 'Battaglia peso + odore', although various later

[85] *The Act of Reading: A Theory of Aesthetic Response* (Baltimore, 1978), 55.

[86] Ibid. 55 f. See also Iser's elaboration of the model in 'Indeterminacy and the Reader-Response in Prose Fiction', in J. Hillis Miller (ed.), *Aspects of Narrative* (New York, 1971), 1–15.

Futurists were to prove that the idiom is capable of even more permutations than Marinetti explored.

'Bombardamento' (as it is called in *Zang tumb tumb*) exists in more forms and covers a far greater period of Marinetti's creative activity than any other work he wrote. There are at least five versions of the work. The most well-known of these are:

1. 'Adrianopoli assiedo orchestra' (*Lacerba* (15 Mar. 1913), 50f.). This largely corresponds to the first pages of the *Zang tumb tumb* version, although, as De Maria points out (*TIF*, p. cxviii), the typography of the *Lacerba* version is less effective. Moreover, there are numerous variations between these two printings, some significant, others marginal.[87]
2. The 1914 *Zang tumb tumb* version', entitled simply 'Bombardamento', which appears in the final section (pp. 693–99 in the *TIF* reprint).

There are three further versions, not itemized in *TIF*:

3. A revised Italian version, 'Bombardamento di Adrianopoli. (Parole in libertà)', published in *I nuovi poeti futuristi* (Milan, 1925), pp. 288 f.
4. A French translation of this, entitled 'Bombardement d'Andrinople: Mots en liberté futuristes' (*I nuovi poeti futuristi*, pp. 308 f.). (Marinetti wrongly dates both the Italian and the French versions here '1911', although the siege itself did not take place until the year after.)
5. A manuscript of the French translation, with a far more experimental layout and various typographical innovations. This manuscript is held among the papers of the Harold Monro Fund at the University of California, Los Angeles.[88]

Looked at as telegraphic writing, this series of reworkings throws light on various features of Marinetti's poetic development.

In the 'Destruction of Syntax' manifesto of 1913, one finds Marinetti

[87] The main body of the *Lacerba* text is in italics, with lines approximately twice as long as those in *Zang tumb tumb*. Not all onomatopoeic words have been printed boldly in *Lacerba*. The spaces between the words-in-freedom there are much shorter and more conventional, with traditional punctuation, whereas the *Zang tumb tumb* layout is much more dramatic, and even includes iconic effects.

[88] Neither the Finch College Museum catalogue *Italian Visual Poetry 1912–1972*, ed. Luigi Ballerini (New York, 1973) nor *Tavole parolibere futuriste (1912–1944)*, ed. Luciano Caruso and Stelio M. Martini, both of which include a facsimile of this MS, dates it. The Department of Special Collections of the University Research Library (University of California, Los Angeles) catalogues the item (745 bx 2) as '*ca.* 1914', although various stylistic features, discussed below, suggest that it probably dates from the same time as—or soon after—the two versions in *I nuovi poeti futuristi*. Caruso and Martini's decision to print the facsimile MS *before* their illustrations from *Zang tumb tumb* creates the impression that Marinetti had already designed a highly experimental layout for the piece that then had to be compromised for the *Zang tumb tumb* printing, which is unlikely to have been the case, even granting that the typography of *Zang tumb tumb* is often more tamely horizontal and conservative than its linguistic radicalism suggests it ought to have been.

advocating the use of 'very brief . . . mathematical and musical signs' (*TIF* 66), a proposal which is elaborated on in 'Geometric and Mechanical Splendour and the Numerical Sensibility'. Here we read: 'My love of . . . essential brevity has naturally given me a taste for numbers.' To substantiate the point, Marinetti suggests that, in contrast to the Futurist utterance 'horizon = sharp bore of the sun + 5 triangular shadows (1 kilometre wide) + 3 lozenges of pink light + 5 fragments of hills + 30 columns of smoke + 23 flames', passéist writing would need 'at least an entire page of description to render this vast and complex battle-horizon' (*TIF* 91 f.). In the various versions of 'Bombardamento' Marinetti fortunately does not adhere to this telegraphic-cum-equational principle in any mechanical way, although there are interesting variations between the versions. Whereas 'Bombardamento di Adrianopoli' has the phrase 'ogni 5 secondi', the French translation changes this to '1 2 3 4 5 secondes'.[89] The first formulation, which also occurs in all previous versions, does not become more telegraphic by being changed to include the numbers 1–5. (Presumably, the insertion of each intervening number for the seconds should suggest the same kind of sequential *Sekundenstil* regularity as Marinetti was aiming at when he wrote the actual word 'regolarità' in the form 're-go-la-ri-tà' in *I nuovi poeti futuristi*.[90]) Elsewhere, there is a general tendency towards abbreviation. The *Lacerba* version writes 'metri' in full, whereas *Zang tumb tumb* uses simply 'm.'; and the abbreviation 'kmq' is also preferred. 'Batterie a tiro rapido' in the first version is reduced to 'batterie tiro rapido', 'i pazzi' becomes simply 'pazzi', and there are other instances of excising articles, prepositions, and other 'redundant' elements. Although the syntax of the *Zang tumb tumb* version has on the whole become more elliptical than that in *Lacerba*, other changes mean that the second version is in some respects not necessarily much more economic. One finds, for example, far more 'free expressive orthography in the second version, with elastic spellings such as 'Timmmpani' and 'suoooooonare'. As a result, the layout becomes even more expansive. The later *I nuovi poeti futuristi* version also injects more 'free expressive orthography' ('sventrrrrare', 'infiniiiito', and similar duplicative forms). Inflation sets in elsewhere too, with '200 shrapnels' now becoming '2000' and with abbreviated measurements being written out in full. Not all changes are improvements, in my opinion. The seemingly impatient sequence 'hallò . . . **hallô hallô**' is now printed with no distinction between the two forms; and exclamation marks, prepositions, and articles removed after the *Lacerba* publication now creep back in. Only some of these effects (like the 're-go-la-ri-tà') can be accounted for by assuming that the telegraphic has been subordinated to the claims of the iconic.

[89] This is another instance of a tendency discussed in Chapter 1 on the 'Typographical Revolution': any conflict between the telegraphic idiom and the principle of iconicity, which may necessitate a less abbreviated form, invariably results in a solution suggesting that iconicity is more important to the Italian Futurists than surface economy.

[90] *I nuovi poeti futuristi*, p. 288.

The main thing that the final experimental manuscript layout (fig. 3.4) shows is that the one area where Marinetti found all the previous versions unsatisfactory was in giving a visual sense of the noise and dynamism of battle.[91] Most of the salient innovations in the new handwritten design, in particular the departures from horizontality, relate to this dimension of the battle-experience. There are other shifts in the direction of an experimental use of free-word typography,[92] but they pale into insignificance when compared with the increased iconicity of the final manuscript. Here, the telegraphic idiom becomes the language of gunfire, the rattattatting of the telegraph-key now transferring easily to the sound of rapid machine-gun fire, and with a series of impressive explosions interrupting the linear flow of the narrative.

The title of the original *Lacerba* version of 'Bombardamento'—'Adrianopoli assiedo orchestra'—takes up an analogy between the experience of battle and an orchestral performance already in the *Intransigeant* reports. The 'Technical Manifesto' elaborates this parallel, positing similarities between the 'wireless imagination' and the act of orchestration: 'It is only with distant analogies that an orchestral style, which is at the same time polyphonous, polymorphous and polychromatic, can embrace the life of material things ... one has to orchestrate the images, according to the maximum degree of disorder' (*TIF* 84).[93] The poet is an orchestrator, for not only does he observe and share with us the impressions of combat; he also makes us feel that he is virtually arranging the battle, conducting it even, for our edification. He is its organizer, not just a privileged participant. He guides us through the event with an air of assured familiarity ('ogni 5 secondi cannoni da assiedo sventrare | spazio'), at times he even intervenes to protect us ('**ciaciaciaciaciaak** | su giù là là intorno in alto attenzione | sulla testa'). The figurative elements in his free-word description

[91] Significantly, the page-proofs for the French version of 'Bombardamento' in *Les Mots en liberté futuristes* display corrections relating almost solely to the onomatopoeic element, which Marinetti seeks to intensify. A page of the proofs (in the Yale Marinetti Archive) is reproduced in the catalogue section of *The Futurist Imagination*, p. 117.

[92] Marinetti was particularly proud of the 'vampe' sequence in the *Zang tumb tumb* version: here, the flames, in the shape of the Italian word for 'flames', envelop the nearby words. He even accused D'Annunzio of plagiarizing the technique (*I nuovi poeti futuristi*, pp. 11 f.). The MS contains a number of other inventive effects, for example, the slur signs joining the words 'ZANG' and 'TOUMB TOUMB' (and encompassing the measurement 'ampleur 50 kmq'), and the 'jeu de pendule' passage, which is set out vertically in the shape of a pendulum (see fig. 3.4). Nevertheless, in size, number, and impact, these iconic devices are less important than the acoustic and dynamic aspects of the overall restructuring.

[93] Cf. the following passage on orchestrative descriptions from the introduction to *Les Mots en liberté futuristes*: 'Words-in-freedom are an absolutely free expression of the universe beyond prosody and syntax, a new way of looking at things and feeling them, a measuring of the universe as the addition of forces in movement. These forces intersect in the conscious and creative self which notes them rigorously with all possible means of expression. Free-word poets thus orchestrate colours, noises, sounds, forming suggestive combinations with the material of language and of patois, of arithmetical and geometrical formulae, archaic words, deformed words and invented ones, the cries of animals and the noise of engines etc.'

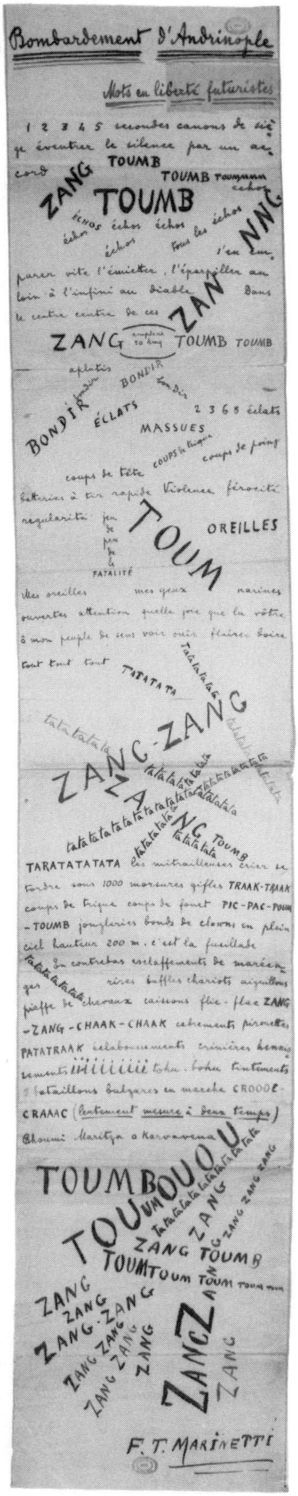

FIG. 3.4. F. T. Marinetti, 'Bombardement d'Andrinople', undated manuscript (University Research Library, University of California, Los Angeles)

('ammutinamento di 500 echi', 'orchestri stagni') are so palpably *his* that it becomes difficult to feel that the battle is anything but his own personally created drama, rather than mediated experience. As the work proceeds, the poet progresses noticeably from being a mere observer ('che gioia vedere udire fiutare tutto | tutto') to the role of exhorter and conductor: 'suoooooonare suoooooonare Graaaaaaande | fragori non cancellare'. 'How magnificent' is his reaction as he sees the flames destroying the positions and in response, as if rising to the occasion, he creates his own comparable destruction of the word for the enemy positions. In the manuscript version, his language explodes, just as the shrapnels do on the field of battle. In this 'teatro ampiezza 300 chilome- | tri', the poet emerges as scriptwriter, director, entire cast, and stage-effects man all rolled into one. But as well as creator and observer of the battle-scene, he plays another role: that of commander. Not only is his vantage-point the Olympian one of a general rather than a lowly participant; he also organizes the bombardment of the reader with devices. (A similar identification with leader-figures occurs in other works by Marinetti: for instance, *Mafarka le futuriste* and *Après la Marne, Joffre visita le front en auto*.) Very often in the poem, words are made to sound like gunfire: 'tam-tumb- | tumb tum-tumb-tamb-tumb | tumb'; elsewhere, especially in the final unpublished version, words explode; they whizz past like sniper-bullets and ricochet off other surfaces. By contrast with all this acoustic commotion, a real telegram is usually a silent form; it is read and silently received, not delivered in this war-equivalent of the singing telegram. It is therefore not surprising to find images like the tapping Morse Code machine, the overheard voice shouting into the field-telephone, or the telegraphed message figuring more frequently in Marinetti's writing than any silent printed messages (the exception being the proof-correcting overture to the entire war cycle). The rhythms of the spoken, heard message play a larger role than the silently read telegraph.

As an attempt at sabotaging the syntax of the passéist literary enemy, 'Bombardamento' displays a predictably Futurist preference for nouns and infinitive verbs (often with a variety of enumerated subjects). One striking feature of the resultant nominal style is the richness, even recherché quality, of the words chosen, something with few analogies in telegraphic language generally, though it is prefigured in Marinetti's *Intransigeant* reports. Zipf's Law in linguistics, which states that there is a correlation between the length and frequency of words, in the sense that the syntagmatic length of units is inversely related to their probability, would seem to suggest that Marinetti is condemned, by dint of the precious vocabulary he uses on this occasion, to produce words which are anything but telegraphic in their structural complexity. Moreover, the removal of the more common words on a principle of redundancy can only ensure that the remaining language is even disproportionately longer on average.

Marinetti's description of the bombardment ranges from abstractions ('violenza', 'ferocia', 'regolarità'), which are the aesthetic equivalent of the general's overview, positivistic registerings of measurements ('ampiezza 50 chilometri quadrati', 'altezza 200m.', 'LENTO DUE TEMPI') to the kind of recording of personal details about no longer noticing how cold one's feet are ('non sento piu i miei piedi gelati') or impressionism that suggests an event experienced at close quarters ('ilari nitriti iiiiii . . . scalpiccii tintinnii'). Clearly, such richness of language extends beyond the registers of the telegram proper; there is more variety here—an amalgam of abstraction and concretion, of rich as well as bureaucratic-cum-military language—than one would expect to find in telegraphese. The main feature of the telegraphic idiom, in this instance, remains a matter of both the syntax and the quality of words chosen.

Since much telegraphic writing tends towards a vertical poetic structure, the extreme being the single-word columns of Khlebnikov, Farfa, and Stramm, Marinetti's habit of extending his layouts horizontally is another feature of this poem worth considering here. When 'Adrianopoli assiedo orchestra' appeared in *Lacerba*, a periodical with parallel columns of print in the manner of a newspaper, there would have been ample opportunity to segment the work into short lines and extend it vertically, as Palazzeschi did with many of his poems. But Marinetti does not choose such a format. Instead, he has the poem printed at the top of the page in such a way that it spreads across more than one column, with the result that it becomes wider than it is long. And, although the lines of the *Zang tumb tumb* version are about half as short as those in *Lacerba*, one still has a visual impression of expansiveness. The main reason for this horizontality, I would suggest, is because Marinetti wants to bombard his reader relentlessly with word-missiles. The longer the line, the more seemingly interminable the attack becomes, whereas short lines would have offered more 'typographical breathing spaces', as Donald Davie calls them.[94]

Another unexpected repercussion of this long line-structure is its fortuitous word-splitting, a feature of both 'Battaglia peso + odore' and *Zang tumb tumb*, one connected with, though not adequately explained by, the fact that Marinetti retains a justified right-hand margin to his poem. A comparison of the various versions of the poem shows that, with lines of varying length in the different printings, different words are split from version to version. The resultant fragmentation of words into such largely non-morphemic forms as 'bizz- | zzarie' and 'impen- | arsi' appears to have no particular semantic significance. The process neither foregrounds important words nor suggests word-plays, of the kind that emerge when Mayakovsky splits words across lines. The main results of the random splitting are: to take the reader on from one line to the next at such a pace as to suggest, in an equivalent to *enjambement*, that the line-ending is of little significance and the poem really consists of one

[94] *Articulate Energy* (London, 1955), 126.

breathless continuum; an impression of arbitrariness, as if the poem was attempting to impose an artificial and weak order on chaotic experience; and a sense of the sheer density of sensations during battle. Despite the economies created by telegraphic style, Marinetti's 'Bombardamento' ultimately gives the impression of being a poem overflowing with events and sensations.

The only major Futurist to comment on the relationship between telegraphic poetry and typographical layout is Mayakovsky, although he has a very different effect in mind to either the Futurist word-column or the uncontained horizontality of a work like Marinetti's 'Bombardamento'.

In 'How to make Verses' ('Kak sdelat′ stikhi'), Mayakovsky links his stepped layout (the *lesenka*)—which we already encountered in the quotation from 'Khorosho' at the beginning of this chapter—with the cult of the telegraphic: 'our condensed, economical structure of verse often forces the excision of inessential words and syllables; if one does not make a stop after these syllables, often one larger than that between the lines, then the rhythm is broken.'[95] The mid-verse pause becomes as important, and possibly even more so, than those at the end of lines. Hence, the *lesenka* is exploited as a way of indicating new orders of rhythmic break within a form of less syntactical, less punctuated poetry where some other method of signalling pauses is required. Here, and in other areas of Futurist poetry, in *Zang tumb tumb* and in the work of Stramm and the *Sturmkreis* poets,[96] layout takes over the role traditionally played by punctuation.

The Polish Futurist Stanisław Młodożeniec's poem 'XX wiek' ('The Twentieth Century'), in his 1921 volume *Kreski i futureski*, brings out the relationship between layout and telegraphic rhythm even more clearly than Mayakovsky's use of the *lesenka* does.

> zawiośniało — latopędzi przez jesienność białośnieże.
> — KINEMATOGRAF KINEMATOGRAF
> KINEMATOGRAF . . .
> słowikując szeptolesia falorycznie caruzieją.
> — GRAMOPATHEFON GRAMOPATHEFON
> GRAMOPATHEFON . . .
> iokohama — kimonooka cię kochają z europy
> — RADIOTELEGRAM RADIOTELEGRAM
> RADIOTELEGRAM . . .
> espaniolę z ledisami parlowacąc sarmaceniem
> ESPERANTISTO ESPERANTISTO
> ESPERANTISTO . . .

[95] *PSS* xii. 114. This essay was written in 1926, but the points also relate to Mayakovsky's earlier Futurist practice.

[96] I have explored some of the implications of this for one *Sturmkreis* poet in 'Aspects of Typography and Layout in August Stramm's Poetry', in J. D. Adler and J. J. White (eds.), *August Stramm: Kritische Essays und unveröffentlichtes Quellenmaterial aus dem Nachlaß des Dichters* (Berlin, 1979), 47–68.

odwarszawiam kometuję dosłoneczniam
AEROPLAN AEROPLAN . . .
zjednoliterzam paplomanię.
— STENOGRAFIA . . .[97]

it's sprung—summerrushes through autumnness whitesnowing.
—CINEMATOGRAPH CINEMATOGRAPH
CINEMATOGRAPH . . .
nightingaling whisperofforests calorwavely are carusoing.
—GRAMOPATHEPHONE GRAMOPATHEPHONE
GRAMOPATHEPHONE . . .
a kimonoeyed yokohama love you from europe
—RADIOTELEGRAM RADIOTELEGRAM
RADIOTELEGRAM . . .
espaniel with laydis parleying with sarmatian
ESPERANTISTO ESPERANTISTO
ESPERANTISTO . . .
I dewarsaw I comet I sunreach
AEROPLANE AEROPLANE . . .
I uniletter the babblomania.
—STENOGRAPHY . . .

The poem, a quasi-manifesto, suggests that the avant-garde poet is availing himself of up-to-date technology, and an up-to-date idiom, to signal his message to the waiting world. The staccato style of a transmitted telegraphic message is effectively reinforced by seeming to come from such a source. Technology not only suggests being up to date; it also suggests power.

As both this work and a poem like 'Bombardamento' demonstrate, the cult of a telegraphic poetic style is very much bound up with issues of rhythm, lay-out, and even punctuation, as well as the central question of how much conventional syntax is retained. Although he himself did not realize it, Marinetti's style in *Zang tumb tumb* represents just one form of telegraphic lyricism. It may have had a strong influence on other Futurists, but few of them actually imitated it slavishly. If the Italian's impact on telegraphic writing was substantial, its results nevertheless are to be found in a whole range of heterogeneous forms of telegraphese.

Even within the area of proposed grammatical reforms, there is quite a variation among individual Futurist theoreticians. In his 'Declaration about the Futurist Theatre' ('Deklaratsiya o futuristicheskom teatre'),[98] Vadim Shershenevich pleads for the removal of both the verb and the adjective from

[97] Quoted from Bogdana Carpenter, *The Poetic Avant-Garde in Poland, 1918–1939* (Seattle, 1983), 46 f.

[98] Shershenevich's 'Declaration' was first published in *Nov'* on 26 Apr. 1914.

Futurist language, and is in this respect close to the prescriptive radicalism of Marinetti. But there were others, even within the Italian Futurist camp, who found Marinetti's proposals too extreme. Auro D'Alba, for instance, felt that Marinetti's axiomatic rejection of the adjective went too far. In 'Pure Lyricism in the Futurist Sensibility' ('Il puro lirismo nella sensibilità futurista'),[99] he voices misgivings about what he calls 'absolute precepts', by which he mainly means the idea of the 'abolition of grammar'; and he puts in a plea for the tactical retention of the adjective. The substantive alone, according to D'Alba, is skeletal. The adjective is needed to give power to the sentence. This theoretical article is immediately followed by an illustrative poem, 'Simultaneità', which duly flaunts its retained adjectives.

Recording his impressions of a Futurist soirée in Berlin, Rudolf Leonhard stated categorically that the new Marinettian telegraphic style he heard there would only be appropriate to a limited number of subjects: the roar of a motor, the sounds of battle, possibly the extremes of ecstasy and violence, and the end of the world, a subject which was, of course, dear to the hearts of German Expressionists at this time.[1] Leonhard's remarks are conceivably not without an intentionally satirical, over-dismissive cutting-edge, but one can easily imagine how someone whose familiarity with the idiom was largely via Marinetti's various public declamations could arrive at such a dogmatic conclusion. Indeed, Marinetti's own theoretical pronouncements about 'essential brevity' also link the style with particular subjects ('intense zones of experience') and with certain effects. But, in fact, telegraphic poetry had the ability to be both thematically and stylistically a much more variegated phenomenon than a familiarity with Marinetti's poetry and theories alone might suggest.

In many respects, one of the most faithful adherents to the letter of Marinetti's telegraphic precepts was the Futurist poet Luciano Folgore. Folgore reiterates many of his mentor's ideas in his own 1914 manifesto 'Synthetic Lyricism and Physical Sensation' ('Lirismo sintetico e sensazione fisica'),[2] in particular those concerning the pruning of the traditionally grammatical sentence. Ideas, he suggests, are to 'leap from the writer's genius in their essential nudity'. The new 'rapid and *telegraphic form*' (there is now no hint of the remote analogies of the 'wireless imagination') will register the physical sensations of the object world. Folgore's preferred term for the requisite telegraphic

[99] *Lacerba* (1 Mar. 1914), 69 f. The poem 'Simultaneità' appears on pp. 70 f.

[1] 'Futuristenabend in Berlin', quoted from *Literatur-Revolution 1910–1925: Dokumente, Manifeste, Programme*, ed. Paul Pörtner, ii. *Zur Begriffsbestimmung der 'Ismen'* (Darmstadt, 1961), 88: 'This technique, the "Telegrammstil", familiar in our literature before Marinetti, is adequate for three themes: the roaring . . . of engines, the clash of battle and possibly the heights of ecstasy and the end of the world.' This reaction to Marinetti was first published in *Die Bücherei Maiandros*, 4–5 (1 May 1913).

[2] *Lacerba* (1 Jan. 1914), 1 f.

idiom is 'synthetic lyricism', a generic title which he had given to various poems published before his manifesto,[3] and one that also prefaces the poem 'Avanscoperta' which is printed immediately beneath 'Synthetic Lyricism and Physical Sensation' in *Lacerba*. In his volume *Ponti sull'oceano: Versi liberi e parole in libertà* (Milan, 1914), Folgore in fact divides the poems into two sections: 'Versi liberi: Lirismo sintetico' and 'Parole in libertà', as if the 'synthetic' were still merely a matter of free verse, rather than words-in-freedom. But it is with words-in-freedom that Folgore's poetry really develops in the direction of extreme ('synthetic') reduction. And it is clear, despite the difference in terminology, that 'synthetic lyricism' is in most respects identical with Marinetti's 'telegraphic lyricism'.

Having credited Marinetti with the reform of traditional orthography and syntax, in particular with the abolition of the adverb and the adjective, Folgore goes on to advocate that the cult of the telegraphic be taken one step further—by eliminating the verb and leaving only the noun. As a result, it is argued, poetry will become even more tangible. In other words, once again a single reform of the grammatical system is being naïvely expected to have a series of profound thematic and qualitative implications. If verbs are abolished, according to Folgore's reasoning, physical sensations will inevitably dominate poetry. With the noun recognized as the 'skeleton' of the sentence, what D'Alba called 'the new material poetics',[4] that is to say, the poetics of matter and physical sensation, would be able to evolve unrestricted by passéist syntax.

Folgore's poems in both sections of *Ponti sull'oceano* (*Bridges over the Ocean*) tend to use an almost exclusively nominal language of the kind described in 'Synthetic Lyricism and Physical Sensation', and it was only with the later volume *Città veloce* (*Speedy City*) of 1919 that he returns to a more conventional syntax. This could be connected with the fact that, as Luciano De Maria suggests (*FF* 477), he is now really 'widening his range of topics' from those treated in his earlier more party-line Futurist position.

The following extract from the poem 'Sobborgo' ('Suburb') is fairly representative of the effects Folgore aspired to achieve with his new brand of 'lirismo sintetico':

Sobborgo: gerani
capelli castani
affacciati,

[3] e.g. 'Caffè notturni' (*Lacerba* (1 May 1913, 87 f.), 'Città ferma. Sciopero generale' (*Lacerba* (1 July 1913), 138), 'Cena' (*Lacerba* (15 Aug. 1913), 177). A further poem in the same idiom, 'Moka' (*Lacerba* (1 Dec. 1914), 332), even sports the generic sub-title 'sensazione fisica'. Marinetti's remarks in 'The Futurist Synthetic Theatre' ('Il teatro futurista sintetico') make clear that 'synthetic' is being used here in the special Futurist sense, which has a direct bearing on the cult of the telegraphic: '*Synthetic*. That is, very brief. To compress into a few minutes, into a few words and gestures, innumerable situations, feelings, ideas, sensations, facts and symbols' (*TIF* 98 f.).

[4] 'Il puro lirismo nella sensibilità futurista', *Lacerba* (1 Mar. 1914), 69 f.

penduli,
fiori di verde e di carne;
cuffiette di nonna,
fazzoletti a brandelli di seta,
macchie di calze—qua e la—
un'ombre di gonna.[5]

Suburb: geraniums
brown hair
looking out of windows,
hanging,
flowers of green and of flesh;
grandmothers' bonnets,
handkerchiefs with torn pieces of silk,
stains of stockings—here and there—
the shadow of a skirt.

This kind of registering style differs considerably in tone from the strident battle-inventories of *Zang tumb tumb*. A largely critical consciousness seems to be observing an actual suburban setting and picking out trivial, yet symptomatic details that strike the eye—the geraniums, a girl's brown hair, grandmothers' bonnets, etc. Alternatively, these associations could come to mind when the idea of a 'sobborgo' is rehearsed. This asyndetic, essentially impressionist nominal style is one to which the telegraphic idiom evidently also lends itself. Rather than effectively suppressing the subjective element, as Marinetti had once encouraged (*TIF* 44 f.), or merely cataloguing the physical world, as Folgore was inclined to predict it would, such a style makes the reader very conscious of the registering, controlling sensibility holding together all the impressions in the poem. In such a handling, the telegraphic style approaches stream-of-consciousness writing.

In 1934, in one of his more crassly opportunistic attempts at retrospective colonizing, Marinetti claimed Joyce as a major beneficiary of the Italian Futurists' language reforms. In 'Joyce and Words-in-Freedom ('Joyce e le parole in libertà')[6] Marinetti takes up the question of interior monologue raised in a recent book on Joyce by Aldo Sorani. He points out to his Italian readers that the technique, far from being Joyce's invention, in fact stems from Edouard Dujardin's *Les Lauriers sont coupés*. But this move to deny Joyce any inventor's rights over the technique is less in the interests of historical accuracy than a tactic to introduce another pedigree: a Futurist one. Indeed, the very terms which Marinetti uses to describe interior monologue—'absolute sincerity, naked and raw . . . a bold and rapid verbalization . . . of spontaneous thoughts,

[5] *Lacerba* (15 Sept. 1913), 199.

[6] *Futurismo*, 2, 61 (15 Feb. 1934). The article is available in a French translation in Giovanni Lista, *Marinetti et le futurisme: Études, documents, iconographie* (Lausanne, 1977), 54.

appearing to be born prior to organized logical discourse'—are all preparing the way for the eventual Futurist annexation:

Do we not here [Marinetti asks disingenuously] have an example of the most Italian of free-words with their free expressive orthography, invented twenty-five years ago by the Italian Futurists . . .? Almost all free-word experiments . . . also contain what is called 'interior monologue', that is to say, the absolute and alogical authenticity of the conscious and unconscious.

While I acknowledge Joyce's originality, it remains beyond contention that he is indebted to Edouard Dujardin for the interior monologue and to the Italian Futurists for words-in-freedom.[7]

Despite the characteristic vagaries of Marinetti's argument (his initial association of the technique with the unconscious, then with both the conscious and the unconscious, his first implying that words-in-freedom contain examples of interior monologue almost by definition, but then suggesting confidently that Joyce must nevertheless be indebted to the Futurists for a free-word-*style*), one can detect Marinetti spelling out certain things about the telegraphic mode in 1934 that he had not taken much cognizance of in his early programmatic writings.[8] The problem would appear to be that in his initial theorizing Marinetti relied on an inadequate model of the psyche. In retrospect, his contrast between the constraints of conventional logic and syntax and the freedom of the 'wireless imagination' cries out for discussion in terms of layer psychology, relating the two modes to conscious and unconscious aspects of the psyche; but Marinetti then seemed unaware of Freud (whose name does not figure once in any of his writings) and of the depth psychological implications of many of his literary proposals. Only now, in the Joyce essay, does he begin to acknowledge the distinct psychological possibilities of telegraphic writing, though still without thinking the idea through.

Some of Marinetti's contemporaries were, however, quicker to spot the link between telegraphic writing and various forms of psychological impressionism. Writing in *Noveyshaya russkaya poeziya: Nabrosok pervyy: Viktor Khlebnikov* (originally published in Prague in 1921), Roman Jakobson quotes a substantial passage concerning 'zones of intense experience' from Shershenevich's translation of the 'Destruction of Syntax' manifesto and adds: 'here we see the cult of

[7] Marinetti makes a similar claim in 'The Italianness of Modern Art' ('Italianità dell'arte moderna'), *Il giornale d'Italia* (24 Nov. 1938) (also in *Artecrazia*, no. 118), arguing that the modernism of Eliot and Joyce would be 'impossible without Italian Futurism' (specifically: without 'our poetic revolution of words-in-freedom'). The article has been reprinted in Enrico Crispolti, *Il mito della macchina e altri temi del futurismo* (Trapani, 1969), 737.

[8] Some connections are admittedly posited between the style of words-in-freedom and the perceiving psyche; e.g. 'the infinitive can provide a sense of life and the elasticity of the intuition that perceives it' (*TIF* 41); 'matter whose essence must be grasped by strokes of intuition' (*TIF* 44); and 'we are entering the untrammelled domain of free intuition . . . the characteristic gift of the Latin races' (*TIF* 47). But the remarks owe more to vitalist philosophy than to depth psychology.

pure impressionism, a pendant, as it were, to the "telegraphic style of the soul" which Peter Altenberg had envisaged.'[9]

In fact, among German-speaking writers, the telegraphic cult seems to have a second source-line, one which began independently of Futurism but by the Expressionist period becomes difficult to disentangle from the Futurist influence. The Naturalist poet Arno Holz claimed to have invented the term 'Telegrammlyrik' long before the *Sturmkreis* Expressionists or the Futurists had thought of it;[10] and by common consent the plays of the leading German Expressionist Carl Sternheim are also written in a form of clipped *Telegrammstil*, with again little likelihood of an Italian influence.[11] And, as Jakobson reminds us, not only had the Viennese Impressionist Peter Altenberg cultivated a similar telegraphic laconicism years before; he had expounded its theory in a variety of contexts.

A number of Altenberg's literary prefaces contain statements very reminiscent of Marinetti's manifestos. For example: 'One can write a two-hundred-page novel, and it can be splendid. One can say the same thing in three pages, and it can be just as splendid. The whole thing is a matter of saving time. Nowadays, there are a great many otherwise industrious people who just don't have the time to read two hundred pages. One gives these people a three-page digest.'[12] Or (the particular passage which Jakobson no doubt had in mind): 'I admit: I love the technique of abbreviation, the telegram-style of the soul. I would like to depict a character in a single sentence, a spiritual experience in a single page, a landscape in a single word.'[13] Even if they do not entail the same cult of the ultra-modern sensibility, such sentiments are not far removed in spirit from the declaration, in Marinetti's 'The Synthetic Futurist Theatre' ('Il teatro futurista sintetico'), that '*It's stupid to write a hundred pages, when one would do*' (*TIF* 100). One thing that nevertheless separates both of these forms of dramatic minimality from the telegraphic idiom which is being explored here is that the focus, for both Altenberg and Marinetti, is on the macro-, not the micro-structure. 'Five-minute' plays and Futurist *sintesi* are, in all but the rarest of cases, not written in pared-down language. The quest for brevity finds its expression in the telescoped nature of the total play, not the language at

[9] *Texte der russischen Formalisten*, ii. 29.

[10] *Das Werk*, x (Berlin, 1925), 618.

[11] Hermann Kunisch refers to 'the soberly expressive pathos of its metaphor-free diction, which is often reduced to the level of telegraphic writing' (*Kleines Handbuch der deutschen Gegenwartsliteratur* (Munich, 1967), 506). Albert Soergel and Curt Hohoff rightly identify yet another association of the style when they observe that the 'social *Telegrammstil*, which in Wilhelmine Germany had entered into a grotesque alliance with a dashing military style of speaking, became literary in Sternheim's plays' (*Dichtung und Dichter der Zeit: Vom Naturalismus bis zur Gegenwart*, ii (Düsseldorf, 1963), 229 f.).

[12] *Bilderbogen des kleinen Lebens* (Berlin, 1905), 164.

[13] *Was der Tag mir zuträgt* (Berlin, 1901), 6. Richard Hamann and Jost Hermand find Altenberg's *Depeschenstil* 'reminiscent of the tachiste style of the Impressionist painters' (*Impressionismus* (Epochen deutscher Kultur von 1870 bis zur Gegenwart, 3; Munich, 1960), 266).

individual-sentence level, which is why the Futurist *sintesi*, although the genre shares certain assumptions about the need for brevity, does not come under investigation in a chapter on the Futurist cult of the telegraphic. Altenberg's whimsical project for a series of 'Fünfminutenszenen' (five-minute plays) resembles the Futurists' theatrical *sintesi*[14] in underlying rationale, even if Altenberg never aspired to the absurdism of Futurist performance. And both forms, one needs to bear in mind, are not so far removed from such modern jokey eccentricities as Andy Mayer and Jim Becker's *One-minute Classics* (London, 1986).

What this dual pedigree to telegraphic writing, and the different forms it takes in Italy and German-speaking countries, reveals is that Marinetti's thinking on the subject was constructed on such a narrow theoretical and experiential basis that it obscured his view of many of the further possibilities of the innovations he was advocating.

Fillia's poem 'telegrafo', published in *I nuovi poeti futuristi*, also exploits the psychological potential of the telegraphic, but this time by using the telegraph itself as an image for the brain. The result is not so remote from the 'dramas of humanized matter' that Marinetti once claimed (*TIF* 44) were anathema to the Futurists.

> le parole sono i ritmi interni dei cervello che ca-
> pisce e non afferra il suono
> esasperazione di una lucidità regolare immo-
> bile freddamente applicata instancabile
> morte del senso udito.[15]

> words are the internal rhythms of the brain which un-
> derstands but does not grasp the sound
> exasperation of a regular lucidity immo-
> bile coldly applied indefatigable
> death of the heard sound.

Fillia here uses a form of sparse writing to suggest a parallel between telegraphic communication and the receipt of language. This time, instead of focusing on the telegraph as the *transmitter* of signals, the interest is on the receiver-analogy (the 'heard sound'). The sense of linguistic scepticism underlying the 'coldly applied' lucidity and the 'death of the heard sound' makes the machine image into something itself cold and reductive. The poem begins with a bare, but nevertheless fully grammatical series of utterances and becomes progressively more telegraphic from line to line, as if to mirror the growing breakdown in communicating the problem of communication. The rhythm here is not the

[14] These have been considered in some detail in Michael Kirby, *Futurist Performance* (New York, 1971) and Mario Verdone, *Teatro italiano d'avanguardia: Drammi e sintesi futuriste* (Rome, 1970).

[15] *I nuovi poeti futuristi*, p. 148.

kind of ecstatic crescendo to be found in so many of Marinetti's battle poems, but a relentlessly self-undermining momentum. Such an exercise in the telegraphic has little to do with the predominantly acoustic, staccato free-word rhythms of early Futurism, but edges towards a series of truths in a measured, halting manner.

Alberto Viviani's poem 'Stati d'animo di un suicida' illustrates a similar process of making the telegraphic idiom psychological (or perhaps one should say, in the light of Impressionism, returning it to a psychological function). The interior monologue of a man on the verge of ending his life, it is the very opposite of almost everything that Marinetti has to say at the beginning of 'Geometric and Mechanical Splendour and the Numerical Sensibility' about the Futurist sensibility's need to get away from introspection and subjectivity.

Nebbia Nebbia
Oriente—
Orrore spasimo
brancolamento tra le folgori—
Fuggire fuggire
Ubriacarsi
VUOTO
da cui si può cadere
in una via deserta
albe invernali
suoni bisbetici
pieni di braccia movimentate
voci oscillanti
lumi miagolanti
Assalto di rumori
vermi sul cuore
sulla gola
nel ventre fra i ginocchi
Gesti magici
resuscitanti cadaveri
mani invisibili
che scompongono misteri
pazientemente edificati
Rumori Rumori
Muraglie
gocciolanti umidi spasimi verdi
nomi di donne
amalgamati di nebbia e di velluto:
Leonora—Berenice
Morella—Beatrice
Svolazzamento di baci intorno
ventose di zucchero

e di musica
SBARAGLIAMENTO: crepuscoli
salotti—amori
smaniosità—pesantezza
brrrrrrividi
Morte.[16]

A disconnected, reduced style is here used to convey the sensations of someone in the process of losing his grip on reality. We are offered a mixture of brief impressions and discontinuous, hazy thoughts ('Horror agony | groping between the flashes of lightning | To escape to escape | To get drunk | VOID'). Boccioni's description of the method of his *stati d'animo* paintings as offering a synthesis of what one remembers and what one sees,[17] a fusion of interior and external reality, in fact comes close to the monologic quality of this poem. The lines 'Fog fog | Orient | Horror agony' appear to be the sense-impressions of this disturbed mind on the verge of suicide. On the other hand, 'Women's names | Mixed with mist and velvet | Leonora—Berenice | Morella—Beatrice' resemble blurred recollections from the past, possibly even, given the number of names, a series of failed relationships that account for his present depression. There are other lines, such as 'Sounds Sounds | Walls | dripping humid green spaces', which resist a clear-cut contextualization in either the character's past or his deranged present. Similarly, the concluding word *Death* ('Morte') could possibly refer to the final act, the end of a life as well as a poem, or it may simply be something he is still contemplating. Viviani not only exploits the syntactical discontinuities of the telegraphic to suggest the vagaries of a particular kind of warped mental interior; he also harnesses such a reduced style's ability to create pronounced gaps, here creating uncertainties about what is past or present, real or imagined, and concerning the actual significance of certain phrases. In the right hands, the telegraphic idiom is as well suited to evoking introspective 'states of mind' as it is to battle-descriptions or the dramatic depiction of any other 'zone of intense experience'.

Viviani's 'Stati d'animo di un suicida' also bears witness to another protean quality to the form: the ease with which it can be accommodated to a predominantly vertical structure, with each discontinuous (telegraphic) phrase or even single word appearing in a new line, in contrast to the more horizontally extended layout encountered in Marinetti's 'Bombardamento'. On the whole, German and Russian poets—notably Stramm and his *Sturmkreis* emulators, as well as Mayakovsky, Kamensky, Khlebnikov, and Kruchenykh—were quicker to perceive this possibility than were Marinetti and the Italians. But the second

[16] *Lacerba* (15 Jan. 1914), 28.

[17] See Boccioni's comments on 'stati d'animo' paintings in Chapter 2 on Futurist collage. Attempts were also made at this time to apply the 'stati d'animo' method to visual poetry: see Giuseppe Steiner's *Stati d'animo disegnati* (Milan, 1923).

major anthology of Italian Futurist poetry, *I nuovi poeti futuristi*, does contain a number of word-columns with only one word per line.

One factor behind a preference for more vertical telegraphic structures is that such word-columns tend to isolate the individual parts of language and create a situation in which the elements are far more divorced from any traditional sentence-matrix. A classic example of this is Mayakovsky's 'Utro' ('Morning') where the effect is used, not only to isolate the individual words, but to introduce rhymes between the end of one line and the beginning of the next. The poem begins:

Угрюмый дождь скосил глаза.
А за
решеткой
четкой
железной мысли проводов —
перина.
И на
нее
встающих звезд
легко оперлись ноги.
Но ги-
бель фонарей,
царей
в короне газа,
для глаза
сделала больней
враждующий букет бульварных проституток.

The gloomy rain squinted its eyes,
And behind
a clear-cut
grille
of the iron thought of wires—
is a featherbed.
And upon
it
lightly leaned the feet
of rising stars.
But the de-
mise of streetlamps,
tsars
in a crown of gas,
has made more painful
for the eye
a quarrelling bouquet of boulevard prostitutes.

A similar structure is used in Mayakovsky's poem 'Iz ulitsy v ulitsu' ('From Street to Street'), where the device is exploited even more radically to advance beyond single-word lines in the direction of an inventive interplay between individual morphemic (and quasi-morphemic) sub-units:

У-	The
лица.	street.
Лица	Faces
у	of
догов	Great Danes
годов	are sharp-
рез-	er
че.	than years.
Че-	A-
рез	cross
железных коней	iron horses
с окон бегущих домов	from windows of running houses
прыгнули первые кубы.[18]	jumped the first cubes.

Here word-splitting facilitates a number of morphemic games: the words *ulitsa* ('street') and *litsa* ('face') are brought into a relationship which suggests that the one is contained etymologically within the other. The phrase *u litsa* suggests both 'by the face' (i.e. preposition and noun) and 'street' at the same time. Various inversions—*dogov* ('of Great Danes') and *godov* ('of years'); *rezche* (sharper) and *cherez* ('around')—are also made possible by the highlighting, splitting, and re-ordering of morphemes in a way which this vertical structure makes easier.

Susan Compton has argued that a restriction of this kind 'to the single word', or even to the morphemic components of a word, 'avoids external reference' much more, and hence the structure begins to focus attention increasingly on its own medium—what the Russian Futurists refer to as the 'self-contained word' ('samovitoe slovo').[19] This is probably truer of the Russian examples with which she is concerned and of certain *Sturmkreis* experiments[20] (some of which will be looked at in the next chapter) than it is of Italian Futurist word-columns. Many of these, especially Palazzeschi's, remain relatively faithful to conventional syntax and are by and large still referential. But a later Futurist work such as Farfa's 'nervi', which begins conventionally with the syntactical sequence 'in | quale | di genova | avverto | la geografia | scienza che amo meno della poesia', soon resorts to a bolder use of column-structure to generate a-syntactical relationships between words:

[18] 'Utro' (*PSS* i. 34 f.); 'Iz ulitsy v ulitsu' (*PSS* i. 38 f.).

[19] *The World Backwards: Russian Futurist Books, 1912–16* (London, 1978), 15.

[20] For a detailed exploration of one of Stramm's main achievements within this word-column convention, see J. D. Adler, 'Urtod: An Interpretation', in Adler and White (eds.), *August Stramm*, pp. 84–98.

nervi
nervature
nevralge
nervosità
nevrastenie[21]

Here, verticality becomes a convenient grid for generating various words sharing the same etymological root. Given his awareness of the advantages of verticality, one might have expected Farfa later in the poem to set out the four phrases 'Il respiro caffè', 'lo spirito caffè', 'la vita caffeistica', and 'il caffeino impulso' using four separate lines. Instead, he works with a subtler rhythm, allowing the first phrase to be broken across three lines—hence beginning the sequence with a slow, halting phrase. But then, by placing all three further variants on a single line, he accelerates the rhythm to an appropriate excitedness. In Farfa's poem, typography and layout have gone some way towards supplanting traditional syntax, but it has done so only because an essentially telegraphic language predominates; i.e. in a situation where one form of syntax has been pared down, another has been resorted to to perform an analogous function.

In 1924 the Ukrainian Futurist poet Mykhaylo Semenko published his 'Kablepoema za okean' ('Cablepoem beyond the Ocean') in *Kobzar* (*The Minstrel*), a volume of poetry irreverently named after the first book by the national bard Taras Shevchenko. 'Panfuturist Cable-greetings and Radio-greetings were', according to Myroslava Mudrak, 'a direct offshoot of the machine aesthetic and the concept of the artist as engineer.' (Semenko had, in fact, worked as a military cable-operator from 1914 to 1917 in Vladivostok, where he was posted after being mobilized.[22]) What Mudrak goes on to call 'the ethos of a collective global spirit and communication by means of telegraphic cables or radio transmissions'[23] is the Panfuturist equivalent of the sensibility discussed at the beginning of this chapter in connection with the cult of the Eiffel Tower. Semenko also views this new telegraphic dimension to modernity in terms of the need for a new abbreviated idiom: 'We need to incorporate freed titanic thought into required form—a thought which operates in masses with synthesized understandings and not with primordial understandings [the contrast is between the language of Futurism and that of passéism]; a thought which formulates its own grammar and its own rules in uniting sentences.'[24]

[21] *I nuovi poeti futuristi*, pp. 120 f.

[22] Semenko is only mentioned *en passant* in Markov, *Russian Futurism*. A useful general survey of his work can be found in Oleh S. Ilnytzkyj, 'Anatomy of a Literary Scandal: Myxajl' Semenko and the Origins of Ukrainian Futurism', *Harvard Ukrainian Studies*, 2 (1978), 467–99. I should like to record my gratitude to Jaroslaw Balan for answering my many questions about Semenko and for initially introducing me to Ukrainian Panfuturism.

[23] *The New Generation and Artistic Modernism in the Ukraine*, p. 162.

[24] 'Poezomalyiarstvo', *Semafor u maybutnye* (May 1922), 35.

Fig. 3.5. Card One of Mykhaylo Semenko, 'Kablepoema za okean', 1924

For Mudrak, 'Kablepoema za okean' epitomizes the new style which Semenko advocates:

> in its rigid, architectonic appearance and hard-lined dispatch, it imparts a sense of invincible strength and durability. The language . . . is tough. The words selected are firm, often eliciting tough, tactile industrial qualities. Furthermore, the visual impact of the typographical layout . . . reinforces the technological constitution of the poem which is arranged according to a network of bridges and scaffolding. In general, the visual impression is one of utmost simplicity—it is a work absolved of all decoration . . . which emphasizes (literally and symbolically) a galvanized integration of content and form.[25]

Fig. 3.5 shows the first page of this eight-page telegraphic poem, which Mudrak analyses in some detail in her study of Ukrainian modernism. Here,[26] as is also the case with *Zang tumb tumb*, the telegraphic does not necessarily result in an

[25] *The New Generation and Artistic Modernism in the Ukraine*, p. 162.

[26] *Kobzar* was published in two versions (Kiev, 1924 and 1925). Mudrak's illustrations appear to be taken from the first edition, according to Balan.

overall brevity of message. Although Marinetti's war volume is telegraphic, it is nevertheless dozens of pages long. And Semenko's 'Kablepoema za okean' is conceived on a grand epic scale—on eight pages, which each have their effect as separate visual impressions, but add up, as Mudrak's detailed analysis demonstrates, to a complex cycle that takes the reader around the world.

Mudrak sees the first 'card' in the series as a 'frontispiece', 'a kind of introduction to the consecutive cards all divided into two major vertical sections'.[27] The speed with which these various cards take one, via a plethora of geographical names and locations, around the globe, is not only a hymn in praise of modern communication, but above all a tribute to Panfuturism, for it is specifically the poet who is traversing the globe in this way with his work, not the cable-transmitter. Again, as with Młodożeniec's 'XX wiek' ('The Twentieth Century'), the message being transmitted is the gospel of Futurism to the outside world, so the sending of a telegraphic message is equated with the launching of a manifesto-like statement. And once more the poet is able to overcome the limitations of geography in exactly the same way as he liberates himself from those of grammar and syntax.

The passage that follows is a translation of part of Card Five of 'Kablepoema za okean'. The telegraphic idiom is explicitly equated with the 'language of the Future' (i.e. the language of Futurist literature, as well); and the rejection of thinking 'in details' and the constriction this brings about is very reminiscent of Marinetti's rejection of the overburdened passéist sentence.

> listen!
> I pressed the button
> 50 parallel 50
> Moscow–Himalaya–Kyoto
> *Our* language—
> Cities and mountains and seas and a field of snow
> Words will be forgotten
> it is constricting to think in details! . . .
> Learn the language of the Future![28]

We have so far considered horizontal and vertical forms of telegraphic layout as separate types, but Semenko's 'Kablepoema' is something of a hybrid form in this respect. The poet, although employing a predominantly horizontal grid format, in fact manages to work in a number of vertical acrostics at the same time. Thus, on Card Two the word 'EUROPE' is spelt out by the first letter of the words in the right-hand column, and on the facing card (3) 'ASIA' is formed from the initial letters of the first four words, while the word 'COMMUNE' is created by the last letters of these same words and those below

[27] *The New Generation and Artistic Modernism in the Ukraine*, p. 163.

[28] Quoted from Mudrak, *The New Generation and Artistic Modernism in the Ukraine*, p. 169.

them. Since the first card includes the slogan 'Workers of the world, unite!', the acrostic effects serve, iconically, to unite the names of the various parts of the world in one large two-directional unifying pattern. The structure joins different parts of the world, and forms new socialist meanings out of their names, at the same time as the message of Panfuturism is transmitted by the cable-poem.

Semenko's method here—the sectionalizing of the page, the use of segmented word-columns to introduce a vertical form of telegraphic writing—is not so distant from some of Kamensky's 'ferro-concrete' experiments, in the sense that there too one has a combination of containing geometrical shapes and a number of word-columns. For example, the left-hand column in his poem 'Dvorets S. I. Shchukin' ('The Mansion S. I. Shchukin') (from *Tango s korovami* (*Tango with Cows*)[29]) is very similar to the marginal columns in the cable-poem: the division of the page into sub-sections, the use of containing lines surrounding blocks of vertically piled-up words and the Constructivist starkness of the overall configurations are very reminiscent of much of Kamensky's work, with which the Ukrainian poet was familiar. Semenko was to go on, with the 'poezo-paintings', as he called them, of *Moya mozaika* (*My Mosaic*) (1922), to work with various forms of word-column juxtapositions which retained the skeletal form of the cable-poem's outline-blocks filled with words, as on a telegram form, but now without such a conscious exploitation of the telegraphic language component.

As these examples show, telegraphic writing took many forms, from the hurried and staccato to the slow and introspective. It could be exuberant, as Marinetti's war poetry so often is, but it could also be melancholic (which seems to confirm the earlier assumption that what we are dealing with is a change of linguistic repertoire, not something with over-specific thematic or value-based implications). The resultant layout was sometimes predominantly vertical, sometimes horizontal. And even within such structures, as Kamensky's word-columns and Semenko's 'Kablepoema za okean' show, there was still considerable latitude for experiments with layouts that could take the purely verbal component in the direction of iconic 'poezo-painting'.

It would be unfair, in the light of all these diverse possibilities, to single Marinetti out for censure because he lacked the imagination to foresee the rich potential of his new idiom. After all, many commentators who have referred to it treat the telegraphic as if it were a relatively monolithic phenomenon displaying a fixed number of features and only suited to a narrow spectrum of subjects. *The* 'telegraphic style' is often mentioned in contexts which suggest that no reader could be in any doubt about exactly what was meant by the phrase. Hence, Kirby remarks in *Futurist Performance* that various passages from

[29] The poem is reproduced (as fig. 94) in Janecek, *The Look of Russian Literature*, p. 134, and discussed on the facing page.

Depero's *Notes on the Theatre* are 'in the telegraphic style'; Tisdall and Bozzolla, no doubt with their eyes firmly on Marinetti's idiom, equate the 'telegraphic' with a 'barrage of nouns, colours, exclamations and directions', a description that would require much modification to apply to the poetry of Farfa, Folgore, or Viviani. Materer, in his study of Vorticism, writes at one point that, 'even if Lewis were not explicitly mentioned . . . the telegraphic style would show the influence of his *Blast*'; and Hermans observes, on the subject of the nature of the telegraphic, that, 'whereas the cinematic montage technique may be said to bear on the poem's macro-structure, the "telegraphic" style mainly affects its micro-structure'.[30] While these remarks all make sense in context, they betray an underlying assumption that what is being discussed is a general style and not just an individual example of it or a form of repertoire. Yet an idiom which could be associated with Apollinaire's poetry, the language of *Blast*, with Marinetti's free-word war poetry, and with the verbal experiments of the Expressionist *Sturmkreis* must be anything but homogeneous. The macro-/micro-structure contrast, for example, would be more difficult to maintain in the case of Semenko's 'Kable-poema' and Kamensky's 'ferro-concrete' works, although it makes very good sense in the specific context in which Hermans uses it.

A further complicating factor, apart from the sheer diversity of the technique and hence the danger of generalizing on the basis of a small number of types, is the tendency of many writers at the time, even as late as the 1920s, to talk of telegraphic discourse as something still to be attained *in the future*, rather than already in existence.[31] This habit may well indicate just how utopian, how Futurist in the sense of belonging to the future, such a phenomenon ultimately was. But it also had other attractions. While it had the aura of modernity and radical experiment about it, a writer could resort to some form of telegraphic idiom without feeling tied down to too constricting a style or too narrow a spectrum of subjects. It could be adapted to a variety of themes and moods and yet retain its fascination as a self-conscious strategy for poetic renewal within the framework of a future-orientated literary movement. This must largely account for its popularity for almost two decades among such a wide range of poets. What is more, as a stylistic tendency with pronounced machine associations (the transmitter, the telegraph, later Khlebnikov's robot radio-voice), it

[30] These quotations are taken, respectively, from Kirby, *Futurist Performance* (p. 117), Tisdall and Bozzolla, *Futurism* (p. 96), Timothy Materer, *Vortex: Pound, Eliot, and Lewis* (Ithaca, 1979), 113, and Hermans, *The Structure of Modernist Poetry*, p. 70.

[31] It may have been timely, for example, for Buzzi to be stressing the need for 'elementi ideo-telegrafici' in his preface to *L'ellisse e la spirale* (Milan, 1915, n.p.); but he can still be found voicing comparable demands—for 'a poetry that would retain the ticking of the Morse code machine . . . something boundless, perhaps: wire-less'—in 1928 ('I tempi di Poesia', *La fiera letteraria* (6 May 1928), 4). And writing in *Die Neue Rundschau* in 1921, Iwan Goll could still prophesy the coming of 'a poetry that would do justice to modern requirements', while largely reviving arguments from Marinetti's pre-war manifestos.

provided a medium particularly attractive to writers whose modernity often lay in trying to bridge the gap between man and machine, rather than simply depicting the new technological furniture of the world around them.

The appeal of this mutability can be seen in the way that telegraphic lyricism, in one form or another, was received during the Expressionist period in Germany, especially at the hands of the Berlin *Sturmkreis*. Indeed, so strong was the influence of Italian Futurism on this group that it has sometimes been suggested that the *Sturmkreis* is really German Futurism by another name. Although this group's reactions to—and its subsequent appropriation of—aspects of Futurist technique has been an extraordinarily well-documented chapter in the history of modernism,[32] the specific reception of the Futurist cult of the telegraphic has received little consideration.

It is already evident from Döblin's open letter to Marinetti (*Der Sturm* (1913)) that nothing as complex as the theory of the 'wireless imagination' or as subtle as the actual practice of telegraphic lyricism in anything but the poem appended to the 'Technical Manifesto of Futurist Literature' was going to be allowed to play a role in Döblin's dismissal of Futurist syntactical reforms. Just as happened in the case of Leonhard's 'Futurist Soirée in Berlin' ('Futuristenabend in Berlin'), Marinetti is reduced to the status of purveyor of a trite *Telegrammstil*. Having begun with a schoolmasterly 'I will show you just what—and how little—the telegraphic style can achieve',[33] Döblin feels able to move on quickly to the conclusion that over-condensation of this kind is limiting rather than constructive. 'You tried to compress everything to such an extent that your glass retorts broke into pieces during the condensing process, and you now have to show us the slivers of glass as the samples of your art. Ecce rubbish . . .'.[34] In fact, as Arnold and Chiellino have demonstrated, Döblin then went on to produce a series of experimental novels (*Berge Meere und Giganten* (*Mountains, Seas and Giants*), *Wallenstein*, *Die drei Sprünge des Wang-lun* (*The Three Leaps of Wang-lun*), and his better-known *Berlin Alexanderplatz*) frequently employing a rhythmic, nominal, syntactically skeletal style strongly influenced by what is being rejected in Italian Futurism's 'lirismo telegrafico'. Arguably, Marinetti himself presents his own telegraphic writing within a very narrow, largely prescriptive framework, but then, Döblin in turn chooses to create a caricature

[32] Armin Arnold, *Die Literatur des Expressionismus: Sprachliche und thematische Quellen* (Stuttgart, 1966) is the pioneering work on the subject. There is also much valuable material in Carmine Chiellino, *Die Futurismusdebatte: Zur Bestimmung des futuristischen Einflusses in Deutschland* (Frankfurt, 1978), Peter Demetz, *Italian Futurism and the German Literary Avant-Garde* (The 1986 Bithell Memorial Lecture; London, 1987), Dorothea Eimert, *Der Einfluß des Futurismus auf die deutsche Malerei* (Cologne, 1974), and Kurt Möser, *Literatur und die 'Große Abstraktion': Kunsttheorien, Poetik und 'abstrakte Dichtung' im 'Sturm', 1910–1930* (Erlangen, 1983).

[33] *Aufsätze zur Literatur*, p. 13. The background to this letter is well sketched in Judith Ryan, 'From Futurism to "Döblinism"', *German Quarterly*, 54 (1981), 415–26.

[34] *Aufsätze zur Literatur*, p. 14.

of this position in order to be able to pretend to explain and reject it, while at the same time working constructively within a related idiom.

It is hardly surprising, given the extremely dismissive contemporary German image of the Futurists' telegraphic *Worttechnik*, that the admirers of August Stramm, the leading *Sturmkreis* poet, felt incensed at the suggestion that his poetry was written in *Telegrammstil*—which, of course, in the wider sense of the word, it most certainly was (even if it had little to do with Marinetti's own declamatory version of the fashionable style). The tendency generally at this time was for the *Sturmkreis* Expressionists to operate with a telegraphic idiom that was in various respects indebted to Italian Futurism but at the same time to ridicule the experiments of Marinetti and his followers as mere *Telegrammstil*, thus enabling them to present their own innovations as far more serious forms of experiment.[35] And although they learnt more from the radically telegraphic techniques of the Italians than they usually cared to admit, the *Sturmkreis* poets actually did take the search for a new abbreviated style in a number of creatively fresh directions.

Marinetti's approach to brevity was largely a reductive one. That is to say, certain grammatical categories were *removed* from the repertoire. And, even if this method was in reality far richer in ambiguities and less straightforwardly decodable than the telegram-analogy would imply, it could easily become somewhat mechanical and acquire the surface characteristics of an all too imitable group mannerism. It is the potential rigidity of Marinetti's language reforms that Döblin (probably unaware of the disagreements between Folgore, D'Alba, and their mentor) was inveighing against in his open letter, as he writes:

> you are monomaniacal, when you tackle syntax . . . I find your generalization horrendous . . . In a full sentence there are various valencies: various sentence functions dominate at different times, now the subject, now the verb, now the adverb; you can increase the power of a word, you can reduce it, you can shorten sentences, you can spread yourself in paragraphs, or you can put a single word, a noun, an adjective, a verb, or an adverb on its own—this is the way in which you can get close to reality.[36]

[35] In her introduction to *Dein Lächeln weint* (Darmstadt, 1956, 11), Inge Stramm, the poet's daughter, quotes the first three lines of a poem by him ('Der Morgen') and declares them to be written in *Telegrammstil*. She then ventures the opinion that such an idiom derives from Stramm's professional activities ('he worked for the German Post Office, he was familiar with this style'). Soergel and Hohoff also suggest that Stramm's work brought him into contact with telegrams and codes of various kinds (*Dichtung und Dichter der Zeit*, p. 173). The trouble with such an explanation is that it is likely to lead to an underestimation of the differences between the language of actual telegrams and what we have found to be the quality of telegraphic poetry. Stramm's admirers, however, have often objected to the term on the more general grounds that it is a pejorative term. Peter Michelsen writes that Stramm's poetry 'is not "telegram style", it is words expressed from the speaker with a sound like the death-rattle of a dying man' ('Zur Sprachform des Frühexpressionismus bei August Stramm', *Euphorion*, 58 (1964), 300).

[36] *Aufsätze zur Literatur*, p. 12.

Futurist theory was, however, not only being taxed with rigidity; it was also being found lacking in creative potential by many *Sturmkreis* commentators.

Lothar Schreyer's 1918 essay 'Expressionist Poetry' ('Expressionistische Dichtung') begins by largely recapitulating the main innovations that the *Sturmkreis* poets, especially Stramm and his emulators, had introduced into the telegraphic idiom, thus making it even further removed from any literal telegram code. Schreyer draws an important distinction between *creative* modes of brevity and simple forms of mechanical reduction ('die einfachen Satzverkürzungen') such as one finds in a telegram and in basic forms of Futurist poetic 'Konzentration'. ('Konzentration' and 'Dezentration' are, in Schreyer's view, two of the main structural principles of the new *Wortkunst* practised by the *Sturmkreis* poets.) 'In simple forms of sentence-reduction everything that is superfluous to the utterance is omitted. But these are not yet examples of real concentration.'[37] Schreyer feels that this represents only a half-way stage and that more genuinely poetic forms of reduction and expressive concentration operate on something more subtle than a literary equivalent of a real telegram's redundancy principle.

There were, of course, at this time a number of Expressionist works which did operate on a principle of 'simple sentence reductions' without this in any way vitiating their impact or making them necessarily inferior to the kind of poetry Schreyer was seeking to defend. One thinks of Georg Kaiser's plays, *Die Bürger von Calais*, *Gas*, and *Gas 2*, for instance. So one has to be wary of slipping into the contemporary evolutionary way of thinking about such experiments in reduction, according to which the latest twist to the theory is invariably presented as an improvement or a refinement on the previous position. What Schreyer is at best contemplating is no more than one further alternative mode, not *the* only conceivable advance on an earlier position.

Truly expressive 'Konzentration' for Schreyer is more than a matter of (Marinetti's) mechanical reduction: it is a focus on the compressed word, ideally neither conjugated nor declined (for 'Biegung = Beschränkung', conjugation means limitation, in Schreyer's scheme of poetic values). The most effective poetic brevity, Schreyer claims, with his attention firmly fixed on the recent achievements of the *Sturmkreis*, is attained by bold shifts in grammatical categories, not just by a simple quasi-lipogrammatical process of removing certain parts of speech. (It is in claiming this that Schreyer is implicitly accusing the Futurists of not having gone far enough.) 'From the shortening of words emerge the changes in words. These changes in words lead to the formation of new words.'[38] And for this reason, because concentration of this kind becomes a matter of neologisms created out of a supposed shifting of grammatical

[37] *Literatur-Revolution 1910–1925: Dokumente, Manifeste, Programme*, ed. Paul Pörtner, i. *Zur Ästhetik und Poetik* (Darmstadt, 1961), 440. The essay was first published in the *Sturm-Bühne*.

[38] Ibid. 438.

categories, Schreyer can declare: 'the grammar of everyday usage is of no importance for literature.'[39] Every work creates its own individual grammar as a consequence of taking the possibilities of telegraphic writing further than the Italian Futurists had done.

August Stramm's poem 'Krieg' ('War') illustrates the kind of poetic creative principle Schreyer has in mind:

Wehe wühlt
Harren starrt entsetzt
Kreißen schüttert
Bären spannt die Glieder
Die Stunde blutet
Frage hebt das Auge
Die Zeit gebärt
Erschöpfung
Jüngt
Der
Tod.[40]

Just how much a work of grammatical and semantic ambiguities (and hence how distant from Schreyer's first category of 'simple sentence reduction') such a poem is becomes clear if one looks at Patrick Bridgwater's valiant attempt at rendering Stramm's 'Krieg' into English:

Grief graves
waiting stares aghast
labouring shatters
giving-birth tenses the limbs
the hour bleeds
question raises its eye
time gives birth
exhaustion
re
vives
death.[41]

Stramm's telegraphic 'laconicism', as the Italians would call it, reduces the evocation of war to a limited constellation of words between which more than one set of semantic relationships obtains, and even the omissions in the repertoire seem to be less a matter of dogmatically excluding certain parts of speech than was often the case in Italian Futurism. Hence, the translator's dilemma. The second line, for example, which in Bridgwater's English translation is interpreted as noun + verb + adverb ('waiting stares aghast') could even

[39] Ibid. 439.
[40] *Das Werk*, ed. René Radrizzani (Wiesbaden, 1963), 80.
[41] August Stramm, *twenty two poems*, trans. Patrick Bridgwater (Wymondham, 1969), n.p.

be noun + verb + verb ('waiting stares scares'). The poem's final sequence is in fact an inversion (i.e. of 'death revives exhaustion') but it would be unhelpful to recast Stramm's word-order into a normal English pattern, for much of the poem's final effect derives from our not knowing until the penultimate word just what the revitalizing force is—and even possibly assuming that 'Erschöpfung' is going to be the subject of 'Jüngt', only to arrive at the psychologically plausible idea of the fear of death giving strength back to the tired soldiers.

Stramm's poem exploits indeterminacies of this kind in a more complex and less mechanical way than Marinetti's 'Bombardamento' did, not simply because of any dependence on neologisms—although, as Schreyer's theory leads one to anticipate, a number of substantivized verbs do make unusual subjects ('Kreißen schüttert', 'Harren starrt entsetzt')—but because of both a lack of clear division between constituent phrases and also, sometimes, a variety of syntactically governed interpretive possibilities. Is the reader to assume the *enjambement* of 'Die Zeit gebärt | Erschöpfung' or one idea ('Die Zeit gebärt') with no object and hence an indefinitive sense of something being born, and 'Erschöpfung' as the subject of the new verb 'jüngen' (normally used with the prefix 'ver-'), here endowed with a strangely transitive sense? Even real telegram-ellipses risk giving rise to the occasional ambiguity, which is presumably why the conventional word 'STOP' was introduced to signal the boundary between one group of immediate constituents and the next cluster, to prevent just such indeterminacy obscuring the message. Stramm, on the other hand, employs poetic *enjambement*—and sometimes merely the suspicion of it—to push poetry in completely the opposite direction. Hence not only does his form of telegraphic lyricism tend to appear semantically richer; it also becomes more verbally concentrated. Rather than simply being systematically shorn of redundant elements, even if the preliminary omission of definite articles, prefixes, and adjectives does operate on a general redundancy principle, it radiates new ambiguities. In addition, whereas many of Marinetti's poems operate on the basis of linguistic reforms formulated outside the poetry and applied consistently from work to work, each of Stramm's poems creates its own individual system.

Underlying the differences between Marinetti's and Stramm's handling of the telegraphic idiom is a completely different attitude—not just to the role of language, but to the world itself. For in Marinetti there lingers an old-fashioned belief in the total communicability of experiences[42] linked with an assumption that pruning the medium will simply lead to greater efficacy. For Stramm, by contrast, the world is evidently more mysterious, thus language can only add to

[42] Shirley Vinall contrasts Bergson's belief that 'words betray reality' with Marinetti's confidence that 'the word can capture the essence of an object' ('Italian Avant-Garde Attitudes to French Culture: Marinetti, Soffici, Papini and Prezzolini, 1897–1915', unpublished Ph.D. thesis (University of Reading, 1975), fo. 439).

this feeling of disorientation. For him, telegraphic lyricism, of a very different kind in any case from Marinetti's, simply compounds, and finds syntactic equivalents for, a sense of confusion and danger. In this respect, the Italian Futurists look distinctly passéist, for all their technological fetichism. By contrast, Stramm emerges as a markedly post-Symbolist poet.

In the proposals outlined in the untitled manifesto—presumed to be by Khlebnikov, Burliuk and Mayakovsky—which introduces *Sadok sudey 2*, there are various resemblances to Marinetti's telegraphic theorizing. 'We have ceased to look at word-formation and word-pronunciation according to grammatical rules. . . . We have shaken syntax loose', the first item reads (*Manifesty*, p. 52). The abolition of punctuation, the rejection of standard orthography, and the focus on the role of prefixes and adjectives would all at first seem to suggest something akin to the ideas in 'Destruction of Syntax', even if the telegraphic metaphor itself is not used in the argument. (But then, much of the discussion of language abbreviation in Russia is carried out without reference to the telegraph metaphor: Shershenevich, for example, prefers the term 'catalogue of images' ('Katalog obrazov' is the name of a poem in his *Loshad' kak Loshad'* (*The Horse as Horse*)); but, as Nilsson has shown in his discussion of the poet's 'verbless poetry',[43] there are strong similarities to the Italian cult of the telegraphic.) Nevertheless, there is one fundamental difference between the 'shaking loose of syntax' in Russian Futurism and the general cult of the telegraphic. The brevity of telegraphic lyricism is primarily a matter of accelerating the delivery of an idea; the removal of redundant and decorative elements facilitates the transmitting of the utterance speedily. But in Russian Futurism the focus is on the individual word. Indeed, by reducing syntax in order to focus on the individual components of a sentence, the poet arguably slows down the pace of delivery; the act of defamiliarization is at the same time an act of retardation.

David Burliuk's poem 'Razdol'e Vremya' ('Boundless Time'), which in layout may at first appear to resemble some of the examples we have looked at so far, in fact displays a concern with what the Futurists in Russia refer to as 'the word as such', not some overall utterance which has to be delivered at a frenetically Futurist pace.

Слово	Word
Олово	Tin
И	And
Камень	Stone
Растопить	Flame
Способен	Of the heart
Пламень	Of the sun
Сердца	Can

[43] See Nilsson, *The Russian Imaginists*, p. 74. Nilsson himself prefers the term 'montage' for the style under discussion.

Солнца	Melt
И	And
Горна	That furnaces
Что застелила	Spring
Весна,	Has covered
Что раскрыла все цветы	That it has opened all the flowers
Что согрела высоты	That it has warmed the heights[44]

Unlike the situation in Stramm's poem, there is no semantic ambiguity as a result of destabilizing syntax here. What is foregrounded is a pattern of parallelisms between various individual words: 'slovo' | 'olovo'; 'kamen′' | 'plamen′'; 'serdtsa' | 'solntsa'. In one sense we are nearer the style of the telegram proper, for language has been logically pruned of its decorative elements, without any of the grammatical and interpretive uncertainties which occur in *Sturmkreis* poetry. But telegrams are still sentence-generated and are reconstructible as full sentences, which means that most forms of the telegraphic still bear some relationship to the sentence-paradigm. The effect in Burliuk's poem, however, is not so much of the single word being shaken loose from its syntactical matrix as of column constructions being *generated* outwards from the single word. The next logical progression from this is the single-word pillar which has each individual word standing on its own and where the constructive principle is very much that of generation, rather than reduction. The 'Blagovest umu' ('Parade of -um') part of Khlebnikov's *Zangezi* contains a section beginning

Проум
Праум
Приум
Ниум
Вэум
Роум
Заум
Выум
Воум
Боум
Быум
Бом! (SP iii. 334)

This sequence and similar generative structures will be considered in our discussion of the 'linguistic' aspect of literary Futurism, since they are not, in any direct way, related to the telegraphic reduction of language. Here, there is no longer a sense of having been reduced from the excesses of conventional, passé-

[44] The poem and English translation are taken from Folejewski, *Futurism and its Place in the Development of Modern Poetry*, pp. 216f.

ist syntax. Rather, the utterance appears to have been generated—out of nothing, as it were—on the basis of word-formation principles.

One of the obvious dangers for the radicalness of a stylistic cult like that of the telegraphic was that it could quickly become a mannerism. Marinetti concludes one of the sections of 'Destruction of Syntax' with the caveat: 'we should at all costs avoid rhetoric and banalities expressed telegraphically' (*TIF* 62). In the German Expressionist context in particular, where the telegraphic fashion took an even greater hold than it did among the Futurists, what had begun as a serious renovation of poetic language fairly rapidly degenerated into self-caricature.

In a remarkably early discussion of the role of *Telegrammstil* in German Expressionist poetry, Richard Samuel and R. Hinton Thomas make a distinction between 'the apparent sobriety and objectivity of the "Telegram Style"' and the 'highly emotional quality' which other forms of reductive style went on to create in its wake[45] (although this would be a more difficult distinction to make in relation to Italian literature, even if it has some attractions in the German context). Yet even *Telegrammstil*, they argue, was gradually abused: 'the "Telegram Style", having become a mode, was degraded further into a mere stammering and babbling when exploited by radical political lyricists.'[46] They cite examples from Johannes R. Becher (e.g. *Panzerwagenballade* (*Ballad of an Armoured Car*)), whose early work is greatly indebted to Italian Futurism,[47] and Ernst Toller (*Die Mutter* (*The Mother*)) to indicate that 'this deterioration of the "Telegram Style" led to a language of verbal skeletons and inarticulate cries'. Certainly, modishness had its dangers here. One only has to consider the difference between the following two passages, one from Behrens and the other from Werfel, to see how telegraphic writing was always teetering on the brink of becoming a parody of itself.[48]

Franz Richard Behrens's array of cabled verses in *B= C. Der Roman der Lyrik* offers a whole epic panorama of world history, told in telegram form:

> Cable Transfers
> 350.50
> Blinder Bettler Homer hungert über die Dörfer
> Baltimore und Ohio

[45] *Expressionism in German Life, Literature and the Theatre* (Cambridge, 1939), 159.

[46] Ibid. 162. This position is partially undermined by the fact that Samuel and Hinton Thomas are willing (p. 160) to see the dramatist Georg Kaiser as a purveyor of this category of 'Telegram Style', when the highly rhetorical, skeletal language of his plays is on the whole anything but sombre and objective.

[47] The subject is explored in Norbert Hopster, *Das Frühwerk Johannes R. Bechers* (Bonn, 1969).

[48] Samuel and Hinton Thomas consider a parody of the style in Fritz von Unruh's play *Platz* (*Expressionism in German Life, Literature and the Theatre*, p. 162).

> 47.97
> Hesiod ermordet ins Meer geschmissen
> Canada Pacific
> 127.87
> Archilochus erdolcht
> Chic Rock Island
> 38.00
> Der Bruder erschießt Anacharsis . . .[49]
>
> Cable Transfers
> 350.50
> Blind beggar Homer hungers around the villages
> Baltimore and Ohio
> 47.97
> Hesiod thrown murdered into the sea
> Canada Pacific
> 127.87
> Archilochus stabbed
> Chic Rock Island
> 38.00
> Brother shoots Anacharsis . . .

As befits its use of stock market elements, *B* = *C* ends with the lines 'Dollars kauft Dollars | Liliencron spuckt aus und geht | Weiter' ('Dollars buy dollars | Liliencron spits and goes | On his way').

Behrens's experiment is unique and unrepeatable. The trouble was, many forms of the telegraphic lent themselves too easily to mimicry and were all too identifiable as the mode of the moment. The following passage from Franz Werfel's *Spiegelmensch* (*Mirror Man*) represents very much (even if in satirical form) the effect of the ubiquitous style's 'rhetoric and banalities expressed telegraphically' that Marinetti had cautioned against:

> Radiogramm sogleich an Times und New York Herald,
> An jedes Sternbild, das die Ohren herhält,
> 'Welterlösung promptest zu erhoffen . . .
> Gott hier 12 Uhr persönlich eingetroffen'.[50]
>
> Radio telegraph immediately to Times and New York Herald,
> To every star that is listening
> 'World salvation shortly to be expected . . .
> God personally arrived time 12.00'.

In general, however, the fate which the telegraphic idiom suffered during the later years of the *Sturmkreis*'s activities was not excess, be it parodistic or unintentional, but sheer ossification. After the early bold achievements of Stramm,

[49] Behrens, *Blutblüte*, pp. 109–25.
[50] Franz Werfel, *Spiegelmensch: Magische Trilogie* (Munich, 1920), 138.

Behrens, and Schreyer, there was a tendency for later Expressionist writers to imitate the compressed idiom of these writers too slavishly. The periodical *Der Sturm* and much of the related poetry and drama published by Heynicke, Knoblauch, Liebmann, and Runge show how easily the telegraphic idiom could lose its ability to vitalize conventional language and become a superficial mannerism.[51] It was only with the more radically minimal, lipogrammatical exercises of David Burliuk's poems 'Bez *R*' ('Without *R*' and 'Bez *R* i *S*' ('Without *R* and *S*') (in *Dokhlaya luna* (*Croaked Moon*)) and Otto Nebel's 'rune-fugues' (considered in the next chapter, on Futurist 'linguistic' literature) that telegraphic writing took a new direction which enabled it to avoid the hazards of modishness.

One thing that clearly emerges from this exploration of various forms of telegraphic writing is just how rich the idiom is. I cannot agree with Helmut Meter's suggestion that, where Marinetti rejects syntax and prunes down the number of grammatical categories in his language, the result is nevertheless a retention of an underlying 'mental syntax'.[52] There are already sequences within the 'Bombardamento' section of *Zang tumb tumb* where the ambiguities of the telegraphic idiom make any notion of a decodable telegram-structure seem unacceptable as an approach to the style. And once one looks beyond Marinetti's personal practice, the device reveals itself as being even richer in possibilities, of both syntactical restructuring and layout, than Marinetti's own theorizing would suggest it could be. It could open the door to an impressionist registering of sensations, to forms of the monologic, and to a variety of moods and registers. It could offer a provocative way of flaunting one's modernism by refusing to use the conventional language that the canonical poets of the past had used. For Marinetti, telegraphic lyricism was a strategy for rejecting the intellectualism of passéist syntax, a means to more intuitive expression, and a way of narrowing the gap between man and machine. It also had a poetological value; in 'Destruction of Syntax', the new 'telegraphic lyricism' is presented as something 'with no taste of the book about it but, rather, as much as possible of the taste of life' (*TIF* 66). While there was a measure of disagreement among the Italian and Russian Futurists about the requisite degree of radicalism to such a new abbreviated idiom, they were generally working on the assumption that the telegraphic was to be arrived at by means of some form of rigorous expunging of certain grammatical categories from the repertoire. However, one only has to compare their practice with the poetry of the Imagists to

[51] The process is analysed in M. S. Jones, 'The Second Generation. August Stramm: *Der Sturm*'s Hero and his Imitators', in his *Der Sturm: A Focus of Expressionism* (Columbia, South Carolina, 1984), 137–70.

[52] 'Maschinenkult und Eschatologie: Zum historischen Ort des italienischen Futurismus', *Sprachkunst*, 17 2 (1986), 90.

appreciate that extreme brevity can be attained by more flexible means. F. S. Flint's exhortation 'use absolutely no word that does not contribute to the presentation' and Ezra Pound's 'use no superfluous word, no adjective which does not reveal something'[53] indicate theoretical standpoints which, although influenced by Italian Futurism,[54] by no means seek to legislate about which parts of speech are to be retained and which should be jettisoned in order to create the new telegraphic idiom. Something that has to be borne in mind—and it will concern us more as we now turn to some of the linguistic aspects of literary Futurism—is the fact that the Italian Futurists were labouring under a handicap with which their Russian and German counterparts did not have to contend. What I have in mind is the fact that the very structure of Italian imposes severe limitations on the forms of telegraphic brevity which could be achieved in the language.

[53] 'Imagisme' (*Poetry* (Mar. 1913)), quoted from *Imagist Poetry*, ed. Peter Jones (Harmondsworth, 1972), 129; 'A few Don'ts by an Imagiste' (ibid. 131). See also Pound's letter to Harriet Monroe of Jan. 1915: 'Language is made out of concrete things. General expressions in non-concrete terms are a laziness. . . . "Epithets" are usually abstractions. . . . The only adjective that is worth using is the adjective that is essential to the sense of the passage, not the decorative frill adjective' (ibid. 142).

[54] On this influence, see Giovanni Cianci, 'Futurism and the English Avant-Garde: The Early Pound between Imagism and Vorticism', *Arbeiten aus Anglistik und Germanistik*, 7 (1981), 3–39.

4.

Linguistic Literature: Neology, New Syntax, Word Games

ARDENGO SOFFICI's poem 'Studio', which was mentioned in the preceding chapter in the context of the Futurists' preoccupation with forms of 'telegraphic lyricism', boasts the striking compound adjective '[cabina] radiotelefantastica'. It is a portmanteau neologism, the middle part of which suggests an amalgam of two existing Italian words: the one (*telefono*) of relatively recent provenance, the other (*fantastico*) with an obviously much longer pedigree. A cognate adverb *radiotelefonicamente* appears to have been already in circulation at the time when Soffici was writing. No doubt, it had the attraction of flaunting the fact that its user was *au fait* with the possibilities of such an ultra-modern means of communication (Marinetti uses it in this spirit in *L'Aeropoema del Golfo della Spezia* (*TIF* 1031)). But Soffici's tripartite adjective is, by contrast, a freshly invented word.

Even comparatively straightforward neologisms of this order are, nevertheless, rare in the writings of the Italian Futurists, although we shall, admittedly, later on encounter another related word—'telegrafonoradio'—in a poem by Luigi Pennone. In Italian Futurism the stress is predominantly on language-*modification*—that is to say, on the streamlining of the conventional grammatical sentence, rather than on language-*creation* in a more fundamental sense. Whilst Marinetti and his followers were, as we have seen, most enthusiastic about revolutionizing language at the level of syntax-reduction and thus creating a new hyper-economical sentence-structure, they invariably give the impression of being quite content to construct their free-word works by using the traditional components of the words already available in the Italian repertoire. It is not until Nelson Morpurgo's *Il fuoco delle piramidi* of 1920 that one comes across any more thoroughgoing attempt at radical word-creation on the basis of inventive morpheme-montages and a series of idiosyncratically semanticized phonemes. In the poem 'Colloqui notturni' ('Nocturnal Conversations'),[1] for example, Morpurgo creates the following effect:

te
pian
gono

[1] *Il fuoco delle piramidi* (Milan, 1920), 17.

merging the verb 'to cry' (*piangere*) with the plural of 'plant' (*pianta*) to give 'plants cry', with the suggestion that the two words share a 'false' (poetic) root. But with experiments of this sort, Morpurgo establishes himself as something of a maverick. His various word-dissecting games are on the whole untypical of mainstream Italian Futurism, and one is tempted to conclude that, as an Egyptian, the poet must have retained an especially defamiliarizing relationship to the Italian language with which he was working.

On the issue of what they saw as the anachronism of conventional syntax, the majority of Italian Futurists would probably have been inclined to endorse the position adopted by their Russian counterparts. These from the outset paraded their 'insurmountable hatred for the language in existence before them' (*Manifesty*, p. 51). Likewise, it is evident that most of the Italian writers in Marinetti's circle would have had good cause to sympathize with Khlebnikov's description of Futurist literature as 'the enemy of the bookish ossification of language' (*SP* v. 233). But from here on they would in all probability have parted company fairly rapidly. One deciding factor would have been that Khlebnikov's chief panacea for such linguistic ossification lay in the area of intensive 'word-creation': 'Just as modern man stocks the depleted waters of rivers with schools of fish, so language-making gives one the right to stock the impoverished waves of language with new life, with extinct or non-existent words' (*SP* v. 234). This is hardly a view that many Italian Futurists would be disposed to echo. Indeed, generally in Italian Futurist circles, the deep-seated dissatisfaction with the current state of the language that also prevailed did not pave the way to the principal corollary that one finds in the founding Russian manifesto 'A Slap in the Face of Public Taste' ('Poshchechina obshchestvennomu vkusu' (1912)). There, the Futurists' disenchantment with the contemporary state of the medium is deemed to imply, as a logical consequence, an emphatic defence of 'the *rights* of poets ... to enlarge vocabulary in its scope with arbitrary and derivative words (creation of new words)' (*Manifesty*, pp. 50 f.). Kruchenykh can, in the light of such a standpoint, claim with some justification that the Russian Futurists 'were the first to say that for portraying the new and the future *completely new words and new combinations* are necessary' (*Manifesty*, p. 75). The Italians, on the other hand, would seem to have regarded themselves as already more or less satisfactorily equipped to portray 'the new and the future—and more generally, to write Futurist literature—on the basis of the linguistic resources of the present. The poetically 'striking newness' which, according to Kruchenykh's 'New Paths of the Word' (Novye puti slova'), was guaranteed to 'come through the combination of words according to their own immanent laws revealed to the poet and not according to the rules of logic or grammar, as has been the case hitherto' (*Manifesty*, p. 75) was not destined to play any major part in the Italian Futurists' extensive programme for language-reform. It was by no means seen as a *sine qua non* of the syntactical innovations being introduced in their words-in-freedom. In fact, their much-trumpeted 'destruction of syntax' invariably had the side-effect of leaving the individual word intact, if one discounts the putting of verbs into the infinitive (which hardly involves the creation of new words so much as a grammatical solecism).

In this respect, the Italian Futurists can also be seen to differ from the majority of German poets of the *Sturmkreis*, that outpost of Central European Futurism, for these too laid a great deal of programmatic stress on the importance of neologism, particularly within poetry. Indeed, it could be argued that, even when the German poets did behave like good Futurists and put their verbs in the infinitive, their tendency to place that infinitive in isolation in word-columns and write it with a capital letter created a new form somewhere between a gerund and a conventional infinitive or something resembling the use of the German infinitive for imperative purposes, so that Marinetti's reform attains less of an innovative quality there than it had in the Italian.

The necessity of *creating* new language could also, as some of the *Sturmkreis* writers saw, be expressed as the need to *destroy* old language (and not just in respect of syntax, either). The movement's 'word-artists' (*Wortkünstler*, a German compound which tellingly has no equivalent in Italian Futurism), should, according to Paul Hatvani, 'first destroy, after which they must bring about an absolute homogeneity in their material. . . . Only then can their creative work begin.'[2] Clearly, what we have here is a position which differs fundamentally from that of the Italian Futurists, despite the strong influence that the latter have on the whole been recognized as having had on the German *Sturmkreis* poets.

Nevertheless, there is one exception to this overall pattern (a pattern which makes the language of the Italian Futurists seem, at word-level, relatively conservative when compared with the neologisms of the *Sturmkreis* and the Russian Cubo-Futurists): that is in the area of essentially additive word-formations as a strategy for revitalizing language.

When it came to the question of names, the Italian Futurists frequently displayed an uncommon penchant for verbal inventiveness: not only in decking out their own children, as Marinetti and Balla did, with such ostentatiously Futurist first names as Elica (Propeller) and Luce (no doubt, *electric* Light) or adopting modish names like Volt (Vincenzo Fani-Ciotti) or *Dinamo* Correnti, but also in the very structure of their various pseudonyms. In the interests of the new *laconicità*, Bruno Corradini streamlined his name to 'Bruno Corra', Cristoforo Mercati became the Futurist poet 'Krimer', Osvaldo Barbieri became 'Bot', Arnaldo Ginanni Corradini was known as 'Arnaldo Ginna', Vittorio Tommasini as 'Farfa'; Luigi Colombo for similar reasons took the

[2] 'Spracherotik', *Der Sturm*, 3 (1912/13), 210.

pseudonym of 'Fillia'—all, in other words, shortening their names as if according to some onomastic version of the telegraphic principle, and in some cases actually returning their name to what might look like a portmanteau or root-form (instead of simply assuming a completely new one). More significant from our point of view, though, is the converse pattern. Balla, for example, frequently chose to promote himself as 'Futurballa'. The title of one of his works from 1917 consists of a similar compound noun: 'Futurrealtà'.[3] In equally augmentative fashion, the running-head to the individual pages of Cangiullo's 1916 collection of poems, *Piedigrotta*, reads 'Cangiullopiedigrotta' on the left-hand pages and 'Piedigrottacangiullo' on the right-hand ones. In fact, rather than the telegraphic process of name- and word-shortening, it is this second, comparatively straightforward, *additive* practice, the formation of simple poetic compounds by the joining together of existing Italian words, that one comes across repeatedly in the one main area of Futurist neology. When it comes to the length of the new coinages, Italian Futurist neologisms seldom display the 'laconic' or 'synthetic' quality one might have expected. The resultant new words are often exceptionally long, looking more like hyper-inflated German compounds than Italian ones.

'Unter den Linden', one of the sub-sections of Ruggero Vasari's linguistically polyglot Berlin cycle, includes the sequence: 'DISCONTO-GESELLSCHAFT: americani inglesi francesi italiani tuttilmondo tuttelelingue' ('DISCOUNT-COMPANY: americans english french italians alltheworld alllanguages'),[4] as if the metropolitan melting-pot of language, the place where large crowds assembled, was to be reflected in this amalgamation of separate words into new telescoped collective nouns. Another city-poem, by Farfa, bears the compound title 'Newyorkcocktail', again a simple attempt at making the word itself into as much of a mixture of various ingredients as the actual city.[5] One of Loris Catrizzi's contributions to Marinetti's *I nuovi poeti futuristi*[6] boasts the neologistic title 'Stradaseragennaio' ('Streeteveningjanuary'), and contains a number of other new coinages: *tangheggiamento* ('as if dancing the tango'), *cristalleriche* ('the sound of crystal'), *violinio* ('the sound of a violin'), *sinfonia-agrandeorchestra* ('a symphony played by the full orchestra'), *manisenzaguanti* ('gloveless hands'), and *legnobagnato* ('wetwood'). Emilio Mario Dolfi's 'Wisky [*sic*] and soda' (also in *I nuovi poeti futuristi*,[7] a richer hunting-ground for neologisms than any prior Futurist anthology) includes the multi-compound *calandosigiuperlesartiedifuoco* ('letting itself down the burning shrouds'). Luigi

[3] For details, see Giorgio de Marchis, *Giacomo Balla: L'aura futurista* (Turin, 1977), 61.

[4] *Venere sul Capricorno* (Naples, 1928), 19. The poem originally appeared in *Der Sturm* in 1922.

[5] The work, first published in Farfa, *Noi miliardario della fantasia* (Milan, 1933), is here quoted from Glauco Viazzi's anthology *I poeti del futurismo, 1909–1944* (Milan, 1983), 572f.

[6] *F. T. Marinetti presenta I nuovi poeti futuristi* (Rome, 1925), 23–8.

[7] Ibid. 69.

Pennone's 'Gli affari del primo porto mediterraneo: Genova' comes up with *macchinacattedrala d'acqua* ('machinecathedral of water') as a description of the port, and goes on to employ a large number of further additive neologisms. The following is an extract from the second part of the poem, with both Pennone's additive neologisms and the new hyphenated compounds of the Italian original indicated in the English by being placed in small capitals:

genova calatadestra mediterraneo moloitalia con faroroma
approdo alle genti del mondo moltiplica
nave × mille mille mille

ciclopico tempio mare legato terra per assalto
cielo frantuma pugno K.O. la crisi mondiale

navi navi navi di tutte le bandiere-tonnellaggi-carichi
sbarcoimbarcano cereali caffè carbone cotone lana
carni pelli spezie legno salnitro vino macchine metalli
denaro uomo strano incrociante linguaggio
sapore dei grandi fratelli vicinolontani

nord
traffico arrivopartenza da per occidente-oriente
sud

in ridda di miliarditonnellaggi miliardilire tumultuanti
negli scagni di banchi borsa delle borse cervello motore
superalimentato tra bancosangiorgio avo banche italia-estero
e antrobuchetrattorie meraviglie sottoripa

quotazioni noli spedizioni carichi scarichi avarie assicura-
zioni stallie controstallie battaglie campali a colpi di
telegrafonoradio[8]

genoa RIGHTQUAY mediterranean PIERITALY with LIGHTHOUSEROME
landing place for the people of the world multiplies
ship × thousand thousand thousand

cyclopean temple sea tied earth for assault
sky shatters punch KO the world crisis

ships ships ships of all COLOURS-TONNAGES-LOADS
UNLOADLOAD cereals coffee coal cotton wool
meat leather spices wood saltpetre wine machines metals
money man strange crossing language
taste of the great brothers far away and yet close

north
traffic ARRIVALDEPARTURE FROM FOR west–east
south

[8] Quoted from Viazzi (ed.), *I poeti del futurismo*, p. 641.

> in a turmoil of THOUSANDMILLIONTONS THOUSANDMILLIONLIRE tumultuous / at the counters of banks purse of the Stock Exchange driving intelligence / OVERFED between SANGIORBANK ancestor banks ITALY-FOREIGN / and DIVESCHEAPRESTAURANTS marvels seafront
>
> quotations freight shipping cargo discharge damages insurance lay-days demurrage pitched battles by telegraphonoradio

Some of Pennone's neologisms would appear to have been invented specifically to perform new metaphorical functions. Italy, for example, whose shape has so frequently invited figurative interpretation, here becomes a gigantic pier ('moloitalia'), with Rome serving as its lighthouse ('faroroma'), as the ships head north on their way to the port of Genoa. ('Rightquay' '*calatadestra*' is presumably a logical description, if one is looking out to sea, i.e. southwards, from the port of Genoa itself, though it might mean 'the quay on the right', if viewed the other way around.) Other new verbal collocations create a lively picture of the maritime setting, in particular the telegraphically compressed 'ships of all colours–tonnages–loads', an idea which would otherwise have required setting out in three separate adjectival strings. But some of the more effective neologisms transcend the compounding capabilities of conventional language in order to capture the unique excitement of a modern port. Words like *arrivopartenza* ('arrivaldeparture'), *sbarcoimbarcano* ('unloadload'), and the pairing *da per* ('from for') give a sense of the bustling to-and-fro of the harbour, with *antrobuchetrattorie* ('divescheaprestaurants') suggesting collectively the kind of mosaic of dubious eating-places to be found in the Genoese dock-area.

In trying to highlight the new words in this poem I have assumed that 'telegrafonoradio' is also a neologism. But at a time when new technological inventions were often spawning more modish terms than were able to survive long enough to be immortalized in the dictionary (the history of early cinema and phonograph terminology bears ample witness to this), it is often difficult to be sure. (For example, a poem by Nicolas Beauduin, published in the *Mercure de France* in January 1914, under the title 'La Poésie de l'époque', contains the phrase 'la phonocinématographie instantanée', where the words for 'phonograph' and 'cinematography' are formed into a portmanteau which looks plausible, yet is not recorded in dictionaries of the period, and which I am convinced is a poetic invention.) In any case, one of the other striking linguistic features of Pennone's piece, apart from its recourse to neologism to communicate the feel of a modern commercial seaport, is the introduction of a large number of technical maritime and industrial terms into the Futurist description. So 'telegrafonoradio' could belong to this category.

The industrialization of the port of Genoa is reflected in the registers of vocabulary used to describe it, words relating to classes of ship, business transactions, types of cargo, and so on. For once, it is the idiolect of early twentieth-century technological modernity, rather than the phenomenon of battle as the Futurist experience *par excellence*, that gives rise to a whole plethora of invented words.

Never a leader to be outdone by his footsoldiers, Marinetti was also soon to be observed trying his hand at a whole series of comparable additive formations. In the 'Dune' section of *Zang tumb tumb* (*TIF* 709), for instance, there occur(s) the word(s) *sei-* | *statuneroe*, translated by Gisuti as 'you havebeenahero', and, by Pioli, with a for him uncharacteristically conventional 'you are a hero'. Actually, if one really wishes to reflect the fact that the original has a line-ending phrase-splitting hyphen (not indicated in Gisuti's text), a more faithful rendition would be a fragmentation of the single word 'youhavebeenahero' into the *enjambement* of the split formation: 'you- | havebeenahero'.[9]

Marinetti's post-war works also contain a number of additive neologisms. *8 anime in una bomba* includes the sequence *torrida profumeria acidacrestomachevole* ('torrid acidsoursickening perfumery' (*TIF* 757)). In the same volume, one also encounters *grande carezza verdombrardente del bosco* ('great green-growing-shadow caress of the woods' (*TIF* 731)), and the totally meaningless *Liroramina-procozazifrufru* (*TIF* 761). And at a number of junctures in *L'Aeropoema del Golfo della Spezia*, the aero-poet employs similarly structured additive neologisms, among them: *oliovinbiancozenzero* ('oil-white-wine-vinegar' (*TIF* 1016)), *Gocciolamento* ('drippingly' (*TIF* 1019)), *occhieggianti* ('eyeing' (*TIF* 1023)) and the sequence *Acciaia autotemprante del mare* ('self-tempering steel of the sea' (*TIF* 1020)). In addition, a number of Marinetti's personal poetic contributions to the 1925 anthology *I nuovi poeti futuristi*, which he edited, display a similar interest in the possibilities of compound or hyphenated neologism. For example: the leitmotif *panpantomimepan* and the single verb *friggeride* ('frylaughing') in 'macchina lirica' ('Machine Lyric'). (The fact that the French version of this poem, 'machine lyrique', contains the substantially more successful word-play 'frire rire frirerire' makes one wonder how often, in fact, Marinetti worked towards his Italian effects from the starting-point of a still greater competence in French.) Other examples in the same anthology include the reference to an *odore caldogrodolce* (in 'ritratto olfattivo di una donna'), which becomes an *odeur chaudaigredouce* (meaning 'a warmsweetsour smell') in the French version, 'portrait olfactif d'une femme', and the phrase *vento-di-cuoio-gelato* ('wind-of-frozen-skin') in 'Con Boccioni a Dosso Casina' ('With Boccioni on the Casina Ridge').[10]

[9] Cf. the English translations in Franco Gisuti, *Italian Futurism as a Sociolinguistic Phenomenon* (Ann Arbor, 1981), 153, and Richard J. Pioli, *Stung by Salt and War: Creative Texts of the Italian Avant-Gardist F. T. Marinetti* (Reading Plus, 2; New York, 1987), 93.

[10] These poems appear on the following pages of *I nuovi poeti futuristi*: 'macchina lirica' (p. 283),

One suspects that it was often the particular challenge of finding verbal equivalents for sense-impressions that gave rise in these poems to a more experimental use of adjectival constructions, in particular. The dog-subject of 'paesaggio d'odori del mio cane-lupo' ('my alsatian's landscape of smells'), on the other hand, seems incapable of rising to the occasion of such a word-creating challenge, despite the overall richness of descriptive vocabulary at his disposal and for all his undoubted capabilities as the first canine *mot-libriste* in both Italian and French.

However, the new word-formations cited above should not be overestimated. They represent a thin harvest by comparison with the work of the *Sturmkreis* poets and the Russian Cubo-Futurists. On the whole, they amount to no more than local bravura effects to be found in a sprinkling of Italian Futurist poems mainly of the post-war period, limited sometimes to just one per poem, and even then, mainly to no more than a few works, the majority of them being by the movement's leader himself. Sometimes, one has the impression that the introduction of neologisms has been reserved for the kind of 'zone of intense experience' where the earlier Marinetti would in all likelihood have resorted to a 'designed analogy' (these appear to have fallen out of favour in the later works). Moreover, a crudely quantitative ethos often informs these neologisms, something which is reminiscent of the deployment of numbers and quantitative measurements to be found so often in Italian Futurist descriptions and in the language of the manifestos.

A far more inventive form of neologism comes into play when Marinetti sets about imitating the rhythm of the telegraph-key tapping out messages in Morse Code, in his 'Il cracracracranio della notte' ('The Croakcroakcroakcranium of the Night', the pun being on the two words 'crac' and 'cranio').[11] The first part of the poem reads:

Il cracracracranio della notte
Polifonia oceanica di rumori-scatti-tonfi
RANE ROSPI
Officina filanda di scricchioliii organizzatissimi
cracracracracracracra tutti i rospi della terra
lavorano al tornio 30000000000 granate nere per
bombardare di nerotradimento terrorironia la
troppo bianca spasimante Luna fiduciosa
La fattoria mano nera calda fende col suo tetto-prua
il dilagaaante fresco oceano di rumori odori argentei
fieni acri
ssssssssssss
sssssssssssssssiiilenzio lunare

'machine lyrique' (p. 303), 'ritratto olfattivo di una donna' (p. 299), 'portrait olfactif d'une femme' (p. 302), and 'Con Boccioni a Dosso Casina' (p. 255).

[11] *Poesia*, new ser., 1 (1920), 22 f.

La luna soffia a gote bianche piene nel suo megafono
di raso bianco
sssssssilenzio che vuole imporre
Vieeeeni! Vieeeeeni!
La Notte è attentiiisssssima a tuttiii iii rumoriii
Via LATTEA muta rivolta di caratteri tipografici
Polvere di rumori già uditi spinpinpinpingere dipinpingere
punpunpunpungere
punpunteggiaturaaaa
Ritmo di telegrafo Morse
Stelle parole telegrafate d'amore-guerra 200000 uomini
di riserva! Presto ad ogni costo   Chiudere il varco
del fronte e del cuore che muore

The croakcroakcroakcranium of night

Ocean polyphony of noise–jerks–thuds
FROGS TOADS
Workshop spinning mill of very organized creaking
croakcroakcroakcroakcroakcroakcroak all toads on earth
are turning 30000000000 black grenades
to bomb with blacktreason terrorirony the
trusting overwhite languishing Moon
The firm black hand cleaves with its roof-prow
the spreading fresh ocean of noise silver smell
sharp hays
sssssssssss
moon sssssssssssssssilence
The moon blows with full white cheeks with its
white satin megaphone
sssssssssilence which it wants to impose
Coooome! Cooooome!
Night is extremely attentive to all noises
The MILKY Way mute rebellion of typographical letters
Dust of noises already heard pushpushpushpaint depipict
punpunpunpunctuate
punpunctuatiooooon
Rhythm of Morse telegraph
Stars telegraphed words of love-war 200000 reserve
men! Soon at any cost   To close the passage
through the front and through the dying heart

The dominant analogy structuring the imagery here is between the croaking of frogs at night and the new nocturnal activity of the Morse Code machines tapping their staccato messages out into the void.

The more such poems come to depend on sound-patterning and inventive word-play for their effects, the weaker any translation of them is likely to seem in comparison with the original. In general, translations of linguistically innovative poems of the period are condemned to be reduced to reflecting little more than the sense in places, rather than being able to offer equivalents of the experiments being made in the original language.

More significant, in this case, than the mechanically additive formations which Marinetti's poem also contains (e.g. *bombardare di nerotradimento terrorironia* ('to bombard with blacktreason terrorirony')), are the new rhythmic duplicating structures, such as the title's 'cracracracranio', or *spinpinpinpingere* (pushpushpushpaint), *punpunteggiaturaaaa* (punpunctuatioooon), and *dipinpingere* (depipict). On the whole, though, even in the majority of his post-war works, Marinetti still appears to favour hyphenated forms of semi-compounding, such as the constellation *note-lagrime-stelle* ('notes-tears-stars') in *Gli indomabili* (*TIF* 892); or the following groupings from *8 anime in una bomba*: *amore-bile-invidia-speranza* ('love–bile–envy–hope' (*TIF* 746)), *mio sangue bromo-salso-jodico-donnaiolo* ('my blood bromine–salt–iodine–ladies' man' (*TIF* 759)), and *nel buio-noia-attesa della carne nemica* ('in the darkness–weariness–waiting of the enemy flesh' (*TIF* 770)). As soon becomes clear, in such contexts the hyphen is no longer reserved as a copula for bold 'wireless analogies', as was the case in much of the earlier poetry. Its function is now more directly associative than juxtapositional.

A more extended passage displaying a genuinely innovative compounding technique comes with the following rhapsodic extract from Marinetti's later *Poema non umano dei tecnicismi* (Milan, 1940):

> O bella ansia di luce di quali nomi vuoi che
> t'incoroni Aggancianuvole Ingoiastelle Bevincoro
> Tracannacielo Oppure se vuoi Scannatenebre
> Strizzabuio Antifogna Crepinvidia Preferisci con
> me Liquidagioia Acquedargento Perledifonte
>
> (*TIF* 1060)

Marinetti's Italian sequence travels badly between the two languages, as the following approximation into English may well merely serve to underline:

> O beautiful anxiety of light with what names do you wish
> to be crowned Claspclouds Swallowstars Drinkinchorus
> Imbibesky Or if you wish Unwinddarkness
> Squeezeobscurity Antisewer Crackenvy You prefer to be crowned with
> my Liquidjoy Silverwaters Fountainpearls

Yet the inventiveness of the Italian original is bought at a high price. Even if one has the relief of finding the occasional successful coinage (e.g. 'Liquidagioia', 'Perledifonte') among this string of desperately innovative compounds, such a rigidly additive technique of word-formation as Marinetti employs here has a tendency to generate inordinately clumsy sequences.[12]

It was, in fact, not Marinetti, but one of the early Futurist painters who was the first to contrive a lighter variation on the additive neologistic procedure.

In the 1913 draft of his 'Futurist Manifesto of Men's Clothing', Balla tells of the time when

> all around us we shall find acrobatic blocks of colour set out like the following word-shapes:
>
> Coffeecornhou Rosegreebastocap transpomotocar
> legcutshop blueblackwhitehouses aerocigarend
> skyrooflifttyellight anomoviesphot barbebbenpurp
>
> Human beings, until now, have dressed (more or less) in black mourning.

So runs Balla's complaint. Hence, 'WE MUST DESTROY ALL PASSÉIST CLOTHING.' The result, we are promised, will be 'hap-hap-hap-hap-happy clothes'.[13]

If Balla's comparison uses two kinds of language to contrast the dowdiness of conventional clothing with an intimation of what the new flamboyant style will be like, Khlebnikov and Kruchenykh, in their turn, resort to sartorial imagery to contrast two kinds of language: in 'The Letter as such' ('Bukva kak takovaya') they ask, concerning the passéists, 'why do they clothe the letter in the grey tunic of prison? You have seen the letters of their words—set out in straight lines, cut short, shorn and all equally grey and colourless—not letters but brands of infamy!' (*Manifesty*, p. 60). Much of the imagery of Soffici's *Primi principî di un'estetica futurista* also plays with contrasts between the letter as the 'mortal clothing of Poetry' and the 'naked idea' (*Archivi*, i. 687).

If one compares Balla's fusion of foreshortened verbal elements with the dazzling kind of Futurist clothing he—and Fortunato Depero—envisaged (a series of illustrations of which can be found in *FF* 324 f.), then it becomes clear

[12] In his survey of augmentatives in Arno Holz's poetry, in particular in *Phantasus* (1898), Bernd Helmut Neumann makes a useful distinction between simple 'lexeme *combinations*' and the more creative neologistic process of 'lexeme *composition*'. Rather than mere combinations of lexemes, Holz's compound words are shown to involve 'language games, games in which sememes are brought into new relationships with one another and give shape to complex poetic fields of meaning. . . . Their semanticization is only interpretable as a complex of various possible meanings' (*Die kleinste poetische Einheit: Semantisch-poetologische Untersuchungen an Hand der Lyrik von Conrad Ferdinand Meyer, Arno Holz, August Stramm und Helmut Heißenbüttel* (Cologne, 1977), 197). Neumann's discussion of various inventive *Phantasus* neologisms, including such complex compounds as 'koständelkicherneckt', 'Tanzpossenklimbimkaschemme', 'Schlingkrautstrebepfeilerstützen', and 'schlankkrankkegelig', in the pages preceding this conclusion, makes it clear that in Holz's case one is largely dealing with something semantically far richer than most of the lexeme-combinations of Italian Futurism. On the other hand, the few instances he gives of more perfunctory 'lexeme combination' in Holz's work (e.g. 'Kraftstolzfreude', p. 185: 'strengthpridepleasure') are sadly reminiscent of the Italians' augmentative effects.

[13] *Futurist Manifestos*, ed. Umbro Apollonio (London, 1973), 132.

that such a pyrotechnical verbal display as this has managed to hit upon a striking linguistic analogy for the disruption of conventional outline, the countermanding of static shape and form-orientated patterning to be achieved with the new garments. Indeed, much of what Germano Celant has to say in the Venice catalogue about the use of 'optical intersections' and the 'kinetic look' of the new Futurist clothing could just as well be applied to the synthesis of truncated verbal forms that Balla has turned to in his manifesto-draft in order to describe the revolution in clothing (with the proviso that Balla's word-montage retains a linearity which imposes severe limitations on his language, whereas his clothes-designs were not constrained by a single dimension in this way).

clothing began to exist as an object and an event, something to be removed from a mainly static conception and made mobile.... Balla proposed visual breaks in the shape of the jacket and the neck-tie, arranged on the body to create extremely original optical patterns ... an irregular, active disorder and dispersal. ... The interrelation of the shapes created unstable spiral effects almost as if the function ... was to multiply the colours and volumes of its movements rather than organize them into formal rigidity ... (*FF* 453)

There is a certain irony to the fact that Balla the visual artist is doing no more than resorting *en passant* to a convenient linguistic metaphor for the kinds of effect he ideally wishes to create with his radically Futurist clothing and yet in doing so he puts many of the writers' neologisms in the shade. For, although the result may still be a long way from the neologistic sophistication of a *Finnegans Wake* or even of Arno Holz's *Phantasus* cycle of poetry, it is nevertheless noticeably more inventive than many of the formulaically additive word-formations which Marinetti was content to indulge in during the following two decades.

It was not until the relatively late 'Manifesto of Aero-poetry' ('Manifesto dell'aeropoesia' (first published in the *Gazzetta del Popolo* of 22 October 1931, in connection with the Pesaro Gallery exhibition)) that Italian Futurist literary theory finally began to place any explicit programmatic emphasis on the formation of new compound words. Originally called simply 'Aero-poetry' and claiming to be specifically 'addressed to Poets and Aviators', Marinetti's manifesto was in due course inserted into the introductory section, headed 'Decollaggio', of *L'aeropoema del Golfo della Spezia* (Milan, 1935; cf. *TIF* 1005–8). But this is a misleading caption for the new context: the term *décollage* would have been an accurate description of Balla's technique of deconstructing words, but it hardly describes Marinetti's process of word-formation in this poem.

The manifesto's fourth proposal (of twenty-two) is that the Futurist aero-poet should attempt to 'destroy time by fusing words together, such as "battle-riverbridgewoods"'. That is to say: just as the rapidly shifting landscape below him has the effect of fusing into a new a-temporal visual synthesis for the pilot as he flies at great speed across it, so too the language of the new Futurist aero-poetry should be made to create analogous sensations by no longer linguistically separating the individual, sequentially experienced topographical elements of a normal journey. Speed was to create new compounds, in other words, just as aviation was now successfully synthesizing what the Futurists regarded as the passéist's sentimentalized landscape. Nevertheless, there are severe limitations to the cumbersome multiple compound suggested, the original of which is the uninspiring Italian neologism 'Battagliafiumeponteboscо'. Not only does it suffer from a structural ungainliness; it also fails to compensate for such inelegance through any substantial gain in semantic richness. (It is, in Neumann's sense, no more than a crude 'lexeme-combination', rather than any aesthetically more productive 'lexeme-composition'.) The resultant word does nothing to take its various components creatively in the direction of a portmanteau effect. Rather than the fusion of two or more conventional words to create a richer entity with *new* associations, all we are offered here is the mechanical accumulation of a series of rather obvious, scarcely reconstituted conventional words. In semantic terms, the result adds up to no more than the sum of its individual nouns. But in all probability that is exactly as Marinetti intended. The manifesto-context makes it clear that this kind of limited verbal experimentation is not being resorted to primarily in order to enrich language semantically, in the way that, say, Holz's and Khlebnikov's rich reservoirs of poetic neologisms are.[14] Marinetti's new formation remains, above all, mimetically governed. It pretends to be no more than a means of communicating the actual experience of flight, just as Balla's verbal deformations are in the final analysis no more than serviceable manifesto metaphors for the 'hap-hap-hap-hap-happy clothes' which he envisaged creating.

There was a later twist to the idea that compound neologisms would occur in aero-poetry. In the 1930s some of the Futurists also created the related genre of parachuting poem. One such work by Ennio De Concini, entitled simply 'Paracadute', includes the following lines *Da terra | la sinfonia aereggia | in un vortice | dicasepalazzicittamonti* ('From earth | the aerial symphony | in a vortex | ofhousespalacescitymountains').[15] With such a work, we re-enter the same area of mimesis to which aero-poetry had condemned Futurism (after collage had already emancipated it from such concerns). The underlying assumption of De Concini's poem is the idea that the experience of falling can be reflected in a series of linguistic deformations matching the visual impressions of the

[14] On the language of Holz's *Phantasus*, see Gerhard Schulz, *Arno Holz: Dilemma eines bürgerlichen Dichterlebens* (Munich, 1974). Khlebnikov's neologisms have been systematically analysed in Ronald Vroon, *Velimir Xlebnikov's Shorter Poems: A Key to the Coinages* (Michigan Slavic Materials, 22; Ann Arbor, 1983). Further material on the subject is to be found in Willem Gerardus Weststeijn, *Velimir Chlebnikov and the Development of Poetical Language in Russian Symbolism and Futurism* (Amsterdam, 1983), and V. P. Grigor'ev, *Grammatika idiostilya: V. Khlebnikov* (Moscow, 1983).

[15] *Aeropoesie futuriste di bombardamenti* (Rome, 1941), 44.

fast-accelerating parachutist. This is a potentially fruitful idea—not that literature should be swamped by parachuting poems, but in the sense that the principle could be transferred to the description of other visual impressions gained by motorists, train-passengers, or aviators while travelling at speed. But unfortunately, as is the case with much of Marinetti's aero-poetry, such new linguistic horizons are not explored with any marked enthusiasm or verve.

Both quantitatively and qualitatively, as the above selection of examples suggests, the art of neology plays a relatively minor role in Italian Futurism, whereas it is central to Russian Futurism and also to much of the poetry of the German *Sturmkreis*.

To appreciate the nature and understand many of the limitations of the Italian Futurists' verbal experiments, it is necessary to bear two main factors in mind. First, one must take account of the ingredients of their poetic ideology: that is to say, what they wanted their literature to depict and how. And second, an issue of equally vital importance, one must take note of the characterizing features of the particular language with, and within, which they found themselves obliged to work. Both factors have a strong bearing on the circumscribed verbal innovativeness of the Italian Futurists, when it comes to neology, and on some of the major differences between the linguistic concerns to be found in their writings and those of Futurists in other countries.

In 1922 the French Dadaist Georges Ribemont-Dessaignes published a preface to the French translation of Ilya Zdanevich's *lidantYUfAram* (*LedentYU as a beAcon*): *Ledentu le Phare. Poème dramatique en 'zaoum' d'Ilia Zdanevitch dit Iliazd. (Précédé d'une préface en français de Georges Ribemont-Dessaignes)* (Paris, 1922).[16] In this, he drew attention to the inherent appropriateness of engaging in certain 'transrational' linguistic experiments in the Russian language. This preface is a particularly important document from our present point of view. It draws attention to a largely underestimated aspect of literary Futurism—namely, the relevance of the particular language used by a writer to the practicability of the linguistic reforms demanded by Futurist theory. It indicates, in other words, the kind of effects that could realistically be expected to be achievable. It is also of some significance because it is now generally assumed that many of the passages in the pamphlet originate from Zdanevich (alias 'Iliazd') himself—which is quite probable, since the poet had moved from Russia to Paris in the early 1920s.[17]

It is in the following excerpt that Ribemont-Dessaignes seeks to relate

[16] Although Vladimir Markov (*Russian Futurism: A History* (Berkeley, 1968), 440) attributes this piece to the year 1947, it was actually first published in 1922. For details, see *Georges Ribemont-Dessaignes. DADA. Manifestes, poèmes, articles, projets (1915–1930)*, ed. Jean-Pierre Begot (Paris, 1974), 172.

[17] Zdanevich's involvement in the writing of this material is discussed in Markov, *Russian Futurism*, p. 357, which explains Markov's decision to include what on the surface looks like the work of a French Dadaist in his subsequent Futurist *Manifesty* anthology.

radical language-orientated poetic experimentation (he is talking specifically about the semantics of 'transrational language' (*zaum'*)) to the kind of verbal repertoire the poet has at his disposal for such an enterprise:

> *Zaum'* is a language with a Russian appearance whose words and whose onomatopoeias are such that they lend themselves to being enriched semantically by various acoustically related words. Invented by the Russians, it has an inevitably Russian appearance. But there could be a French *zaum'*, a German *zaum'*, or an English one, or even an international one, something which would, however, diminish its accessibility. Russian is perhaps *the most supple language, the one richest in interferences, in opportunities for sonal and semantic games*. It seems to soak up associations and germs of meanings coming from outside itself. That is to say that the shape of its sounds is infinitely more expressive in its representation of the meaning of a word and of the verbal object in the Russian language than is the case with French or English, for example. (*Manifesty*, p. 181, my italics)

Unfortunately, Ribemont-Dessaignes does not elaborate on what specific linguistic features are implied by the 'suppleness' of the Russian language, but clearly its segmentability, its facility in building new words by the addition of affixes and the process of compounding, and, not least, the readiness of its users to do just that, are all characteristics which noticeably distinguish Russian from the rather strait-jacketing Italian which Marinetti and his followers had at their disposal.

In *Futurism: The Story of a Modern Art Movement. A New Appraisal*, Rosa Trillo Clough makes a similar point about the divergent nature of the various languages of Futurism when she compares the defamiliarizing quality of certain linguistic innovations in Armando Mazza's poem *Altare* (*Altar*) with her English translation of them. The original reads:

TRAMONTO ALTARE
tutte le torce e tutti gl'incensi accesi
spiumío ventilante di fiati di campane lontane
fumea cerula globulare zuccherina che si fonde
lievezza di spume nuvolari
lubricità sfrangiata d'aniline
cigolío del sILENZio brunito
come un immenso fermaglio che si chiuda[18]

The English rendering of the poem which she offers runs as follows:

sunset altar
all the torches and all the incense aflame
windswept foaming of breath of distant bells
blue spherical sweet vapors which dissolve
lightness of cloudy spray

[18] *Firmamento* (Milan, 1920), p. 43.

unravelled slipperiness of blue
creaking of the silence burnished
like a huge closing brooch.

She notes that her version scarcely does full justice to Mazza's language because 'compound words, which are not unusual in English, are extremely contrary to the nature of the Italian language'.[19] The difference between the phrases 'Tramonto altare' and 'sunset altar' is the kind of thing she presumably has in mind.

In her *Geschichte des Futurismus*, Christa Baumgarth remarks at one point 'that the Italian language proves to be highly unsuitable for these kinds of experiments', that is the sort of linguistic reforms which Futurism seemed to be advocating, but which she finds more fully in evidence in the poetry of Stramm, Behrens, and the Russian Futurists than in Italian literature.[20] Both German and Russian, it is implied, are by dint of certain structural features of word-formation better suited to particular kinds of neologistic experiment than the Italian language. And as even a rudimentary formal comparison of the various languages of Futurist literature shows, there is much truth in this. They do not all lend themselves to the same kinds of reform or experimentation.

The languages of the world can be taxonomically divided into three main types: *isolating*, *agglutinating*, and *inflecting*.[21] In an isolating language, of which Vietnamese is an even purer example than the usually cited Chinese, all words are invariable. Consequently, the number of morphemes in such a language is virtually the equivalent of the number of words in the entire repertoire. This is another way of saying that the language cannot form compounds easily and that individual words are not created from morphemic sub-units.

In an agglutinating language, such as Turkish, words are typically composed of a sequence of morphs with each morph representing one morpheme. However, as far as I can see, this grouping has little bearing on the linguistic interests of the Futurists (indeed, it might be argued that, from their point of view, such a system already represents a lesser degree of redundancy than prevailed in the languages they were seeking to reform in the name of poetic expressiveness). Rather, it is the category of inflecting languages which is vital to an understanding of literary Futurism, for it is the difference between them and non-inflecting ones which accounts for the difficult passage of Futurist poetic theory between one culture and another.

Inflecting (or 'fusional') languages cannot normally be segmented into morphs, or, as John Lyons puts it, 'only . . . at the price of arbitrariness, incon-

[19] *Futurism: The Story of a Modern Art Movement. A New Appraisal* (New York, 1961), 162.

[20] *Geschichte des Futurismus* (Reinbek, 1966), 141.

[21] See John Lyons, *Introduction to Theoretical Linguistics* (Cambridge, 1968), 187 ff.

sistency and the proliferation of allomorphs'.[22] This is why the modern linguist has some difficulty in accounting for the various declensional endings of, say, Latin words as if they were 'allomorphs'. But this is not the issue which is of primary relevance to literary Futurism in its various national forms. Rather, it is the sheer difference between the kinds of innovative poetic behaviour possible in an inflected language and not possible in an uninflected one. Much Futurist theory would appear to be above all targeted upon the reform of inflecting languages. Indeed, it might well be argued that some of the movement's main manifesto-proposals, such as putting verbs in the infinitive or dropping redundant affixes, were really attempting to shift naturally inflecting and segmentable languages in the direction of isolating ones. (It has even been suggested that a familiarity with Chinese may have influenced Marinetti in this respect, as it most certainly influenced Ezra Pound.) There is a paradox here, in the sense that many of the proposals emanating from Marinetti's Italian Futurist camp either do not relate very closely to the actual conditions of the Italian language or at best effect in it reforms which look like no more than compromises with radicality when compared with what is possible in more segmenting languages.

There are, in fact, two main structural features to the various languages of literary Futurism that are of major relevance to the question of which adapt themselves more readily to innovative procedures. First, there is the question of *inflection*. As an inflected language, Russian is, as Baumgarth points out, inherently more able to lend itself to certain kinds of truncation in the cause of reforms aimed at destroying syntax, since in such a context syntax/grammar is largely registered through inflectional endings. (This was why MacNeice expressed the opinion that Auden was making things difficult for himself by trying 'to attain in English, an uninflected language, effects that would be comparably easier in other languages and would produce more linguistically sophisticated results'.)

Tellingly, when Shershenevich attempted to offer the Russian readers of his *Futurizm bez maski* (*Futurism without a Mask*) a digest of Marinetti's 'Founding and Manifesto of Futurism' ('Fondazione e Manifesto del Futurismo' (1909)), he attributed to the Italian the words 'Down with words in the nominative case and in the infinitive'. As Anna Lawton points out, this passage only makes sense if viewed as Shershenevich's 'Russification of Marinetti's declaration, inasmuch as in Italian there is no nominative case'.[23] Certainly, it is an indication of the

[22] Ibid. 189.

[23] *Futurizm bez maski* (Moscow, 1913), 47. Anna Lawton considers Shershenevich's mediation of Italian Futurism to contemporary Russia in Chapter Two of her *Vadim Shershenevich: From Futurism to Imaginism* (Ann Arbor, 1981). She follows her observation about Shershenevich's 'Russification of Marinetti' (p. 49) with the point that the former in any case totally misrepresents Marinetti's position on verbs in the infinitive (something which he in fact advocated). This is taken to imply 'lack of familiarity', on Shershenevich's part, with the Italian manifestos in which the issue was discussed.

fact that the Russian language is open to linguistic effects which might, in a sense, be thought of as capable of out-Marinettiing Marinetti himself.

The second factor, apart from inflection versus non-inflection, is that of the comparative *segmentability* of the individual languages of Futurism. It is clear that German and Russian display a higher degree of segmentability than is to be found in Italian—or English, for that matter. One only has to compare the original Russian of what is the most famous Russian Futurist poem of all—'Zaklyatiye smekhom' ('Incantation by Laughter')—with the ungainly English coinages resorted to by its various intrepid translators, in order to realize that Khlebnikov was operating in a language where word-formation was a relatively protean activity, something that is less the case in English.

Khlebnikov's poem, seen by many as the foundation-stone of Russian Futurist poetry, reads as follows in the original:

> О, рассмейтесь, смехачи!
> О, засмейтесь, смехачи!
> Что смеются смехами, что смеянствуют смеяльно,
> О, засмейтесь усмеяльно!
> О рассмешищ надсмеяльных — смех усмейных смехачей!
> О, иссмейся рассмеяльно смех надсмейных смеячей!
> Смейво, смейво,
> Усмей, осмей, смешики, смешики,
> Смеюнчики, смеюнчики.
> О, рассмейтесь, смехачи!
> О, засмейтесь, смехачи!
>
> (*SP* ii. 35)

In his commentary on the derivational structure of the poem, Raymond Cooke remarks: 'It is clear . . . that the building of words from [such] roots could not take place without . . . the suffixes and prefixes which abound in Russian.' (The general pattern in Russian, it might be added, is that prefixes are very frequently attached to verbal roots to create new words, whereas suffixation is less pronounced, no doubt partly because the end of the verb, the element following the root, is usually reserved for conjugational and tense-indicating signifiers. With nouns, on the other hand, suffixation is another dominant means of creating cognates, to which declensional endings then still have to be added.) 'This [word-building] is evident to a large degree in "Incantation by Laughter" . . . where one can find nominal (*smekhachi*, *smeyunchik*), verbal (*issmeysya*, *usmey*, *smeyanstvuyut*) and adjectival/adverbial (*nadsmeynykh*, *smeyvo*) examples.'[24]

[24] Raymond Cooke, *Velimir Khlebnikov: A Critical Study* (Cambridge, 1987), 71.

The process of attaching one of the most common adverbial endings ('-no') to the root, or a prefixed version of it, creates new words with easily inferable meanings, such as 'laughingly' (*smeyal'no*) or 'mock-laughingly' (*usmeyal'no*). The ending *-khach*, signifying someone who does something, creates a neologism for 'laugher'; and if the root is modified by adverbial affixes (e.g. the infix '-anstv-' of *smeyanstvuyut* functioning as an intensifier), then clearly various highly differentiated connotations to the general idea of laughter will result. Khlebnikov's neologisms are judiciously distributed within the poem so as to frame the new coinages within a context of familiar words. Thus, there is only one neologism (the noun for 'laugher') in the first two lines. Even after this, the interweaving of neologisms, based on standard formative principles and familiar words, means that the poem never becomes opaque, even though only about half the words it uses belong to the repertoire of conventional Russian.

Compared with the simple examples of neology in Italian Futurism, where formations of the additive kind largely result in the neologisms thus created staying within the same grammatical category as their sum parts, here neologism often becomes a matter of shifting grammatical categories. This is in itself a phenomenon which surely suggests that the 'destruction of syntax', one of the Italians' declared principal aims, remains incompatible with the kind of word-creation Khlebnikov is engaged in. For effects of this kind to be meaningful, the reader has to be aware of the grammatical origins of the constituent morphemes and at the same time their new syntactical status, as indicated by the new affixes.

It is an instructive exercise to set this classic of Russian Futurism within a linguistically comparative framework by juxtaposing it with the various attempts to translate it into English, a language of a structurally different kind. What emerges is evidence of the strong causal relationship between the type of effects achieved by Khlebnikov and the structural properties of the Russian language. Attempts at translating the work from an inflected and segmentable language like Russian into English (an uninflected and relatively 'unsupple' language, when it comes to word-formation) readily highlight operative differences.

Markov includes the following English translation of the poem in his *Russian Futurism*.

O you laughniks, laugh it out!
O you laughniks, laugh it forth!
You who laugh it up and down,
Laugh along so laughily,
Laugh it off belaughingly
Laughters of the laughing laughniks, overlaugh the laughathons!
Laughiness of the laughish laughers, counterlaugh the Laughdom's laughs!
Laughio! Laughio!

Dislaugh, relaugh, laughlets, laughlets,
Laughulets, laughulets.
O you laughniks, laugh it out!
O you laughniks, laugh it forth!

The Proffers' *Ardis Anthology*[25]—presumably because of the sheer problem of virtual untranslatability—offers a number of English versions of the poem and suggestions for the generation of further ones. This is how two of the various *Ardis Anthology* versions begin:

O laugh it round, you laughsters!
O laugh it up, you laughsters! . . .

O guff it out, you guffsters!
O guff it up, you guffsters! . . .

Since, once the dominant root has been settled upon, the process of working through the neologistic incantatory pattern becomes somewhat mechanical, they then offer a series of further suggestions, by doing no more than indicating how the first two formulaic lines might run:

O gig it out, you gigglers! . . .
O chuck it out, you chucklers! . . .
O chort it out, you chortlers! . . .

Even when the chosen English words are evidently all constructed upon the 'laugh'- root, as is the case with such coinages as the 'laughniks', 'laughathons', 'laughiness', and 'laughulets' offered in Markov's translation,[26] the registers and tones of these derivational forms have an on the whole awkward feel to them, when set alongside the wittier Russian neologisms they are intended to render. (However, it should be borne in mind that the fundamental problem rests, not with the translators, who have on the whole shown remarkable ingenuity in their attempts at putting the poem into English, but with certain intractable language differences.) Whereas the entire Russian poem is constructed upon formations based on the morpheme *smekh*, a Russian root,[27] an unbounded

[25] *The Ardis Anthology of Russian Futurism*, ed. Ellendea Proffer and Carl R. Proffer (Ann Arbor, 1980), 21. Cf. Nils Åke Nilsson, 'How to Translate Avant-garde Poetry: Some Attempts with Xlebnikov's "Incantation by Laughter"', in Nilsson (ed.), *Velimir Chlebnikov: A Stockholm Symposium* (Stockholm Studies in Russian Literature, 20; Stockholm, 1985), 133–50.

[26] *Russian Futurism*, pp. 7 f.

[27] Victor Erlich expresses the consensus view, when he describes the neologisms of 'Incantation by Laughter' as 'newly-coined derivatives from the root *smex*' (*Russian Formalism: History—Doctrine*[3] (The Hague, 1969), 46). Yet it is noticeable that Khlebnikov has in places shrunk the root more radically to the shorter form *sme*. In other words, we could be dealing with an example of what Ronald Vroon in other contexts calls a 'pseudo-morpheme' (*Velimir Xlebnikov's Shorter Poems*, pp. 160, 166). Looking at the structure another way, and one which is on the whole preferable to Erlich's formulation, Peter Urban declares that 'the poem consists of a series of regular Russian derivations from the root "sme" or the word-stems "smekh", "smei" and "smesh"' (*Velimir Chlebnikov: Werke*, i. *Poesie*, ed. Urban (Reinbek, 1972), 394).

morpheme without any distinctly verbal or nominal associations, its counterpart, the English word 'laugh', looks less like a root than already a self-contained verb, and hence something radically different from Khlebnikov's starting-point. Anything added to such a verbal element, to make it into a noun or an adverb, to change it into another verb with a different meaning, or to make it adjectival, is bound to result in a very different kind of English coinage from the ones in the Russian original.

An added difference stems from the fact that there is a greater eccentricity to the English formations than the Russian ones. Certain English words in these tentative translations are evidently formed by a process of morphological analogy, just as Vroon and Cooke have demonstrated many of Khlebnikov's neologisms to be modelled on other pre-existing formation patterns.[28] But the problem lies in the fact that some of the English models can hardly be deemed 'canonical', in Vroon's sense. Thus we have 'laughnik', presumably on the analogy of 'beatnik' or 'beachnik' ('refusenik' had yet to enter the English language at the time of this translation); 'laughathon' based on 'marathon'; and 'laughadors' on the analogy with 'matadors' or 'torreadors'. To match Khlebnikov's inventiveness, in other words, the unfortunate translator may find himself compelled to journey far afield—to such idiolects as surfer slang, loan-words from Spanish, and false etymologies of a far more stilted kind than the poet's own ones. Some of the translations suggested also engender reader-response difficulties not presented by the original. For example, no English reader of the first line would be able to understand the meaning of the exhortation 'gig it up' or of the neologism 'chort' in isolation; the implications of the invented verb will remain unclear until one gets to 'gigglers' or 'chortlers', which will then supply the root-meaning retroactively. On the other hand, the Russian root, already clearly identified in the poem's conventional use of the verb 'rassmeytes'' will give a palpable semantic derivation to all of the neologisms in a much more effective unifying way.

I have not been able to locate any Italian translations of this poem, and, tellingly, it was also one of the few works quoted in full not to be translated in Jean-Claude Lanne's *Velimir Khlebnikov, poète futurien* (Paris, 1983), the major study of the poet's work in a Romance language, i.e. an uninflected language less rich in affixial constructions. But the various versions in Peter Urban's German parallel-text edition[29] show how able German is, as another highly

[28] Thus, in the case of the poems 'Ploskost'' ('Flatness'), 'Lelyanoyu nochi . . .' ('Singing Nights . . .'), 'Eto parus rekacha . . .' ('That is a Sail of the Stream . . .'), and 'Morskoy bereg' ('Sea Shore') (*SP* iii. 341, 73, 202, 281), Vroon shows the way in which a whole series of nouns ending with *-avl'*/*-avel'* are modelled on the word *zhuravl'* ('crane'), also the title of one of Khlebnikov's best-known poems. See also Cooke, *Velimir Khlebnikov: A Critical Study*, pp. 96 ff.

[29] Cf. the various German translations of the poem by Hans Magnus Enzensberger, Rolf Fieguth, Franz Mon, Klaus Reichert, and Gerhard Rühm, in *Velimir Chlebnikov: Werke*, ed. Urban, i. 19–23. In contrast to the flexibility of Russian and German, French shows the same resistance to inventive

segmented language much closer to Russian in its word-forming abilities than either French or English, to reflect the structuring principles of Khlebnikov's Russian original.

If in Point Three of *Sadok sudey*, some of the leading Russian Futurists are able to claim: 'the role of . . . suffixes has clearly been made conscious by us' (*Manifesty*, p. 52), then this is in no small part due to the influential role played by suffixes and prefixes in the particular language with which they are working. Given, by contrast, the nature of Italian, it is difficult to imagine how Marinetti or any of his followers would ever have been able to make such a claim, had they even tried to take their linguistic experiments more decisively in this direction. On the other hand, the German language allows almost as much 'flexibility' for neology as Russian does, and the self-styled *Wortkünstler* (word-artists) of the *Sturmkreis*, in many respects a counterpart to Cubo-Futurism, were able to be as innovative as their Russian opposite numbers.

Just as Markov describes 'Incantation by Laughter' as drawing on structures 'inherent in the [Russian] language',[30] one could say the same about the neologisms coined by the German *Sturmkreis* poet August Stramm in his poem 'Freudenhaus' ('Brothel'), for the poet is here creating similar effects, again ones only possible in certain languages:

Lichte dirnen aus den Fenstern
Die Seuche
Spreitet an der Tür
Und bietet Weiberstöhnen aus!
Frauenseelen schämen grelle Lache!
Mutterschösse gähnen Kindestod!
Ungeborenes
Geistet
Dünstelnd
Durch die Räume!
Scheu

compounding as Italian. In 'Les Caractères du "Futurisme français" à travers l'œuvre de Pierre Albert-Birot' (*Présence de F. T. Marinetti: Actes du Colloque International tenu à l'UNESCO*, ed. Jean-Claude Marcadé (Lausanne, 1982), 160), Giovanni Lista discusses a poem by Albert-Birot in which he detects marked signs of Marinetti's influence (his journal *SIC* was, after all, instrumental in publishing various Futurist writings in French translation at the time). The poem, 'Derrière la fenêtre' ('Behind the Window') (*SIC*, 3 (Mar. 1916)), contains a number of simple compound neologisms of the kind to be found in Italian Futurism: such formations as *Pluiedegrisdepluie* ('Rainofgreyofrain') and *Untramwayéléctrique* ('Anelectrictram'). In the sixth issue of *SIC* (June 1916), Albert-Birot asks: 'What can syntax do, above all in an analytical language? It can connect and bind one's utterance and express it in a complex way, but it creates a complete lack of harmony between the movement of the utterance and that of the underlying thought. One can mentally grasp two or three images in a second, but it would require thirty seconds to express just one in normal language.' Lista sees Severini as the main Futurist inspiration for such ideas, linking notions of dynamism, perception, and a new poetic language.

30 *Russian Futurism*, p. 7.

Im Winkel
Schamzerpört
Verkriecht sich
Das Geschlecht!

Patrick Bridgwater's English rendering of this linguistically highly innovative poem reads:

Lights wench out of windows
contagion
straddles at the door
and offers moaning whores for sale!
Women's souls shame shrill laughter!
Mothers' wombs yawn child-death!
Unborn things
spook
mustily
through the rooms!
Aghast
in a corner
shame-morrified
cowers
sex![31]

While *DU/Liebesgedichte*, the collection in which this poem originally appeared, was being prepared for publication, Stramm commented in some detail on his neologism 'schamzerpört' (l. 13). The word is a portmanteau formation, perhaps more accurately segmentable into 'shame-de-gnation' (a truncated version of 'shame/de[stroy]/[indi]gnation'), though generally well captured in Bridgwater's hyphenated 'shame-morrified', with the second part's overtones of 'mortified' and 'horrified' (the residual hyphen no doubt a gesture to the limitations of English word-formation, even in an experimental context).

Stramm has the following to say about this particular word in a letter of 22 May 1914 to his publisher Herwarth Walden: 'I should especially like to draw attention to the penultimate line [i.e. in the proof-version, not in the final printed one] in which the word "schamzerpört" has been altered by the printer to "schamzerstört".' Stramm's complaint is that the printer's substitution (meaning 'destroyed by shame') is a relatively conventional German compound, compared with the poet's coinage, with the result that such a well-meant offering tends to trivialize the bold effect that Stramm was aiming at. Stramm's

[31] August Stramm, *Das Werk*, ed. René Radrizzani (Wiesbaden, 1963), 14. August Stramm, *twenty two poems*, trans. Patrick Bridgwater (Wymondham, 1969), n.p.

reaction to the dilution of his innovative coinage is to remark, presumably disingenuously (though one can never be too sure with German writers):

I don't know whether this can be attributed to an erroneous reading on the printer's part or whether it was a result of his feeling for language. Be that as it may, my word 'schamzerpört' says more to me than this other word. Shame [*Scham*] and indignation [*Empörung*] are struggling with one another, and shame crushes [*zerdrückt*] the other concept. Even 'schamempört' does not manage to express this; in any case, my intuition is that the essence of the word 'empören' does not lie in the prefix 'em-', which at best has an explanatory value for the grammarian, my feeling is that the idea of sensing indignation really lies in the 'pören' part, or more simply, ultimately in the single sound 'pö'. If you remove the two dots above the sound, the whole idea will collapse! On these grounds, I consider 'schamzerpört' to be the only combination that really expresses all that. I don't trust printers who think![32]

In this letter, Stramm gets very close to the kind of linguistic, or perhaps one should say, pseudo-linguistic, theorizing that is often to be encountered in Kruchenykh's and Khlebnikov's etymological reflections concerning the language of poetry. In fact, the latter frequently attributes semantic significance to minimal parts of words which do not, by normal linguistic criteria, appear to have a morphemic value. What Ronald Vroon[33] refers to as Khlebnikov's progress towards the 'total semantization' of the phonological components of language is revealed in many of the poet's commentaries on word-formation. Indeed his conception of 'internal declension' (one of the pillars of the linguistic theory underlying much of his more inventive word-formations and also Mayakovsky's) is based on just such a practice. In *Uchitel' i uchenik* (*Teacher and Pupil*) we are presented with the argument that 'the declension of a stem should give the resulting words meanings which are opposite'. The minimal pair *bobr/babr* is offered as one example of this eccentric notion of 'declining' the stems of nouns, or rather of viewing morphemic contrasts existing between certain similar words in modern Russian as the result of a supposed earlier act of declension.

bobr [beaver] and *babr* [Siberian tiger], signifying a harmless rodent and a terrifying beast of prey, and formed from the accusative and genitive cases of the common stem 'bo',[34] show by their very structure that the *bobr* is to be chased, to be hunted as game, and the *babr* is to be feared, since here man himself may become the object of the hunt

[32] *Das Werk*, p. 427.

[33] Vroon, *Velimir Xlebnikov's Shorter Poems*, p. 18.

[34] Whether one views 'internal declension' as a matter of morphological analogy, in the linguistic sense, or poetic derivation is essentially a matter of interpretive framework. Clearly, Khlebnikov believed that he could legitimately posit an analogy between the meanings expressed by vowel changes in nominal declensions and the vowels at the core of certain, usually monosyllabic, words. Thus, 'if the genitive case answers the question "whence", and the accusative and dative cases the questions "whither" and "where"' (*SP* v. 171), vowel changes in minimal pairs which echo these declensional endings will, according to this theory, also reflect oppositional meanings.

on the part of the beast. In this instance, the most simple form, by changing its case, changes the meaning of the verbal construct. (*SP* v. 171)

Clearly, Stramm's ruminations on the potential morphemic significance of the two phonemes *p* and *ö* operate on the same level of 'pseudo-derivational procedures' (to use Vroon's phrase) as Khlebnikov's 'internal declension' theory or his notion (discussed later) that the first letter of a word already determines its meaning. Both are engaging in what in strictly linguistic terms might be thought of as an act of over-semanticization. Obviously, the sound *pö* on its own possesses no conventionalized meaning in German. (Even without its umlaut, it would not mean what Stramm wants it to mean!) In one sense, Stramm here goes even further than Khlebnikov when he attributes ultimate semantic value to the phonological marker of the two dots of the umlaut itself: 'if you remove the two dots above the sound, the whole idea will collapse!'

Yet by contrast to the fairly radical theoretical assumptions being hinted at in Stramm's letter to Walden, the actual neologism in his poem 'Freudenhaus' still reveals a relatively conventional structure. The first part of the triple compound 'schamzerpört' scarcely triggers off innovative expectations at all. In such an example, one merely encounters a familiar German use of the noun as first part in a compound, functioning with adjectival connotation. Numerous formations of this kind are permissible (*schamrot*, *schambedingt*, *schamentleert*); indeed, Dürsteler gives a whole host of examples of such formations in his study of Expressionist coinage. But, even though he rightly presents them as examples of neology, such effects are less radical than, say, Khlebnikov's creations—for instance, his unorthodox attaching of an adverbial suffix (*-no*) or a nominalizing one (*-khach*) to a word that cannot normally bear it, or his creation of the Russian word *budetlyanin* (meaning 'Futurist') from *budet* (third person singular of the future of the verb 'to be') and the suffix *-lyanin* (designating a person according to geographical or social milieu), hence the translation 'future-dwellers' to suggest the connotations. By contrast, if *schamrot* is possible in German, then *schamgrün* is not difficult to construct, or, for that matter, the printer's *schamzerstört*.

What a comparison between German *Sturmkreis* and Russian Cubo-Futurist word-creation makes very clear is that many new words may, in absolute terms, be innovations, and yet they may lack the boldness of some other ungrammatical forms of neologism. Thus, while many of Dürsteler's examples (*angstallein*, *frühlingsmüde*, *augenlodernd*, *faustgeballt*, *wehlachend*, *gähnmüd*) may certainly be new, in the sense that these are the first attested uses of such combinations, they are formed on the basis of such widely accepted and followed principles of morphological analogy that this allows them to integrate easily into the German language, which at least suggests that there is much less of a defamiliarizing or language-enriching effect to such innovations.

In his *Sprache: Strukturen und Funktionen*, E. Coserik makes a useful distinction between 'norm' and 'system' which may help differentiate the various kinds of neologisms being considered here. Language change (*Sprachwandel*), according to Coserik, involves 'a reaction against the norm', but one which remains 'permitted by the system'.[35] Many of the Expressionist neologisms discussed by Dürsteler remain permitted by the system, whereas Stramm's 'schamzerpört' and many other *Sturmkreis* and Cubo-Futurist neologisms are in fact a reaction against the *system*, not just *norm*-breaking, *system*-countenanced new formations.

German compounds formed from combinations of noun + adjective, noun + participle, adjective + noun, and so forth are part of a far more accessible field of constructible formations than the affix + noun (or verb) combinations which have such a bold effect in both German and Russian. The fact that, even if they do not hitherto exist, compounds coined by adding the nominal prefix *scham-* are relatively easy to construct in German, is presumably the equivalent of the 'suppleness' that Ribemont-Dessaignes attributes to the Russian language. The really innovative part of Stramm's word only comes with its second and third components: the one commonly a verbal prefix ('zer-') with negative, destructive associations, and the other, looking like a verbal root *pör* (with past participial *t*-ending, giving it an adjectival function). But this is a false impression. The normally reflexive German verb *sich empören* does not consist of a root *pör* with an infinitive ending *-en* and a prefixal *em-*. The verb is formed from the adverb 'empor', with umlauting and verbal endings. Originally, in Middle High German, the word 'empor' itself was written as two words: 'in bore' (i.e. 'in die Höhe', upwards, to the heights), which at least segments the modern morpheme 'empor' down into two further extinct morphemes, in the way that Stramm's reading of his neologism does (though the second one does not relate directly to Stramm's semanticized *pö*). Although he explicitly postulates a prefix *em-*, it is not possible to be sure whether Stramm was familiar with the word's actual Middle High German etymology. But I find it difficult to imagine how an educated German with his markedly linguistic interests[36] could have failed to be aware of the fact that the splitting of 'empor' into two units was not meaningful in modern German. The most plausible conclusion, on the

[35] E. Coserik, *Sprache: Strukturen und Funktionen*[3] (Tübingen, 1979), 55–9. Also 'Sistema, norma e "parola"', in *Studi linguistici in onore di Vittore Pisani*, i (Brescia, 1969), 235–53.

[36] Jeremy Adler has done much to bring out the link between Stramm's neology and the unearthing of older meanings dormant in modern German words. See 'On the Centenary of August Stramm: An Appreciation of *Geschehen*, *Rudimentär*, *Sancta Susanna* and "Abend"', *Publications of the English Goethe Society*, 44 (1974), esp. pp. 22–6, and '*Urtod*: An Interpretation', in J. D. Adler and J. J. White (eds.), *August Stramm: Kritische Essays und unveröffentlichtes Quellenmaterial aus dem Nachlaß des Dichters* (Berlin, 1979), 84–98, esp. pp. 88, 92, and 94. Stramm's essay 'Deutsche Titel' (Adler and White (eds.), *August Stramm*, pp. 153–7), furnishes pertinent evidence of the poet's professional interest in the possibilities of more elegant forms of word-formation in bureaucratic German.

basis of circumstantial evidence, is that he is consciously working with what, from our own contemporary point of view, looks like a false, 'poetic' etymology, in order to create new suggestive effects. But, as is also the case with many of Khlebnikov's formations, what may on the surface look like a neologism at the same time amounts to a returning of language, whether consciously or not, to its earlier roots.[37] In this respect, Futurism is looking back to the remote past rather than forward to the Italian Futurists' future, although the process is janus-headed: the resurrection of past roots and lost etymologies will at the same time furnish the language of later generations.

Whereas Khlebnikov embeds much of his linguistic thinking on word-formation (which, even if we are inclined to see it as false etymologizing, he actually did believe in[38]) in the dramatic *Teacher and Pupil* dialogue or in the actual structures of his poems, Stramm's few observations on his poetic neologisms remain buried in correspondence with his publisher. As a result, in the Russian's œuvre there is an in-built poetology, whereas in Stramm's poem the significance of the neologism must remain more dark.

Something to which no reference is made in the letter to Walden, but which remains an equally important feature of this work, and of much of Stramm's poetry in general, is the way in which many of the principal neologisms are created by shifting words from one grammatical category to another—an effect which is also not easy to achieve in non-inflected and less 'supple' languages than German and Russian. Dürsteler suggests that neology by means of shifting words from one grammatical category to another is an invention specifically of the *Sturmkreis* Expressionists,[39] but, as Ronald Vroon's study of the coinages in Khlebnikov's shorter poems shows, it is also a central feature of the latter's

[37] Heinz Peter Dürsteler, *Sprachliche Neuschöpfungen im Expressionismus* (Thun, 1954), 62 ff., which is mainly concerned with the work of the *Sturmkreis* poets, records four main areas of intensive language renewal during this period: (1) genuine neologisms (compounded and derivational); (2) modifications of present vocabulary (usually the removal of affixes); (3) recourse to archaisms; and (4) regionalisms. These categories are, however, by no means watertight. Sometimes the third (archaisms) can be arrived at, even fortuitously, by experimentation within the first two areas.

[38] On Khlebnikov's belief in his more speculative theories, see Vroon, *Velimir Xlebnikov's Shorter Poems*, p. 169. In *The One and a Half-Eyed Archer*, trans. John E. Bowlt (Newtonville, Mass., 1977), 56, Livshits declared that it was impossible 'to draw a line between Khlebnikov's poetical work and his philological researches'.

[39] Dürsteler (*Sprachliche Neuschöpfungen im Expressionismus*, p. 45) finds that 52% of new adjectives in German Expressionism are formed from other parts of speech. In particular, he indicates how widespread the formation of abstract nouns is from pre-existent adjectives. Dürsteler also advances the claim (p. 74) that neology by means of shifting words from one grammatical category to another even represents 'a new form of word-formation, invented by the Expressionists'. In the light of the now large body of scholarly work on Russian Futurist language-creation, this can be seen to overestimate the innovativeness of what was happening in Germany, particularly in the *Sturmkreis*. Apart from Vroon's work, Assya Humesky's 'Neologisms of Two Russian Futurists: Igor′ Severjanin and Vladimir Majakovskij', diss. (Harvard, 1962) and *Majakovskij and his Neologisms* (New York, 1964) show how widespread these processes were among both the Ego-Futurists and the Cubo-Futurists.

language experiments. And it is certainly a technique which shows how markedly similar some of the experiments of the *Sturmkreis* writers and the Cubo-Futurists were. Those intent on playing down the Futurist aspect of the *Sturmkreis* tend to emphasize differences with *Italian* Futurism, rather than considering the issue in the light of Russian Futurist writing.

Stramm's poem contains a number of examples of word-creation by shifting words from one grammatical category to another. The verb *dirnen* ('to whore') is the first instance of this procedure, with the new verb being derived from a common German noun. Already here an English translator is likely to encounter difficulties in one respect. For the noun *Dirne* means 'prostitute' nowadays, and, if the translator wants a comparable verb, then the most obvious English choice would be 'whore'. Yet whereas 'whore' in English has the appearance of either a noun or a verb, for there is no available marker equivalent to that in the German to show the difference between the two functions, Stramm's word is unequivocally a *verbal* neologism. The German noun *Dirne* has been made into a verb by the simple addition of an *-n* (in this instance, the sign of the third-person plural conjugation of the present tense), and writing it with a lower case initial letter in order to distinguish the form from the normal plural of the German noun, *Dirnen*. The effect of Stramm's line 'Lichte dirnen aus den Fenstern' is consequently much bolder than the English translation 'Lights whore out of windows' would be. There is a greater boldness, something which Patrick Bridgwater's version nicely captures by employing the noun 'wench' as a verb (although it may well be that because 'to wench' exists in English, this sequence could trigger off different associations than the 'aus den Fenstern *dirnen*' of the original).

Another grammatical innovation in Stramm's poem, not this time the result of shifting words from one category to another, is the transitive use of verbs which would not normally be used in this way in standard German. (Norbert Hopster's study *Das Frühwerk Johannes R. Bechers* shows that this is also a recurrent feature of the style of his Futurist phase, as well). The verb *gähnen* ('to yawn')—here with *Kindestod* ('child-death') as either its object or its predicate—is one such instance. And *schämen*, normally used reflexively with the pronominal *sich*, here receives a new transitive function with *grelle Lache* ('shrill laughter') as its object. In both cases, the grotesqueness of a hostile atmosphere is reflected in an unnervingly disjointed grammatical formation.

A further striking feature to the poem's fifth line lies in the fact that Stramm prefers the relatively rare word 'Lache' to the normal German nouns *das Lachen* or *das Gelächter*. Possibly he has opted for this feminine noun for laughter in order to offer a parallel to the word *Träne* ('tear'), a word which lies by implication behind much of the description. Or it could be that he wants a noun that appears to grow more straightforwardly out of the verb, a more 'telegraphic' noun—or even, thinking more radically, one which looks like the source-noun from which the familiar verb might have been derived.[40] Inevitably, Patrick Bridgwater's English translation, using the normal English noun 'laughter', cannot reflect the implications of this choice. Even when Stramm is not resorting to neologisms, he often employs words with a less familiar feel to them, which means that any English rendering of his poetry is likely to level many of its semantic niceties. And since in this poem many of the linguistic innovations and strange words help to create a sense of the hostile and the grotesque, there is a strong likelihood that such an atmosphere will become diluted in translation.

Stramm may not himself have had anything to say about the phenomenon of neology via changing the grammatical category of words (but then, nor does Khlebnikov); but one of his fellow *Sturmkreis* poets, Lothar Schreyer, discussed such a process of word-formation in some detail at the time. Indeed, one generally finds an intensity of linguistic discussion among the *Sturmkreis* writers (though mainly after their mentor Stramm's death on the Eastern Front in 1915) which is unmatched in the manifestos and occasional poetological commentaries of the Italian Futurists and really only has its counterpart among the work of the OPOYAZ[41] theoreticians and some of the manifestos and linguistic observations of the Cubo-Futurists of contemporary Russia.

In his article 'Drama' ('Das Drama'),[42] Schreyer makes a number of points which are as applicable to the poetry being created by Walden's circle as to their theatrical experiments. 'The grammar of everyday language is unimportant for the new work of word-art [*Wortkunstwerk*]', he declares—and by 'Wortkunst' he largely means the art of language-creation.

> Every word can change its grammatical form in such a work of art. The changed form can assume the role previously played by grammatical relationships. Without the context of grammar the forms of relationship between words become isolated values, whose power to interact with one another extends outwards without a specific controlling direction towards infinity. This power does not move in a particular direction, but in all directions. The effect of the one word overlaps with the effect of others. The resultant words exist in no fixed grammatical interrelationship.

Clearly, even though the *Wortkunst* theoreticians have a habit of resorting to hyperbole on the question of the peripherality of grammar and the relative unimportance of the sentence (this is a consequence of their interest in the single word[43]), what Schreyer is describing here is of great significance for the

[40] Dürsteler (*Sprachliche Neuschöpfungen im Expressionismus*, p. 41) sees abstract substantives as a main area of word-creation, cultivated by the Expressionists 'either because the conventional abstract nouns were too weak, or because the requisite abstractions did not yet exist'. Even Stramm's preference for 'Lache' could be a token of dissatisfaction with the nouns *Lachen* and *Gelächter*, in the sense that he wishes a more abstract connotation than these words will bear.

[41] On the connections between OPOYAZ (The Society for the Study of Poetic Language) and Futurist language-experimentation, see Krystyna Pomorska, *Russian Formalist Theory and its Poetic Ambiance* (The Hague, 1968), 21–32.

[42] *Der Sturm*, 7 (1916), 119.

[43] 'The artist's material is not language, but the word', according to Walden ('Kritik der

linguistic aspect of Futurist literature. It concerns the establishment of new structures of relationship within poetry, once the conventional, controlling grammatical matrix has been weakened (though again not, it must be stressed, *destroyed*[44]). Now, words receive their significance within a new kind of poetic constellation. What this in essence implies is that disruptions of grammar are brought about primarily for the sake of creating new semantic relationships. Neology *per se* remains less important to the *Sturmkreis* than it is among the Russian Cubo-Futurists. The main role of new words is as a means to a further end: that of establishing fresh, more complex interrelationships between the individual words of a poem. To put it another way: whereas neology in Russian Futurism is often a matter of enriching the semantic scope of language by means of the individual new formation and then the placing of the newly constructed words in relatively familiar syntagmatic relationships, with the *Sturmkreis* poets more emphasis is being laid on the way neologism creates new semantic interfaces between the various constituent words of a poem. One may, of course, need the context of other words to understand the connotations of some of Khlebnikov's coinages,[45] but that is still a very different thing from the kinds of ambiguous refraction engendered by the *Sturmkreis*'s loosening of grammar's hold.

In his influential 'Expressionist Poetry' ('Expressionistische Dichtung') of 1918, Schreyer has much to say about the creation of new words by means of shifting grammatical functions or pruning compound words to create simpler forms. The argument is set within the framework of his thinking about 'Konzentration' and 'Dezentration', subtractive and permutational creative structures, which for Schreyer are essentially matters of rhythm.[46] Just as the previous manifesto had set out to emphasize the way in which a systematic reduction in conventional grammar would inevitably focus attention on the individual word and establish a new complex of relationships with the other surrounding words, so here, the focus is on the autonomous word as an entity.

The shape of the individual word concentrates the idea behind it into a single word. This concentration of the word is a concentration of both content and form. To give

vorexpressionistischen Dichtung' ('A Critique of Pre-Expressionist Poetry'), *Der Sturm*, 11 (1920), 11). The point was also made in 'Das Begriffliche in der Dichtung' ('The Conceptual in Poetry'), *Der Sturm*, 9 (1918/19), 66. Similar sentiments can be found in Schreyer's 'Das Wort' ('The Word'), *Der Sturm*, 13 (1922), 125 ff. 'The word rules. The word destroys the sentence', Walden even claims in his introduction to *Expressionistische Dichtungen vom Weltkrieg bis zur Gegenwart*, ed. H. W. and P. A. Silbermann (Berlin, 1932), 12.

[44] On Schreyer's recognition of the fact that, no matter how much the hold of syntax is weakened, a grammatical framework is nevertheless a prerequisite for most of the *Sturmkreis* rhythmic effects, see Neumann, *Die kleinste poetische Einheit*, p. 262.

[45] The point is repeatedly illustrated in Vroon, *Velimir Xlebnikov's Shorter Poems*, e.g. pp. 32, 50, 92, 102.

[46] 'Concentration and decentration are forms of rhythmic structure', *Sturm-Bühne: Jahrbuch des Theaters der Expressionisten*, 6th ser. (Berlin, 1918), p. 1.

shape to the word's meaning with a minimum number of phonetic effects, that's the aim. The result is the abbreviation of words.

This abbreviation of words can be a matter of going back to the root:

Bären for *gebären* [to bear]
schwenden for *verschwenden* [to waste]
wandeln for *verwandeln* [to change]

Declensional and conjugational endings can be omitted. The definite and indefinite articles must be removed everywhere where they are not utterly necessary. Word truncations lead to changes in words. These changes in words lead to the formation of new words. Substantives are created out of verbs, verbs from substantives; for example, from the noun 'Kind' one can form the verb 'kinden'.

Surprisingly, Schreyer does not specify just what this new verb might mean. Is a verb formed from such a noun (*Kind* meaning 'child') to mean 'to be a child', 'to behave like a child', 'to produce children', or does it have some other arcane connotation? We are not told.

Instead, Schreyer goes on to posit what he assumes to be a logical connection between this focus on the pruning of certain conventional elements in the individual word and the creation of an abbreviated 'telegraphic style' at sentence level. Nevertheless, as one can see in the case of Russian Futurism, these are really separate phenomena, not to be confused, and certainly not automatic consequences of one another. Khlebnikov can create neologisms by changing the grammatical categories of certain familiar words by means of attaching affixes. In the same way as he formed the word *budetlyanin*, he creates *vremlyanin* ('time-dweller') and *lyublyanin* ('love-dweller'), in each case forming a new noun from a familiar Russian verb. Yet he employs such new words without resorting to the radical extremes of sentence-structure reduction that interest the poets of the *Sturmkreis*. And, by the same token, the 'telegraphic' pruning of the sentence so popular in Italian Futurism clearly did not necessarily require any telescoping of individual words by reducing them to their simpler roots or removing grammatical indicators. The fact that the two processes are considered in conjunction with one another by certain *Sturmkreis* theoreticians is, in fact, merely coincidental to that context. Indeed, even within the *Sturmkreis*, some of the theoreticians do not bring the two issues into any close relationship with one another. Rudolf Blümner, for instance, talks about the 'dissection of phenomena into their sub-forms', a process which he observes in Cubism and which he, like the Russian Cubo-Futurists,[47] is seeking to transfer to the techniques of *Sturmkreis* literature,[48] but he does not relate this process of language-dissection in the case of the individual word to any form of

[47] David Burliuk's essay 'Kubizm' had already appeared in the founding *Poshchechina obshchestvennomu vkusu* (*A Slap in the Face of Public Taste*) (Moscow, 1912), 95–110. The Cubism–Cubo-Futurism relationship is explored in detail in Juliette R. Stapanian, *Mayakovsky's Cubo-Futurist Vision* (Houston, Texas, 1986).

[48] *Der Geist des Kubismus und die Künste* (Berlin, 1921), 118.

telegraphic sentence-structure. Nevertheless, the fact that even some of the *Sturmkreis* theoreticians can consider at all the possibilities of radically changing the shape of individual words is again a consequence of the language with which they are operating.

Many of Stramm's poems, the models for much retrospective *Sturmkreis* theorizing about the directions the new *Wortkunst* should take, give a concrete indication of the new kinds of interrelationships between words that Schreyer merely intimates in his theoretical writings. For example, the war-poem 'Patrouille' ('Patrol'), from Stramm's posthumously published collection *Tropfblut. Gedichte aus dem Krieg* (1919), is very much an illustration of the new kind of structure Schreyer has in mind.

> Die Steine feinden
> Fenster grinst Verrat
> Äste würgen
> Berge Sträucher blättern raschlig
> Gellen
> Tod.[49]

Patrick Bridgwater's translation of this reads:

> The stones malice
> window grins treachery
> branches strangle
> loomy bushes leafrattling feign
> shrieking
> death.[50]

Even the first two lines here are open to a variety of readings, since the loosening of grammar and the lack of any channelling with the help of punctuation tend to move the interpretive possibilities in the direction of the openness Schreyer describes. The initial phrase 'Die Steine feinden' becomes in Bridgwater's translation 'The stones malice'—a version which offers the same admixture of conventional plural noun ('Steine') and neologistic verb, combining at the same time the concrete object and the abstract threat. Although, in Stramm's German, the new verb might to all appearances look as if it had been created out of the noun for 'enemy' (*Feind*), this is not the only conceivable derivation. There also exists the prefixed transitive verb *anfeinden* ('to show hostility to someone'). Hence, another viable interpretation is that 'feinden' is derived from such an affixed verb by a process of reducing it to its root-form (a pattern analogous to Schreyer's example of the formation of a root-verb 'bären' from the more usual prefixed form 'gebären').

However, it is really from the second line onwards that more substantial

[49] *Das Werk*, p. 86. [50] *twenty two poems*, n.p.

ambiguities begin to arise. The most obvious reading is one assuming a $\{N^{subj.} + VB^{trans.} + N^{obj.}\}$ construction, i.e. 'Window grins treachery'. But there a parallelism could be intended between the first two lines of the poem. 'Die Steine feinden' might well have its equivalent in the two words 'Fenster grinst' (after all, *grinsen* does not normally take an object); and the word 'Verrat' could be an afterthought or a sudden interrupting exclamation. That is to say, if the window is grinning, this impression could be deceptive (a 'Verrat'), since the environment is generally hostile for the soldier on patrol, and what might look misleadingly innocuous, could be part of a trap. The issue of potential relationships (of which parts of the poem are the actual Immediate Constituents) is left even more open by the third line, which returns the poem to the pattern $\{N^{subj.} + VB\}$, but this time with the added complication that 'würgen' would, as a transitive verb, normally be assumed to take an object. The reader's expectation of an object may even be held over to the next line, albeit not to be fulfilled there either. Hence complex relationships and tensions are being established between the various sequences. What is at stake is not the creation of neologism as the primary focus of language-revitalization, so much as an undermining of grammatical security as a correlative to the insecurity felt by Stramm's soldiers on patrol.

This remarkably powerful and disturbing pattern of linguistic disorientation is continued in the final three lines of the poem, again to some considerable extent due to the ambiguities engendered by Stramm's use of grammatically duplicitous neologisms. The meaning of the word 'Berge' has been the subject of much debate, with opinions dividing about whether it is a plural noun or an adjective with a plural 'e'-ending attached. Elmar Bozzetti holds that it is an adjective formed from the verb *bergen* (meaning 'to protect').[51] There are precedents for such a construction in Stramm's work. As Neumann points out,[52] the phrase 'Rauscher Wein' in the poem 'Blüte'[53] had originally been written with the more normal adjective 'rauschender Wein' in a draft version, hence it is plausible to assume a process of creating a new adjectival form by means of the removal of the gerundive ending '-end', rather than by the addition of adjectival endings to nouns such as *Berg* or *Rausch*. Bozzetti has coined the term 'Stammadjektiv' or 'root adjective' for such forms.[54] By contrast, H. Maclean, in paraphrasing the poem, suggests a more obvious nominal function with his

[51] Elmar Bozzetti, 'Untersuchungen zu Lyrik und Drama August Stramms', unpubl. diss. (Cologne, 1961), fo. 60.

[52] Neumann, *Die kleinste poetische Einheit*, p. 269.

[53] Stramm, *Das Werk*, p. 29.

[54] 'Untersuchungen zu Lyrik und Drama August Stramms', fo. 61. Dürsteler does concede that new portmanteau-formations (which he refers to as 'Zusammenschöpfungen') can create semantic ambiguities. Thus, of the word 'kriesseln' in Stramm's poem 'Menschheit' (*Das Werk*, p. 6), he writes: 'it is either to be considered as a fusion of *kräuseln* and *rieseln* or as the dialect influence of Low German: kris(s)eln = to scratch' (*Sprachliche Neuschöpfungen im Expressionismus*, p. 60).

mention of 'stones, window, branches, *hills*, bushes'.[55] In fact, there is absolutely no reason why one should not even posit an *enjambement* from the previous line, with a noun 'Berge' acting as the delayed and somewhat surreal object of the verb 'würgen'; such dislocated visions are common enough both in other Expressionist literature and in the painting of the period. One critic, Herbert Lehnert, goes so far as to insert an exegetical comma between 'Berge' and 'Sträucher', defending such a bold move by arguing that a comma is there by implication, by dint of its very absence. The effect of such a comma, however, would be to make 'Berge' unambiguously into a noun, forming the initial part of a list and thus followed by other nouns.[56]

Most of these readings propose *one* grammatical function only. They contend either that 'Berge' is exclusively adjectival or that it is exclusively nominal. What is more, there is often as much dogmatism about the word's derivation as there is about its grammatical role in the poem, for it is usually assumed to be unequivocally derived either from the verb *bergen or* from the noun *Berg*. It may be that Schreyer's ideas (especially those in 'Expressionist Poetry') concerning the creation of new words by shifting grammatical categories have misled commentators into assuming that Stramm the neologist must always simply be changing verbs to nouns or engaging in some other form of grammatical redistribution. Or it may be that some commentators are more the victims of a passéistically grammatical horizon of expectation, or their own particular language, than the poet they are investigating. In fact, as his practice elsewhere indicates,[57] Stramm's effects are usually more subtle than Schreyer's prescriptions. Rather than grammatical transference or even wilful obscurity, they involve highly poetic and thematically motivated forms of polyvalence, and ones very dependent on the uncertainties created by layout (which so often becomes one further substitute for the conventional function of grammar in *Sturmkreis* poetry). It is only the unfortunate translator who may have to disentangle such nets of neologisms by assigning hybrid words unequivocally to one grammatical category or another.

What these examples from Khlebnikov's and *Sturmkreis* practice clearly indicate is that certain types of language-creation are favoured by certain languages; hence, in their dissatisfaction with the language of passéism, the various Futurisms are bound to be moving towards different solutions. If for Kruchenykh and Khlebnikov the 'work of art' was 'the art of the word' ('The Word as such' ('Slovo kak takovoye'), *Manifesty*, p. 59), for Marinetti language was largely a matter of syntax, and hence Futurist telegraphic lyricism would be

[55] 'Expressionism', in J. M. Ritchie (ed.), *Periods in German Literature*, i (London, 1966), 273 f.

[56] *Struktur und Sprachmagie: Zur Methode der Lyrik-Interpretation* (Stuttgart, 1966), 157.

[57] This is one of the main aspects of Stramm's work explored in my 'Aspects of Typography and Layout in Stramm's Poetry', in Adler and White (eds.), *August Stramm*, pp. 47–68.

a matter of restricting the hold of traditional syntax over the sentence. Without wanting to make it into no more than a displacement activity, I would therefore nevertheless want to suggest that Italian Futurist concern with 'telegraphic' forms of poetic experimentation cannot be divorced from what they could (or could not) do to reform language at individual word-level. What is, by contrast, less important for the encouragement of any such experimentation is the degree of linguistic sophistication at the poets' disposal, on a theoretical level.

Clearly, none of the Futurists or *Sturmkreis* writers had access to the central linguistic concepts of the 'phoneme' and the 'morpheme'—central because it was at these levels that so much of the real language-enriching creativity took place. There are, admittedly, moments when some of the Russian Futurists appear to be groping their way towards the notion of the phoneme. But then, in going on to attribute semantic status to individual letters of the alphabet or even suprasegmentals, they show that there is no rigour, in a strictly linguistic sense, to the way in which they segmentalize their poetic language.

In the Italian context, Enzo Mainardi's potentially useful notion of 'molecules of sound' (in the poem of that name: 'Le molecule del suono'[58]) could in the right hands have led to an exploration of a language-generating model based on something equivalent to the phoneme, and hence to the postulation of the difference between meaningful and non-meaningful phonemes, and thus to the morpheme. But this was not to be; the poet's interest in the minimal units of sound is largely restricted to the phenomenon of *sound* as sound, and not to the more technical issue of language as a meaningful set of phonological elements.

Anna Lawton has drawn attention to a letter by the poet Pino Masnata which also approaches the question of the divisibility of languages via the concept of the molecule. 'I realized', Masnata writes, 'that it is possible to construct a new poetic thought by binding substantives with other substantives by means of adjectives and verbs. The same process which occurs in molecules. I can compare substantives to "atoms".'[59] Unfortunately, by positing a similarity between substantives and atoms, Masnata effectively limits the potential helpfulness of his insight, from a linguistic point of view. For, if in physics the atom is the smallest definable unit which in various permutations forms the molecular structure of various substances, then its equivalent in the context of language cannot be the single word, whether that word be a noun, an adjective, a verb, or any other part of speech. The effect of Masnata's metaphor, not

[58] See 'Le molecule del suono', in Viazzi (ed.), *I poeti del futurismo*, pp. 534 f. According to Markov (*Russian Futurism*, p. 347), the notion of 'molecules of sound' also preoccupied Kruchenykh for some time.

[59] Letter to Giovanni Acquaviva, quoted in Mario Verdone, *Che cose è il Futurismo* (Rome, 1970), 45. The letter is discussed in Anna Lawton's 'Russian and Italian Futurist Manifestos', *Slavic and East European Journal*, 20 (1976), 416.

surprisingly, therefore, is to focus attention once more, in a way not untypical of Italian Futurism's general language preoccupations, on the syntactic characteristics of *the sentence*, not on the morpho-phonemic structure of individual words or their constituent parts. Again, though, the reason for this focus is not simply the lack of any adequate technical framework. After all, the Russian Futurists also lacked a sophisticated methodological framework within which to analyse the meaningful ingredients of individual words (apart from the time-honoured ideas of suffixes, prefixes, and infixes). If Khlebnikov could complain that 'there is no science of word creation' (*SP* v. 228), then this could be understood in two ways: not only that the writers did not have any systematic criteria to work with, but that the study of language-formation in general was still in its infancy. But a number of the Russian Futurists, including Mayakovsky, Zdanevich, Kruchenykh, and Khlebnikov himelf, nevertheless intuitively pushed their exploration of language in the morpho-phonemic direction, the main reason being, as we have seen, that they had a differently structured language to work with, as well as different thematic preoccupations from the Italian Futurists.

'Linguistic literature', by which I mean literature that invents new forms of language for the future or focuses more analytic attention reflexively on its own linguistic medium than on any extrinsic subject-matter, is such a widespread Futurist phenomenon that it would be impossible even to itemize its main achievements here. What I intend to do instead is to single out four specific aspects of this kind of writing, features which are of particular significance, either because they introduce new techniques and perspectives into literature or because they are central to the concerns of the Futurists. These topics are: (1) the interest in artificial languages (*zaum'*, *onomalingua*, and various other invented languages), which I should like to treat more comparatively than has been the case hitherto; (2) the use in Futurist poetry of vertical structures to give the impression of exploring the creation of language within a syntagmatic/paradigmatic framework; (3) the role of what might be termed 'pseudo-linguistic' theories in Futurist language-creation; and (4) the imposition of artificial ground rules, such as those associated with the palindrome or the lipogram, in order to create new forms of meta-grammatical structure in works where conventional syntax has been either reduced or marginalized. Of these, the 'transrational language' aspect of my first category has loomed largest in discussions of the linguistic focus of Futurist writing.

The telegraphic term *zaum'*, an abbreviation of the Russian phrase *zaumnyy yazyk* ('transrational language'), was coined by Alexei Kruchenykh, although he did not invent what the term stood for. In 'New Ways of the Word' ('Novye puti slova'), he presents a series of arguments for placing a high value on language as pure sound. The prefix *za-* of *zaum'/zaumnyy* is important in this respect, for it stresses that the new sound-language has gone *beyond* conventional words, rather than being content with foregrounding the sound-ingredient of everyday words. And, as is usual in Futurist contexts, 'going beyond' is heavily valorized: indeed, in this case, transcending conventional words becomes more than a mere rejection of passéist language (as either an act of provocation or an example of wilfully nonsensical art); it becomes the key to new mystical meanings. For, in Kruchenykh's view, 'new content *can only be born* when new expressive devices, new forms, are achieved' (*Manifesty*, p. 72). The fundamental belief that pure sound (the isolation and reordering of available sounds in non- or partially morphemic sequences) will create new 'transrational' meanings at the same time explains the attraction that earlier forms of glossolalia held for some of the Russian Futurists.[60] In essence much of their *zaum'* is a modernist version of the old 'speaking with tongues'.

'THE WORD IS BROADER THAN ITS MEANING,' Kruchenykh claims. 'The word (and the sounds composing it) is not just foreshortened thought, not mere logic, it is above all the transrational (its mystical and aesthetic components)' (*Manifesty*, p. 66). Pure sound is thus often presumed to be richer in meaning than conventional words, to radiate 'mystical' (transrational) meanings. In *zaum'*, according to Terentyev, one could 'touch unapproachable subjects'.[61] Almost as an article of faith among certain Futurists, apparently nonsensical strings of neologisms and clusters of pure sound become the apotheosis of inspirational writing.

Elsewhere in this study we have looked at other Futurist attempts at revitalizing language: either by reasserting its materiality and rediscovering the iconicity, above all of the printed word, or by reducing the hold of syntax over language (what Shershenevich in *2 × 2 = 5* calls *lomat' grammatiku* ('breaking up grammar')). Much *zaum'* can also be considered as a reaction against the inadequacy of existing language. In Kruchenykh's case, however, one particularly marked feature of the reaction is that language is being viewed above all from an essentially pre-printed perspective. Kamensky and Zdanevich may be concerned with the typographical possibilities of *zaum'*, but Kruchenykh is less so.

The first examples of Futurist 'transrational language' appeared in *Pomada* (1913), including the most famous *zaum'* poem of all: Kruchenykh's 'Dyr bul shchyl'. It is a work destined to receive its true emphasis only as a performed poem, rather than in the form of a series of printed words on the page. The visual effect of 'Dyr bul shchyl' is of little import, except perhaps inasmuch as

[60] For an early recognition of this parallel, see Viktor Shklovsky, 'Predposylki futurizma', *Golos zhizni*, 18 (1915), 6–9.

[61] *Kruchenykh grandiozar'* (Tiflis, 1919), 13.

the spacing between the individual sounds serves as an indication that they should be isolated, not elided in performance:

дыр бул щыл
убещур
скум
вы со бу
р л зз, —
(*Manifesty*,
p. 84)

As a consequence of his concern with the meanings he assumed to be hidden in the sounds of language, Kruchenykh's *zaum′* tends to be either non-morphemic or sub-morphemic. 'Dyr bul shchyl' is constructed from small groups of letters, forming sound-shapes which have virtually no bearing on pre-existing words or morphemes in the Russian language and also which possess no iconic function, yet nevertheless this example of *zaum′* has a Slavic quality. Almost all of its sounds are part of a specifically Russian alphabet and no one would confuse such a sequence with Jabberwocky or a German *Lautgedicht*, although some of the phonemes occur in sequences not nowadays permitted in Russian (*shch*, for instance, would have to be followed by *i*, not *y*). That there has been no lack of postulated meanings for these sounds is in part attributable to their degree of abstraction. For, inevitably, pure sound lends itself to associative responses. In Nilsson's eyes, 'Dyr bul shchyl' can be read according to two main codes: as a Futurist *zaum′* experiment (here the term is being understood over-restrictively to mean pure nonsense, which only some *zaum′* is) or as a primitivist return to an 'intimation of Turkic languages'.[62] Weststeijn attributes another code to the chosen sounds with his suggestion that one can detect in the poem the evidence of 'a polemic with the Symbolists' favouring of word-music', i.e. a predilection for '"ugly" sounds in reaction to the over-cultivation of the melodic'.[63] The claim made subsequently by Khlebnikov and the author in *Slovo kak takovoye* that the poem is 'more Russian than all Pushkin's poetry' (*Manifesty*, p. 55) would certainly suggest that an anti-passéist, provocative element was at work in the construction of this sound-poem.

At times, the ingredients of Kruchenykh's *zaum′* are no more than individual letters, rather than pseudo-morphemic clusters masquerading as the words of an as yet unknown (or long-lost) language. The poem 'Vysoty (Vselenskiyy yazyk)' ('Heights (Universal Language)') from *Dokhlaya luna* (*Croaked Moon* (1913)) falls into this category:

[62] 'Futurism, Primitivism and the Russian Avant-garde', *Russian Literature*, 8 (1980), 481.
[63] *Velimir Chlebnikov*, p. 14.

е у ю
и а о
о а
о а е е и е я
о а
е у и е и
и е е
и и ы и е и и ы[64]

Such a poem would appear to serve as a gloss on one of the various categories of *zaum'* enumerated in Kruchenykh's 'Declaration of Transrational Language' ('Deklaratsiya zaumnogo yazyka'): namely, words for 'when one does not need language—religious ecstasy, love ... affectionate names' (*Manifesty*, p. 180). That is to say, language for occasions when conventional language will not suffice (a more mystical idea than Marinetti's dramatic 'zones of intense experience'). Markov points out that what lies behind this vowel-pattern is 'the Russian text of the prayer "Credo"',[65] which adds a further dimension to our awareness that Kruchenykh is trying to revitalize the language of poetry, to take 'rational' religious language up to the level of ecstatic glossolalia by isolating its transrational elements (in this case, its vowels).[66]

By contrast to the extreme *zaum'* that can be found in Kruchenykh's poetry and some of Zdanevich's *dra*-plays,[67] Khlebnikov's interest in the 'transrational' potential of language tends to be semantic rather than exclusively phonological, with the result that his creative experiments tend to be with morphemic, rather than sub-morphemic units (the rider being that he extends the process of morphemicization to individual letters at times).[68]

In fact, as soon as one turns from the largely nonsensical *zaum'* of works like 'Dyr bul shchyl' and 'Vysoty' to Khlebnikov's *zaum'*, the situation becomes more complex than can be accounted for by any simple contrast between phonology and semantics. For a start, there is not *one* form of *zaumnyy yazyk* in

[64] Quoted from *Modern Russian Poetry*, ed. Vladimir Markov and Merrill Sparks (London, 1966), 360.

[65] *Russian Futurism*, p. 121.

[66] Khlebnikov also posits links between 'the language of the gods' and earlier glossolalia in 'O stikhakh' (*SP* v. 225–7).

[67] Zdanevich's *zaum'* is explored in some detail in Markov, *Russian Futurism*, pp. 350–8, and Gerald Janecek, *The Look of Russian Literature: Avant-garde Visual Experiments, 1900–1930* (Princeton, New Jersey, 1984), 164–83.

[68] Peter Steiner puts forward the view that the rift between Kruchenykh and Khlebnikov over *zaum'* 'corresponds to the conflicting theories of poetic language in the Formalist movement. OPOYAZ's early concern for poetic sound and its emotive qualities betrays Kruchenykh's influence, whereas the Moscow Linguistic Circle's insistence on the meaningfulness of linguistic sound reflects the logocentrism of Khlebnikov's *zaum'*' (*Russian Formalism: A Metapoetics* (Ithaca, NY, 1983), 146 f.).

Khlebnikov's writings, but a number. Using Khlebnikov's own terminology, Ronald Vroon differentiates four main categories:

1. the 'language of the stars' (*zvyozdnyy yazyk*), the projected universal language for the world of the future,
2. the 'language of the gods' (*yazyk bogov*),
3. the 'language of the birds' (*ptichiy yazyk*),
4. 'sound-painting' (*zvukopis'*), the most famous example being his poem 'Bobeobi pelis' guby' ('Bobeobi sang the Lips').

Of these types, the least innovative is the 'language of the birds'. In part of *Zangezi*, we are offered a series of bird-calls, each ornithologically identified by name and given its appropriate sound (*SP* iii. 318 f.). In contrast to such onomatopoeia, what Khlebnikov calls 'sound-painting' is also no absolute innovation, but something akin to the method of Rimbaud's 'Voyelles', except that, in Khlebnikov's case, colour-values are attached to consonantal clusters, rather than vowels. The resultant writing is non-iconic (unlike the 'language of the birds') and depends for an understanding on the availability of the poet's own explanatory glossaries.[69]

The 'language of the gods', on the other hand, is nearer to the pure *zaum'* favoured by Kruchenykh; it is defined by Vroon as 'gibberish' because Khlebnikov 'provides no glossary, no models, no clues which would lend meaning in the ordinary sense of the word to the coinages'.[70] But it is clear from the phrase 'language of the gods' that Khlebnikov attributes supreme status to this form of *zaum'*. For any reader not aware of the values being invested in such non-sense language by its author, however, the result of hearing such acoustic sequences would differ little from the effect of such radically different works as Hugo Ball's 'gadji beri bimba' or Rudolf Blümner's *Ango laïna*, both abstract sound-poems but with less mystical pretensions than 'the language of the gods'.

One of the features distinguishing totally non-morphemic *zaum'* from other forms of Futurist writing, indeed from most other literature, is the fact that the results themselves are often less significant than the way in which they are framed, commented on, or understood. Both Vroon and Cooke make the point that Khlebnikov deliberately contextualizes his *zaum'* in such a way as to make it more comprehensible than if encountered in isolation. In the case of *Zangezi*, stage-directions, commentaries, and extrinsic information enable us to understand the significance of sounds which would be meaningless on their own. But what I have in mind is something less specific to Khlebnikov than this. In many instances of 'transrational' non-sense, it is not a matter of comprehensibility, but the establishing of the status attributable to the technique and the staking out of the terms with which it should be viewed. If one were given 'Dyr bul

[69] See Vroon, *Velimir Xlebnikov's Shorter Poems*, pp. 181 f.

[70] Ibid. 191, 188.

shchyl', some of the 'language of the birds', and the 'language of the gods' in isolation, there would be no way in which one could assess their significance or their success. But much *zaum'* comes equipped with its own commentary and identifies the terms in which it should be understood; in other words more conventional language establishes *zaum'*'s premisses.

In these commentaries, there is a tendency on the part of Futurist poets (one which has since been echoed by commentators) to view *zaum'* on an evaluative scale rising from onomatopoeic neologisms and imitated languages (e.g. the 'language of the birds' in Khlebnikov's *Zangezi* or the spoof 'German' of Zdanevich's *Yanko krul' albanskay* (*Yanko the King of Albania*)) up to pure forms of sound-poetry such as Kruchenykh's 'Dyr bul shchyl' and Khlebnikov's 'language of the stars' and 'the gods'. The way the Futurists presented these absolute sound-poems suggests that the highest value is being placed on total abstraction and that nonsense is being equated with the most powerful transcending of the limitations of rational language (Kruchenykh's word 'Vysoty' ('Heights') for an extreme *zaum'* poem suggests this). But there are contexts in which such abstraction serves less radical purposes.

In two places in the Futurist opera *Pobeda nad solntsem* (*Victory over the Sun*), Kruchenyk uses it for a less elevated effect. In Scene Six, when a Young Man comes in and sings a 'bourgeois song', the result certainly still has all the hallmarks of Futurist *zaum'*:

ю ю юк
ю ю юк
гр гр гр
пм
пм
др др рд рд
у у у
к н к н лк м
ба ба ба ба[71]

But this is neither the language of the gods nor some ecstatic speaking with tongues. Instead, the effect is to satirize the trivial. A similar effect occurs right at the end of Kruchenykh's and Matyushin's opera, when the Aviator sings a war song, beginning 'luh luh luh' and ending 'vida diba'.[72] Clearly, in both of these instances, *zaum'* is being given the opposite function from the ones usually emphasized in Futurist theorizing about transrational language. Here it is employed to lampoon the banalities of everyday language. The passéist language of the bourgeoisie and the military are thereby reduced to nonsense;

[71] Quoted from the facsimile in Gisela Erbsloh (ed.), *'Pobeda nad solncem': Ein futuristisches Drama von A. Kručenych* (Slavistische Beiträge, 90; Munich, 1976), 20f.

[72] Ibid. 23.

and for once *zaum′* non-sense, instead of functioning as 'trans-sense' and opening up a gateway to higher sense, remains at the level of the trivial.

Mario Carli's *sintesi*, *Stati d'animo* (1916), uses something akin to Kruchenykh's nonsense *zaum′* for similar purposes. The scene is an open-air café, with various people sitting around either reading or talking, but not in conventional Italian. The play begins:

L'AFFARISTA (*mordendosi continuamente le unghie*). Astrr ghrr frr magnakalacafù bragimaciucianù astrù gru frusciù micenè micenè.

LO STUDENTE (*con nostalgia e lieve amarezza*). Auflìn bergìn ochipitecio sesisettott orazz birrvin adlgo artù viòoo bacimentosa.

LA COCOTTE (*sbirciando a vicenda l'affarista e lo studente, e sorbendo una menta al seltz*). Chiono chiono psi psi psi marì ninnì marì ninnì miotù tuasì coccolò coccolò coccolò còscela cìcciala momoroguà.

L'IMPIEGATO (*sfogliando un giornale*). Ito rito marito oro coro coloro aso naso rimaso ugo nugo rifugo ito oro aso ugo ugo aso oro ito ito ugo oro aso aso oro ugo ugo ito ito it to o.

IL GIORNALISTA (*parlando animatamente col vicino*). Icc acc biricìch biriciach ciaciabù cinciabò perafàra garatina mitalùra vatrasàca oidonè oidonè psciatt inglanèss farazòna blim blam blumm!

IL DEPUTATO (*con grandi gesti oratorî e voce tonante*). Mirfagò poròfon caralàzz brugùn falà neidor prokolifrogòtipo campogofù tlombô![73]

Only the Poet and the Philosopher speak anything resembling conventional language, the former with his telegraphic 'Shudder mystery wings sunset icily', the latter with his ruminations about 'comparisons forces origins tension . . .'. Yet one suspects that the thrust of the main nonsense parts of the play is to satirize the inane exchanges of philistine café-life, as well as provoking the *sintesi*-audience.

In his discussion of Zdanevich's transrational language, Ribemont-Dessaignes makes the point that 'there could be a French *zaum′*, a German *zaum′* or an English one, or even an international one', though he does in this respect add the caveat that this would on the whole tend to 'diminish its accessibility'. Presumably he means that a language created *ex nihilo* would be inordinately more difficult to comprehend than one constructed in some kind of relationship to an already existent language. That he even makes this remark is an indication of how his view of the subject is dominated by Zdanevich's *zaum′* rather than Kruchenykh's or even some of Khlebnikov's, for in Zdanevich's *zaum′* there are often echoes of other languages and of Russian itself.

Of course, as far as English literature is concerned, it could be argued that a prototypical *zaum′* did already exist in the form of Jabberwocky. In Chapter Four of his *Futuristy* Kornei Chukovsky also proposes—possibly satirically—

[73] Published in the *Teatro futurista sintetico* supplement to *Gli avvenimenti* (2–9 Apr. 1916), 22. René de Obaldia uses a similar technique in *Génousie* (1960).

another candidate for the status of transrational language *avant la lettre* in the shape of some of the more well-known incantatory passages in Longfellow's *Song of Hiawatha*. (The proposal of this conventionally meaningful form of language as *zaum'* suggests that, at the time Chukovsky was writing, there was some elasticity to the concept, capable as it was of applying to such a range of diverse forms, from syntactically conventional incantatory language and onomatopoeic neology right through to abstract sound-patterns and forms of glossolalia.) Nevertheless, the contexts and expectations associated with Futurist *zaum'*-writing hardly make either *Hiawatha* or Jabberwocky serious contenders for such categorization. Otherwise, nonsense poetry and *zaum'* would be isomorphic phenomena, when in fact the frameworks of understanding within which the two types of work occur are quite dissimilar.

Rather surprisingly, given the tendency of Futurist scholars juxtaposing Italian and Russian Futurism to play them off against one another, little note has been taken of the various experiments very much resembling Russian *zaum'* language that occur among the writings of some of the Italian Futurists.

There are, for example, local effects in a number of poems by Aldo Palazzeschi where acoustic effects are employed, not for the sake of imitating real-life noises (as is the case with the sounds of the drops of water at the end of his most famous work, 'La fontana malata' ('The Sick Fountain')), but for a more abstract effect. In 'Le carovane', for instance, the poet is looking out of a window observing the metaphorical 'caravans' of town life going by ('Carovane di ragni | carovane di cani | carovane di piedi' etc.) until there occurs the following sequence:

> Ma che cos'e tutto quel passare,
> tutto quell'andare, sostare e ripigliare?
> Son tutte carovane carovane carovane
> vane vane vane vane vane vane
> eeeeeeeeeeeeeeeeeeeeeeeeeeeeeeee
> e . . . e . . . e . . . e . . . e . . . e . . . e . . . e . . .[74]

As the word 'carovane' breaks down, first into a second morpheme then into its final letter 'e', one has a sense of the sheer assault on the senses of city life passing by. Here, Palazzeschi is not using such fragmentation for acoustically mimetic effects, as he often does, but in order to create a kind of *zaum'* 'designed analogy' for the impression of a dynamic world.

In his 1916 *onomalingua* manifesto, Fortunato Depero describes a new form of 'abstract verbalization' which, according to him, is derived 'from onomatopoeia, from *rumorismo* [one of the words for Futurist musical/sound experiments], and from the brutality of Futurist words-in-freedom'. This new onomatopoeic Futurist tongue is described as 'the language of natural forces:

[74] *Opere giovanili* (Milan, 1958), 95–7.

wind, rain, sea, etc., and of artificial acoustic entities, created by Man: bicycles, trams, trains, cars and all machines'.[75]

The manifesto echoes many of Russolo's ideas in 'The Art of Noises' ('L'arte dei rumori') of three years before ('after being conquered by Futurist eyes, our multiplied sensibilities will at last hear with Futurist ears' (*FF* 562)). There, Russolo associated noise with the modern world: 'ancient life was all silence. In the nineteenth century, with the invention of the machine, Noise was born. Today, Noise triumphs and reigns supreme over our sensibility' (*FF* 560). In fact, response to noise was invariably a good litmus test for distinguishing Futurist from passéist. A Futurist like Russolo can triumphantly declare: 'WE FIND FAR MORE ENJOYMENT IN MENTALLY COMBINING THE NOISES OF TRAMS, BACKFIRING MOTORS, CARRIAGES AND BAWLING CROWDS THAN IN LISTENING AGAIN, FOR EXAMPLE, TO THE "EROICA" OR THE "PASTORAL".' The ultimate arbiter of this new aesthetic is, hardly surprisingly, the pace and overridingly acoustic quality of modern city life:

> If we go through a great modern capital with our ears more alert than our eyes, we can delight in distinguishing the murmur of water, air and gas in metal pipes, the mutter of motors, breathing and pulsing like animals, the throbbing of valves, the thudding of pistons, the screeching of mechanical saws, the jolting of a tram on its rails, the cracking of whips, the flapping of curtains and flags. We can enjoy orchestrating in our minds the crashing of metal shop blinds, the slamming of doors, the muttering, and shuffling of crowds, the variety of din from railway stations, iron foundries, spinning mills, printing works, electric power stations, and ground railways. (*FF* 561)

But neither Russolo nor Depero is content to leave it at such impressionistic 'mental orchestrations'. The one uses his *intonarumori* to create a new kind of noise-music; and the other invents a new poetic *onomalingua* aimed at giving literary reflection to the sounds of the modern world.

The *onomalingua* manifesto as published in *Depero futurista* is accompanied by various illustrations of the new tongue. In 'Tramvai' (fig. 4.1)[76] the poet not only uses a form of *zaum'* to record the language of the tram, but avails himself of the same idiom to speak back to the vehicle. Moreover, Depero's new language was never simply conceived of as an onomatopoeic imitation of the universe of elemental sounds in the outside world (an acoustic sphere which Depero rather unhelpfully widens to include 'the monologues of clowns and music-hall comics'); it also has the kind of utopian hopes pinned to it which one sometimes encounters in the case of Russian Futurist *zaum'*. With the help of *onomalingua* (Depero's term is itself a new coinage), 'it will be possible to talk and efficiently come to an agreement with the elements of the universe, with animals and with machines'.[77] Hence the poet's underlying conviction that his *onomalingua* will become 'a poetic language of universal comprehension', with

[75] *Depero futurista* (Milan–Paris, 1927), n.p. [76] Ibid. [77] Ibid.

Avevo un tale appetito che i sensi mi si elettrizzarono, ed ebbi voracissimo desiderio di mangiare:

Lire 7 e 75 centesimi BOLLITO E MINESTRA

NARICI-PENDOLI | assorbimento
VESCICO-SPANDERE | AUTO -TIK-MATICO

STARNUZIN | PEPA' | PAPE'
ZIO ZIN | TATO' | PIPE'

gipun forchendere

lucitagli, lamafettire, taglizz-minuzz-ire-ire-ire-ire-ire-ire-ire-ire-ire-ire-ire-ire-ire-ire.....
POLITICHILLA nello
ESOFAMAGOSTO

ECCE BOMI–LENTEGIAMA–
CALEICELUCERE
BRILLI–NERVI–CIAIRE
DI.......GE.......RI.......RE........

Ebbi sete tale da volermi sconciamente ubbriacare:

E' OE' OLLA E-I-A ½-1-50-1000
cobria
zamar titù capor ià
malà tem lucar

1-10 ***ancor-infinicobria!***

glo-glu-glo | glu-ampa
glu-spen | glu-buon
glu-reden | glu-bodon

Ah! ci buol **LECE-LEMBO'**
INFINI-COBRIA
litritò tribon vin no soross

ad un tratto mi ricordo d'un appuntamento alle 1.90 e pago il conto 10.50. Vado incontro al TRAM che saluto METALLICAMENTE:

TLEMMO TLA -
ti comm - Vi Emme iiiii
BICAROMM TRAMM
ME COM SAI TI STO?
RISSI COMISSI FUSSI
li to stei mei tleime vroiii sioiii oiii... SOMBÒSOTRÀ

Il tramme ferocemente macchinario mi risponde:

PLOMMO-PLAMMO-
AMMIZIR-
Kin–LISSIN–VUISSI
STROMIRTI-TZING
oogoolanga-buli-tzi
BUM-TOM-SBRE!
COMUGANAL-TZINGHI-LONGA
TAN-KLATAN
KLOGO-SOKRÀ
MAK - TOKRA' - TIKRU' - TUKUISS
QUECC-LUCC
TZAUU-SOKRIPINK
pingh-fuingh-cocipingh
-zifinghi!

AUAAA! è morto!

Un passante venne schiacciato dal Tramvai — eco lontana miagolante del funerale:

Stee sobeom dominum pecoemuu uuuuuuuuuuuu
foneum sabateumm mirobeeeeeeeeeeeeeeee uise
conte gnau BEUMMmmmmm
u u u u u u u
e e e e e e e
prispi-miserere-beatum misericordiam
tuuaaaaa . . . .
. . . .

ora uscito:

„L'ITALIA"
„IL PICCOLO"

ed il grido si disperse in arabeschi vocali:

COLLI TIPICO | COCCHI
CUPOLI | ILLI
TI PI CO

Roma, 1916.

FIG. 4.1. Fortunato Depero, 'Tramvai', 1927

the result that, when it comes to full fruition, no translations will be required any more.

Such sentiments are obviously not all that far from some of the utopian hopes being pinned on Russian *zaum'*. In fact, with this analogy in mind, Glauco Viazzi refers to *onomalingua* as 'a language which is trans-, or infra-, rational'.[78]

In a similar vein to Depero, Kruchenykh, in his 'Declaration of Transrational Language' (*Manifesty*, p. 180), prophesies that 'transrational language' will eventually 'result in a universal poetic language which is born organically, not artificially, like Esperanto'. And Khlebnikov writes in comparably hopeful terms to Depero's of a *zaum'* which 'can unite men'. 'Rational languages', on the other hand, he reasons, 'are already disuniting men' (*SP* v. 236). Elsewhere, he sums up his position with the question: 'What is better, a universal language or universal slaughter?' (*SP* v. 266). (One suspects, though, that Marinetti might have had a different answer to this rhetorical question than Khlebnikov is clearly expecting!)

Depero's new *onomalingua*, a mixture of natural and artificial sounds, has marked similarities to the mimetic kind of *zaum'* that Khlebnikov created with his 'language of the birds'. And another Italian Futurist, Balla, even created

[78] *I poeti del futurismo*, p. 316.

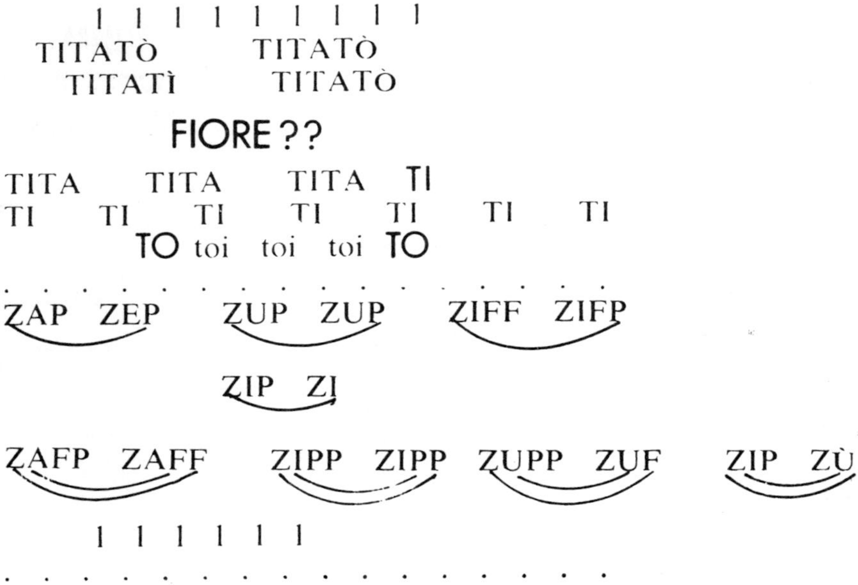

FIG. 4.2. Giacomo Balla, 'canzone di maggio', undated

something far more reminiscent of the language of the birds with his poem 'canzone di maggio' (fig. 4.2).[79] It may be possible, as Viazzi suggests,[80] to construe this work as a totally non-referential sound-poem. But, given that its title refers to the month of May, I am more inclined to the idea that the sounds should be interpreted as birdsong; after all, Balla must have had some reason for including this referential dimension in his title. However, another possibility is that the title refers to the way in which the words are growing up out of the ground (i.e. out of the page) like the flowers also mentioned in the fourth line.

In general, the majority of Italian Futurist equivalents to *zaum′* are not only referential; they tend to be concerned with the sounds made by the 'artificial nature' of an urban, industrialized landscape, rather than the natural sounds (birds, wind, rain, sea) that Depero mentions in his *onomalingua* manifesto. In this respect, Balla's *sintesi* '*onomatopea rumorista Macchina Tipografica*' is more typical.

In this work, which is a cross between a 'happening' and a conventional performance-poem, twelve typists are to repeat the following sequences for a minute, as if taking part in a typing lesson:

[79] Written 1914, quoted from Maurizio Fagioli dell'Arco (ed.), *Ricostruzione futurista dell'universo* (Rome, 1968), no. 58.

[80] *I poeti del futurismo*, p. 314.

Dodici persone
ognuno ripetere per un minuto di seguito
le seguenti onomatopee rumoriste

1° settesettesettesettesettesette
2° nennenennenennenennenennenenne
3° vùùùùmmùùvùùùùùmmùùvùu
4° tè.tè.tè.tè.tè.tè.tè.tè.tè.tè.tè.tè.tè.
5° miaaaaaanavanò,miaaaaaanavanò
6° sta---sta---sta---sta---sta
7° lalalalalalalalalalalalalalalalala
8° ftftftftftftftftftftftftftft
9° riòriòrièrièriòriòrièrièriòriòrièrièriòriò
10° scscscscspsspscscscscspssps
11° vèvèvèvèvèvèvèvèvèvèvèvèvèvèvève
12° nunnnònònunnnònònunnnònònunnnònò[81]

What is not clear from Balla's instructions is whether the figures are to repeat the sounds with their voices or type them on machines. Presumably, the former is the more likely realization of the idea, since it would lead to more acoustic differentiation, yet one can conceive of a performance of the second kind, where a greater fusion of man and machine would be made possible. Then, presumably, different letter-keys would be literally struck by different performers.

One further crucial difference between much Italian sound-poetry and its Russian *zaum′* counterparts can be summed up by saying that, whereas the Italian *zaum′* poets (Russolo, Depero, Balla) were on the whole concerned with *the language of sound*, and hence produced a referential form of onomatopoeia, their Russian counterparts were much more concerned with *the sound of language*.[82]

There were, however, exceptions to this pattern within Italian Futurism. Giovanni Papini's 'Lettere in libertà' ('Letters-in-freedom'), doubtless an attempt to go one better than Marinetti with his *words*-in-freedom, is first and foremost a triumphant evocation of the materiality of the individual letters of the alphabet, as each one is liberated and allowed to shout, as it were, at the top of its voice:

aaaaaaaaaaaaaaaaaaaaaaaaaaaaaaaaaaaaaaaaaaaa
bbbbbbbbbbbbbbbbbbbbbbbbbbbbbbbbbbbbb
cccccccccccccccccccccccccccccccccccccccccccc[83]

The poem works its way through the alphabet, liberating each letter in turn, although it is not clear why two clusters of letters (jkl, wxy) are omitted from

[81] *Ricostruzione futurista dell'universo*, no. 118.
[82] Tellingly, it was Shershenevich, the arch-Marinettian among the Russian Futurists, who objected in *Zelenaya ulitsa* to any simple equation of the word with its sound component.
[83] *Lacerba* (3 Apr. 1915), 112.

the full potential sequence, unless, in a performed version of the poem, they would simply duplicate the sounds already represented—or even be less easy to proclaim. But the result is the liberation of letters from the constraint of the word. The sheer elation with which these individual letters appear to rejoice in their new-found freedom is not far removed from what Kruchenykh has to say (in Section D of his 'Declaration of Transrational Language') about *zaum'* as a language which 'awakens and liberates creative imagination without offending it by anything concrete' (*Manifesty*, p. 179).

Zaum', *onomalingua*, and their various unnamed analogues in other Futurist quarters share with mainstream literary Futurism a dissatisfaction with passéist language—although not always (witness Khlebnikov) a rejection of conventional syntax. In Italian Futurism, experiments tend to be confined within a referential framework, still operating, that is, with the techniques of onomatopoeia. On the other hand, with writers like Kruchenykh and (sometimes) Zdanevich, the ultimate goal would appear to be total abstraction. But this, viewed from an Italian Futurist vantage-point, would amount to depriving Futurist language of the ability to prepare thematically for the Futurist future that was the concern of the majority of Marinetti's followers. Radical *zaum'*, in other words, pushes Futurism towards its frontier with Dada.

	A	
	Adnil	
	Danil	
	Nadil	
	Nalid	Alnid
	Dilan	Aldin
	Lanid	Ildan
Linda	Landi	
Ilnda	Naldi	
Nilda	Dalni	
Indla		
Indal		
Lnida		
Lndia		
Lndai		
Lidna		
Lidan		

FIG. 4.3. Guillaume Apollinaire, from 'Quelconqueries II', 1914

On 15 July 1914, Apollinaire published a permutational poem (fig. 4.3) in *Lacerba*. It appeared under the rubric 'Quelconqueries' ('Odds and Ends')—a self-deprecating gesture on the poet's part, but also an indication of just how marginal permutational experiments of this kind were in the eyes of the Italian Futurists. Yet in a modest way, untypical even of Apollinaire's own *œuvre*, this poem points towards a form of linguistic literature which was far less peripheral in Germany and among the Russian Futurists.

The subject of Apollinaire's poem is essentially the resources of language itself. One could, indeed, view it as an exploration of the word-generating potential of just five letters of the alphabet, using them to create a series of real and invented names, some familiar, but most with an exotic or totally alien ring to them. Yet, although the technique is permutational, it is far from stochastic, in the way that some forms of *zaum'*-creation were.[84] Instead, the poem is a literary equivalent to the well-known game of seeing how many words can be formed from one noun, the added challenge here being that the source-word (or repertoire of letters) is relatively restricted and there is a grammatical control: a focus on proper nouns, and by that token an embargo on other parts of speech. Language is here not only being palpably created out of language; it is primarily being used to reflect upon the medium rather than reflect any world beyond the word.

What is of particular relevance to the line of approach I wish to take in the next part of my discussion is the *generative* aspect of the work, a feature which reveals itself even in the vertical column-structure of the layout. I should point out that I do not mean 'generative' here in the technical sense in which it is used in Generative Grammar or in related transformational contexts. There, the model (derived from mathematics) is a way of describing the ability of grammar to produce the sentences of a language. My interest, by contrast, is on the way in which one line of a poem, or phrase in an often anaphoric sequence, enters into a paradigmatic relationship with those around it, to the extent that one gains the impression that the various utterances have been generated by the use of a literary version of the slot-and-filler procedure. The link between verticality and language-generation that one sees in Apollinaire's poem is a vital one in much language-oriented Futurist poetry. And in the case of this work, one suspects that it is only the constraints of format in *Lacerba* that prevent all the words being laid out in a single column.

If one requires a linguistic model with which to account for Apollinaire's simple generation of a series of names created on the basis of a repertoire of five letters of the alphabet, three consonants and two vowels, then a better framework than that offered by Generative Grammar is the less sophisticated one based on the syntagmatic/paradigmatic relationship between words in an utterance. Words occupying a particular place in an utterance stand in a paradigmatic relationship to other words that could fill that place and at the same time relate syntagmatically to the other words in the sequence. In fact, it is possible to

[84] In his 'Declaration of Transrational Language' ('Deklaratsiya zaumnogo yazyka'), Kruchenykh distinguishes three main kinds of *zaum'*, the third of which is 'the aleatory (alogical, fortuitous, a creative breakthrough) . . . misprints, lapses . . . stuttering, baby talk' (*Manifesty*, p. 180).

spatialize the model by thinking of the syntagmatic dimension as a horizontal one, with the repertoire of words in the same paradigm stacked above each slot, available to fill it, as in those ingenious Baroque poem-creating wheels and cylinders. Apollinaire's word-game shows this model in operation: each place or *slot* in the syntagma of the name to be created can be filled by either a vowel or a consonant, each part of the repertoire is to be used once only, and the words have to be placed in sequences which make for pronouncable names, even though it is clear that some of the words the process produces have unusual consonantal clusters, from a European point of view.

Nevertheless, there are two respects in which Apollinaire's procedure still differs from most Futurist 'generative' poems. First, it is subjected to less syntagmatic pressure: that is to say, there is a larger measure of free variation between where vowels and consonants can occur than obtains in most Futurist poetry's use of the sentence-syntagma. Second, there is its concentration on the creation of single words by reordering individual phonemes, rather than creating longer utterances on the basis of ordering morphemes or entire words.

Generation, in Futurist word-column poetry, habitually takes one of two main forms. The one can be thought of as a simple syntagmatic/paradigmatic structure where each line, sometimes containing no more than a single word, has an equivalence to the preceding line and is built on exactly the same pattern as its successor, with the result that individual lines seem to generate their successors and be generated by their predecessors.

The second form is the poem which increases (or decreases) in size from line to line, in such a way that the lines also appear to be generated from one another—this time by a process of accretion or subtraction. One form in this second category, one with a long pedigree, is known as the 'disappearing poem' or 'Schwundgedicht'.[85] The most well-known (non-Futurist) example is the ABRACADABRA spell, created by removing the final letter from the previous line until one arrives at the final 'A'. (Often the written charm was used to invite illnesses to disappear, just as the poem itself did.)

A B R A C A D A B R A
A B R A C A D A B R
A B R A C A D A B
A B R A C A D A
A B R A C A D
A B R A C A
A B R A C
A B R A
A B R
A B
A

[85] For a discussion of the 'Schwundgedicht', see Alfred Liede, *Dichtung als Spiel: Studien zur Unsinnspoesie an den Grenzen der Sprache*, ii (Berlin, 1963), 266f.

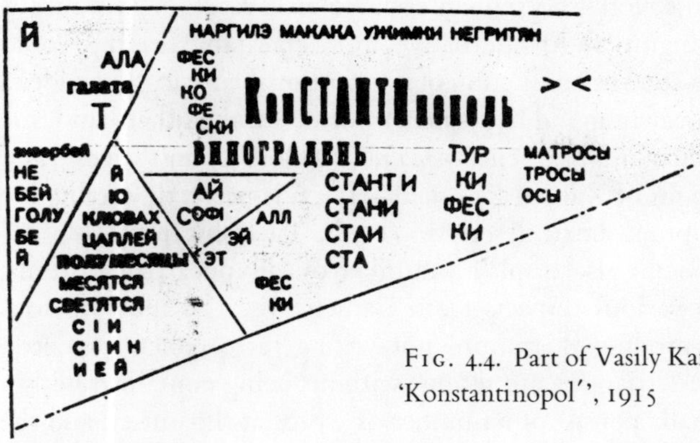

FIG. 4.4. Part of Vasily Kamensky, 'Konstantinopol'', 1915

But, whereas the ABRACADABRA configuration works with a structure created by removing single letters of the alphabet, Futurist poetry operates more at word-level. The following sequence of Kamensky's from *Rykayushchiy Parnas* (*Roaring Parnassus*)[86] operates on such a principle of structural diminution, with the result that, by the end, language has been reduced to zero, as in the traditional 'Schwundgedicht'.

Я	I
Излучистая	Meandering
Лучистая	Radiant
Чистая	Clean/pure
Истая	True/keen
Стая	Flight/flock
Тая	Thawing/concealing
Ая	(f. ending)
Я	I

Disappearing structures of this kind often occur within the constituent segments of Kamensky's 'ferro-concrete' poems. Janecek discusses one such pattern in the main polygon at the top of 'Konstantinopol'' (fig. 4.4).[87] Here, we move from the name 'Konstantinopol'' itself, via 'Stanti' (allegedly the way the poet heard the city's name while there on a visit), to 'stani' (*zaum'*), 'stany', 'stai', and 'sta'. Janecek comments: '*stani* means nothing in Russian, but is close to a number of words (stan [noun]), stany, stan! [imperative]. *Stai* might be the plural of *staya* (flocks) and would therefore be linked to the pigeons and herons to the left. *Sta* could be an inflected form of *sto* (one hundred).'[88] In other words, the challenge of shrinking words from line to line appears to have been

[86] St Petersburg, 1914, 46.

[87] *Tango s korovami: Zhelezobetonnye poemy* (Moscow, 1914), n.p.

[88] *The Look of Russian Literature*, p. 127.

such a constraint that the word-shortenings do not always result in clear meanings (or significant ambiguities). An alternative, more charitable reading would be to assume that the systematic reduction of conventional language provides a fruitful strategy for unearthing hidden (*zaum'*) words. On the other hand, the same poem's further 'shrinking' sequence—*matrosy* ('sailors'), *trosy* ('ropes'), and *osy* ('wasps')—to the right of Kamensky's *stanti*-effect, is perfectly intelligible and pertinent to an impressionistic description of the Turkish capital.

August Stramm's poems also display a number of 'disappearing' patterns, though they differ in various respects from Kamensky's. The most obvious visual difference, a by-product of Stramm's not writing 'ferro-concrete' poetry, lies in the way his word-triangles are set out without being contained within outlines. The lines usually consist of a number of words at the outset, and the triangularity is no more than impressionistic, by comparison with the 'Constructivism' of Kamensky's shapes. Another difference, a more fundamental one, is that Stramm's reducing strategy does not result from a truncation of the previous word by transposing only some of its morphemes on to the next line. Whereas Kamensky can run from 'chistaya' via 'istaya' to 'staya', Stramm's individual lines are shortening, but the constituent words of each line are not constructed from the morphemes (or pseudo-morphemes) of the earlier ones.

The poem 'Verzweifelt' ('Desperate') illustrates these differences clearly enough:

Droben schmettert ein greller Stein
Nacht grant Glas
Die Zeiten stehn
Ich
Steine.
Weit
Glast
Du![89]

Bridgwater translates the poem as follows (although even his faithful rendering produces a work with noticeably different contours):

Over there a glaring stone shatters
night grains glass
the times stand still
I
stone.
Far-off
you
glass![90]

[89] *Das Werk*, p. 23.
[90] *twenty two poems*, n.p.

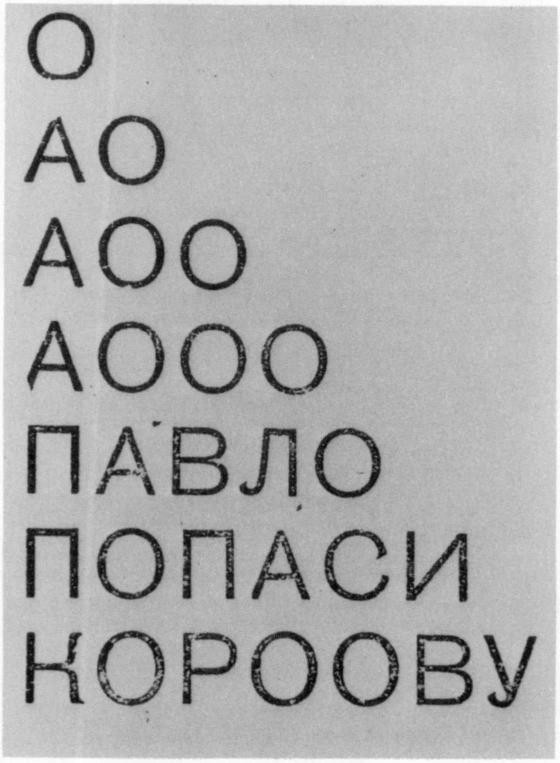

FIG. 4.5. Mykhaylo Semenko, 'Sil′s′kyy peyzazh', 1922

Stramm's layouts owe their triangularity above all to his concern for changing rhythmic patterns. The effect is such that, as the poem progresses, the pauses between words become more pronounced. At first, the end-line pauses separate whole phrases; later they isolate individual words and tend to bring about a retardation in rhythm. Although variations in rhythmic patterning inevitably also result from Kamensky's structures entailing progressive truncation, the main focus there is on shape and meaning: on the extent to which longer words contain shorter ones. Kamensky's poems are more visual in effect, Stramm's mainly acoustic.

The opposite structure is also quite common in Futurist poetry. To consider one example: Semenko's 'Sil′s′kyy peyzazh' ('Village Landscape') (fig. 4.5),[91] originally in *Poezomalyarstvo* (1922), gives the impression of being a conscious inversion of many of Kamensky's effects (with which Semenko was familiar). In English, the poem means simply: 'O, AO, AOO, AOOO, PAVLO, TEND THE COOW'—the elastic spelling of 'cow' taking up the duplicated 'o's earlier in the poem and introducing an onomatopoeic mooing sound.

[91] Mykhayl′ Semenko, *Vybrani tvory*, i, ed. L. Kriger (analecta slavica, 23; Würzburg, 1979), 241.

Establishing 'linguistic' relationships of this generative/reductive kind between words within an overall structure of verticality is most easily achieved within a framework of anaphora, rather than a 'disappearing' structure. Indeed, one of the most fertile areas of literary Futurism's focus on 'the word as such' and on the constituent parts of the autotelic word (as opposed to any syntactical context) comes in the word-columns of the Russian Futurists and the *Sturmkreis* poets.

'Henceforth', Khlebnikov and Kruchenykh write in 'The Word as such', 'a work of art could consist of a *single word*, and simply by a skilful alteration of that word the richness and expressiveness of artistic form might be attained' (*Manifesty*, p. 59).

The most famous example of such 'skilful alteration' within an overall anaphoric structure comes in Plane Nine (the 'Blagovest umu' section) of Khlebnikov's play *Zangezi*. Here the audience is promised the 'parade of all kinds of UM'. The sound of a bell ringing is used as a frame upon which to generate a whole series of neologisms created from the root '-UM'. Various prefixes are attached to *-um* (meaning 'sense' or 'reason', as in *zaum'*) to create an array (or 'parade') of both permissible and invented nouns. Thus, if *no* means 'but', *noum* will suggest an adversary form of reason, and so on. Different shades of connotation are engendered by different prefixes.

'Sound the alarm', Zangezi orders, 'send the sound through the mind! Toll the big bell, the great tocsin of intelligence! All the inflections of the human brain will pass in review before you, all the permutations of "-UM"! Look up and behold. Join us now in song, all of you!' There follows a characteristically Futurist generative word-pillar, with each item appearing as a variation on the meaning of its predecessor and the whole sequence offering examples of words within the same linguistic paradigm.

Тише! Тише. Он говорит!

Зангези: Благовест в ум! Большой набат в разум, в колокол ума! Все оттенки мозга пройдут перед вами на смотру всех родов разума. Вот! Пойте все вместе за мной!

I

Гоум.
Оум.
Уум,
Паум.
Соум меня
И тех, кого не знаю
Моум.
Боум.
Лаум.
Чеум.
Бом!

Бим
Бам!

II

Проум
Праум
Приум
Ниум
Вэум
Роум
Заум
Выум
Воум
Боум
Быум
Бом!
Помогайте, звонари, я устал.

III

Доум.
Даум.
Миум.
Раум.
Хоум.
Хаум.
Бейте в благовест ума!
Вот колокол и веревка.

IV

Суум.
Изум.
Неум.
Наум.
Двуум.
Треум.
Деум.
Бом!
Зоум.
Коум.
Соум.
Поум.
Глаум.
Раум.
Ноум.

(*SP* iii. 334 f.)

While marshalling his 'parade', Khlebnikov favours a grammatical term for the interrelationship between its various participants: they are described as 'inflections of the human brain'. Instead of each word being perceived as a separate linguistic item, these forms—all generated by prefixation from *um*—are not seen just as 'inflections' of that word (in the way that 'internal declension' interprets individual words as parts of some new declensional system); they are even presented figuratively as *declensions of the mind*. The grammatical metaphor makes the various 'declensional' cases (this time, the product of affixation, rather than case-ending) into images for the mind's various states, activities, and relationships to the world around it.

The semantic possibilities of these individual formations have enjoyed much exegetical attention. Originally, authoritative guidance was offered by Khlebnikov's own notes to Plane Nine of *Zangezi* (*SP* iii. 336 f.). These have since been extensively augmented by commentators, especially in recent years,[92] but the tendency has been to focus on the semantics of the individual words, not the generative matrix within which they occur. Yet what this sequence offers is the drama of word-creation in action. The process of adding prefixes engenders a whole new sequence of neologisms, each modelled on the structure of the previous one.

It is a method to which Khlebnikov often returns, usually more restrictively, in many of his poems, and using not single-word lines, but anaphoric sequences of phrases. The effect can vary in mood from the incantatory:

Тебе поем родун!
Тебе поем бывун!
Тебе поем радун!
Тебе поем ведун!
Тебе поем седун!
Тебе поем владун
Тебе поем колдун!

(*SP* ii. 271)

and

Зазовь.
Зазовь манности тайн.
Зазовь обманной печали.
Зазовь уманной устали.
Зазовь сипких тростников.
Зазовь зыбких облаков.
Зазовь водностных тайн.
Зазовь.

(*SP* ii. 287)

[92] See Charlotte Douglas (ed.), *Velimir Khlebnikov. The King of Time: Poems, Fictions, Visions of the Future*, trans. Paul Schmidt (Cambridge, Mass., 1985), 207–9; and *Velimir Chlebnikov: Werke*, ed. Urban, i. 401–4.

to more subtle acoustic and semantic play:

Воздушный воздухан.
Воздухее воздухеи,
Воздухее воздухини,
Сидушистый сидухан,
Сидухее сидухини,
Сидухее сидухеи,
Колышистый колыхан,
Колыхее колыхини,
Колыхее колыхеи,
Едушистый едухан,
Едухее едухеи,
Видушистый видухан.
Видухее видухеи,
Видухее видухини.

(*SP* v. 84)

The *Sturmkreis* writers also produced much poetry of this kind. Franz Richard Behrens's 'Tannentod' contains a comparable (and equally untranslatable) generative sequence, as it describes the axes threatening the fir-trees:

Beile geilen Treten
Sande beten
Sande leben
Sande heben
Weihen sonnen
Weihen binden
Weihen kranzen
Weihen ketten
Rote Tannen irren rote Tode.[93]

Of course, verticality does not necessarily have to be harnessed to such a language-generating function. As we saw in the previous chapter, Mayakovsky's 'Utro' ('Morning') splits words downwards in such a way as to create an *enjambement* even between individual word-parts in a vertical pillar. As a result, there is no 'structure of equivalence' (Jakobson) between individual lines, and hence no sense of various forms being generated syntagmatically from previous utterances. Instead new meanings and relationships emerge from the segmenting of words, rather than from parallelism of the kind we have just been considering. And a poem like Palazzeschi's 'La fontana malata' ('The Sick Fountain')[94] merely arranges brief, conventional sequences in short lines with no sense of marked linguistic innovation to the overall effect. On the other

[93] *Blutblüte: Die gesammelten Gedichte: Werkausgabe*, ed. Gerhard Rühm, i (Munich, 1979), 17.
[94] *Poemi* (Florence, 1909), 123–5.

hand, Giovanni Gerbino's 'Romanza dell'attesa' ('Romance of Expectation') does at least exploit verticality in a way suggesting that each line deliberately varies the sound-pattern of the previous one or is being generated within the same syntagmatic framework.

Cara	Dear
Mara	Mara
quanto	how much
pianto!	crying!
Questo	This
mesto	sad
cuore	heart
muore	dies
a poco	little
a poco.	by little.
Due	Two
tue	words
parole	from you
sole	alone
lo	will help
sprebbero	it
guarire	heal
lo	will make
farbbero	it
gioire.	rejoice.
Cara	Dear
Mara	mean
avara!	Mara!
amara![95]	hateful one!

Yet even this poem, although it isolates its individual words in a manner more reminiscent of the Russian Futurists, is not built on the kind of asyndetic principle that their structures usually embody. It is rhyme, rather than grammatical equivalence, that generates the individual words, with the exception of 'quanto | pianto' and the final 'Mara | avara | amara' sequence, where semantic word-play does come into force.

Nevertheless, such a work is exceptional; in most of the poems of the period verticality combines with the use of parallelism to contribute to the sense of forms being generated according to a linguistic paradigm.

Many of the Futurist poets not only dissected and inventively explored aspects of language in their literary works; they also theorized about language—in manifestos, declarations, essays, dialogues, and treatises. However, many of the ideas they put forward could at best be described as quasi-linguistic or as a form

[95] *Telegrafo e telefono dell'anima* (Milan, 1926), 127.

of poet's linguistics. More often than not, those so engaged were concerned with positing ingenious (false) etymologies, or attributing semantic value to sub-morphemic units, or inventing links between words that were in fact, philologically speaking, unrelated.

The notion of 'internal declension' elucidated in Khlebnikov's *Teacher and Pupil* is one such spurious linguistic theory. Khlebnikov's argument is based on a grammatical metaphor and is arrived at by a process of linguistic analogy, but the analogy is an erroneous one. In itself, 'declension' is, of course, a framework appropriate to Russian and not to many of the other uninflected languages of Futurism. Yet, notwithstanding the fact that Khlebnikov is *not* talking about the inflectional endings of the Russian language, his theory appears to be based on the assumption that a proto-Russian language with suffix-indicated declension might once have had a further system of internal declension, as well.

According to the Teacher, 'if the genitive case answers the question "whence", and the accusative and dative cases the questions "whither" and "where", then the declension of a stem in these cases should give the resulting words meanings which are opposite' (*SP* v. 171). In this wildly speculative spirit, meanings attached to the vowels in declensional suffixes are transferred to those vowels or combinations of vowels when they occur in another context: in the middle of a usually mono- or disyllabic word. But surprisingly, instead of working systematically towards a rigorous series of internal declensions based on the various cases of the Russian noun, Khlebnikov confines himself to contrastive structures. For this reason, Vroon can justifiably refer to 'internal declension' as a 'principle of antonymy'.[96] Yet there is no compelling reason why it should be so, except in the sense that 'declension' is being used merely as a poetic metaphor for contrastive structures. Khlebnikov's key examples are pairs of words with opposite meanings, 'where from' and 'where to' being the main underlying declensional contrast.

There is, in fact, a double process of selection here, first, through the focusing on contrastive pairs of case-endings which help to structure the thinking about antithetical meanings into something approximating Minimal Pairs; and then, at a more pragmatic level, in the selection of individual examples which are intended to serve as evidence of an underlying pattern. Clearly, it is not difficult to locate exceptions to these principles. But then, if Khlebnikov were correct in his claim to be uncovering residues of meaning from the older strata of language, noting exceptions in modern Russian would not be an insuperable obstacle.

In essence, 'internal declension' might be summed up as two things: a pseudo-linguistic hypothesis often believed in by its proponents, and a means of creating contrastive poetic effects. In the second sense, however, it is nothing peculiar to Khlebnikov. As Yuri Tynanyov puts it—here as a general axiom—'if

[96] *Velimir Xlebnikov's Shorter Poems*, p. 164.

you line up in formation disparate, but like-sounding words, they will become cognates';[97] or at least, one should perhaps add, they will appear to acquire such a status. What becomes interesting is the application of such views to the process of verbal creation, the point at which pseudo-linguistic theorizing is converted into *linguistic literature*. No matter how questionable the linguistic foundations of the underlying theorizing, it can still, as we have already seen in Stramm's case, lead to semantically rich poetry.

Khlebnikov's poem 'Ya videl' ('I saw') (*SP* iii. 27) is an example of the principle of 'internal declension' being exploited in such a way that words are made to relate to one another so as to suggest common etymological sources. It starts:

Я видел	I saw
Выдел	(the) separation
Весен	of springs
В осень,	in autumn,
Зная	(I) know
Знои	the glow
Синей	deeper blue
Сони.	than a sleeper.
А ты пяты	And you the heel
На мячике созвездия	on the ball of the stars
Полет бросаешь,	You fly
В изгиб	to a curving
Из губ	of lips
Свернув	having bent
Свой локоть	your elbow
Белого излома.	to the white bend.

(*SP* iii. 27)

Later plays on declension include *put′* ('way') and *pet′* ('singing'), and *letaya* ('flying') and *lataya* ('putting on'). If there is semantic opposition here, as Khlebnikov's theory would lead us to expect, then it often requires a tortuous gloss to bring out some specious notional contrast. Also, the fact that, as Vroon points out, Khlebnikov 'has to transgress word boundaries and resort to oblique cases (e.g. genitive plural) in order to establish the desired opposition—*vesna* ('spring') vs. *osen′* will not do, but *vesen* ('of springs') vs. *v osen′* ('in the autumn') is an ideal reflection of the opposition between the seasons'[98]—suggests that the poet often had to work hard to lay bare his postulated contrastive element. Arguably, this could simply be a sign that it had been buried by the overlay of modern language. But another interpretation suggests that there may be more of the *jeu d'esprit* to such effects than the usual assumption that Khlebnikov was utterly serious about the linguistic pseudo-theories he was putting forward.

[97] *Arkhaisty i novatory* (Leningrad, 1929), 560.

[98] *Velimir Xlebnikov's Shorter Poems*, p. 164.

It is Khlebnikov, of course, who is primarily associated with 'internal declension'. But related effects occur in the poetry of other Russian Futurists: for example, Mayakovsky's line *Nash bog beg* ('Our god is speed'[99]) or his phrase *k ruke reki* ('towards the arm of the river').[1] One might question whether such effects are necessarily based on a theory of internal declension (although the first of these two examples does also occur in Khlebnikov's *Teacher and Pupil*). Certainly, it might be more commonsensical to assume that a memorable effect simply results from the juxtaposition of two phonologically related words, without any linguistic underpinning. But then, despite the well-laid-out theory of 'internal declension', that may also be the reader's response to some of the oppositions in a poem like 'Ya videl'. In such cases, sound becomes (as it always has been in poetry) a form of rhetoric, persuading us that there is a logical connection between the two ideas, because there is one between the sound of the two words.

Apart from 'internal declension', the most detailed (pseudo-) linguistic theory in the writings of the Futurists was the idea that the presence of specific letters in a word, above all in initial position, in some way governs its meaning. Kruchenykh, for instance, felt that the phoneme 'z' suggested connotations of movement, discontent, excitement.[2]

In 'Our Foundation' ('Nasha osnova'), Khlebnikov claimed that the 'first consonant of a simple word governs the whole word' and that 'words beginning with the same consonant are united by the same concept' (*SP* v. 235 f.). It is this second assumption that dominates both his theorizing and the host of poems generated from such a premiss. 'The initial sound', it is stressed, 'is of a special nature, differing from that of those accompanying it' (*SP* v. 191). According to Raymond Cooke, 'By 1912 Khlebnikov had already embarked on the path of "semanticizing" the initial letter of the word.'[3] Over the next few years he was to produce many annotated lists of words all commencing with the same initial letter and therefore presumed to share areas of meaning. Thus, *kh* was analysed as signalling 'types of human structures', examples being *khram* ('temple'), *khlev* ('cowshed'), *khoromy* ('mansion') (*SP* v. 200)—needless to say, the name Khlebnikov is not considered in this context. Often there is a sensuous dimension to the readings, with the posited meanings being seen in terms of physical shapes, trajectories, and geographical associations (with more abstract words figuring infrequently). Thus, the names *Asia, Amerika, Afrika, Avstralia* are said to bear traces of a proto-language in which the letter 'a' signified 'dry land' (*SP* v. 192).

Apart from his tireless collecting of examples in support of this eccentric linguistic theory—his 'Slovar bukvi El' ('Dictionary on the Letter L'), for

[99] 'Nash marsh' ('Our March'), *PSS* i. 75.
[1] 'Koe-chto pro Peterburg' ('A Few Things about Petersburg'), *PSS* i. 121.
[2] *Sdvigologiya russkogo yazyka* (Moscow, 1923), 40 f.
[3] *Velimir Khlebnikov: A Critical Study*, p. 75.

Here is the way the syllable *so* [with] is a region that encompasses *son* [sleep], *solntse* [sun], *sila* [strength], *solod* [malt], *slovo* [word], *sladkii* [sweet], *soi* [clan: Macedonian dial.], *sad* [garden], *selo* [settlement], *sol′* [salt], *slyt′* [to be reputed], *syn* [son].

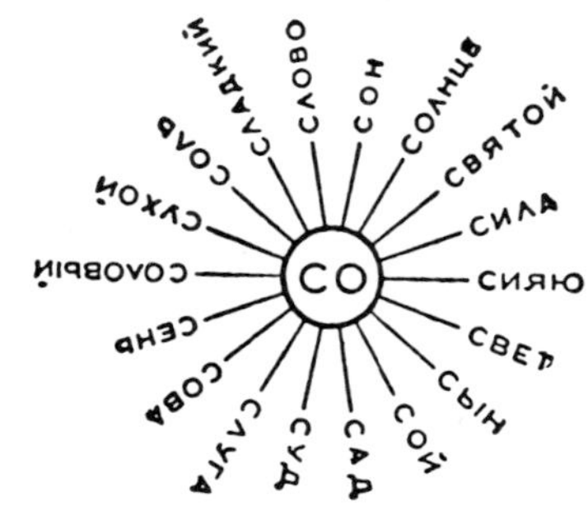

FIG. 4.6. Beginning of Velimir Khlebnikov, 'Kakim obrazom v *so* est′ oblast′ sna', 1912–13

instance—Khlebnikov also produced a whole series of poems involving words beginning with the same initial (e.g. 'Slovo o El' ('The Song of L'), 'Dva i tri' ('Two and Three')). His essay 'How in *so* there is a region' ('Kakim obrazom v *so* est′ oblast′ sna') (fig. 4.6) even includes a chart indicating the shared properties of S-words, in a manner reminiscent of shaped poetry. But this is rare. In general, Khlebnikov neither highlights the semantically dominant letter typographically, nor does he break with linearity in order to bring out any underlying common denominators.

Although it masqueraded as linguistic theory, one might argue that reflection of this sort is no different from the kind of associative values that were attached to certain letters of the alphabet in former times, the most famous example being Rimbaud's 'Voyelles', with its identification of different vowels with different colours. Rimbaud's poem is, in fact, referred to in Nikolai Burliuk's manifesto 'Poetic Beginnings' ('Poeticheskie nachala' (*Manifesty*, p. 79)), and, as Cooke points out,[4] it was well known to the writers of the Hylaea Group. There is even a direct parallel in Khlebnikov's own writing: in his 'sound-painting', with its association of certain consonants with particular colours (*b* with red, *p* as black, and so on (*SP* v. 269)). But in the case of pre-Futurist examples, what one is dealing with is no more than a subjective association, whereas there is a far stronger tendency among the Futurists to advance these associations, not as subjective response, but as part of a linguistic system.

As we have seen already in the case of the adjective 'schamzerpört', August Stramm also attributes semantic value to minuscule (sub-morphemic) parts of language—even to the point of seeing the ultimate significance of this com-

[4] *Velimir Khlebnikov. A Critical Study*, p. 209.

pound as lying in its umlaut. His comments on another compound, 'Laubwelk', further illustrate the same tendency to engage in quasi-linguistic ruminations about certain of his poetic neologisms. His comments, however, in no way privilege the initial letter in a word.

In a letter of 22 May 1914 Stramm sent Walden two versions of his poem 'Untreu' ('Unfaithful'). They differed only in respect of one word. The one version begins:

> Dein Lächeln weint in meiner Brust
> Die glutverbissnen Lippen eisen
> Im Atem wittert Laubwelk![5]

Bridgwater has:

> Your smile weeps in my breast
> the fire-wrought lips ice over
> your breathing weathers leaf-wither![6]

Where this version has *Laubwelk* ('leaf-wither'), the other retreats to the more conventional *Welkes Laub* ('withered leaves'). Although he submits both versions for his publisher's consideration, Stramm leaves Walden in no doubt that he prefers the bolder compound:

> 'Welkes Laub' [i.e. adjective plus noun, instead of the compound noun] has an admittedly softer and more melodious ring to it, but my feeling is that the two words are also less specific, whereas 'Laubwelk' contains more of the idea of the aroma, which was my main concern. Thus [by using the compound] I am able to avoid the duplication of the letter W at the beginning of two successive words, something which I particularly wanted to avoid because I had deliberately repeated the prefix 'ver-' as a trigger for the feeling of transience [*Ver*gehen] and of leaving [*Ver*lassen]. Finally, the repeated combination of the letter 'T' with an 'L' following it awakens in me a sense of gliding, of breath wafting past. All good reasons, then, why I prefer the last version.[7]

Even if, in Stramm's case, the idea is not enshrined in some full-fledged linguistic theory, we once again encounter a desire to attribute morphemic status to the phoneme.

At one moment in *8 anime in una bomba* there is a hint that Marinetti may also be attaching such values to the initial letter of certain words. He refers to 'the lugubrious predominance of the great M's Mountains Mother Moral Malevolence Martyrdom' (*TIF* 980), the Italian words being 'Montagne Madre Morale Male Martirio'. But no indication is given as to whether he seriously associates the initial letter with one particular semantic field.

In general, as this unclear instance indicates, sub-morphemic semanticization usually needs to receive some form of theoretical underpinning, if it is to

[5] *Das Werk*, p. 16. [6] *twenty two poems*, n.p. [7] *Das Werk*, pp. 426 f.

avoid ambiguity. On its own, whether one is thinking of Stramm's interpretation of the umlauted 'o' sound, Khlebnikov's views about the significance of initial letters, or Marinetti's 'M's, the effect risks remaining opaque. It needs a commentary from outside the work or some form of internalized explanation, as is the case in *Zangezi*.

In this context, one has to distinguish between poems resulting from the linguistic assumption that a certain meaning is attached to the initial letter of a word and the more traditional tautogram.[8] Maria Goretti's 'Volteggi acrobatici del mio apparecchio sulla lettera A forse suggeritimi da quell'alto campanile' ('Acrobatic Somersaults on the Letter A by my Device, which was perhaps suggested to me from that High Bell-tower'), is a work composed entirely of sequences beginning with the letter 'A' and employing a large number of further 'A's in the individual incantatory lines. But this is not because some semantic common denominator is being attributed to these words. The poem, a conventional tautogram, at most derives a certain acoustic sonority from the resultant vowel-patterning (hence the bell analogy which Khlebnikov also introduces in *Zangezi*).

Abbrividire
Ansia
Ansia che Affoca Avvampa Abbrucia
Ansia d'Azzurro nell'Abbacinante cielo di stamani
Ansia d'Azzurro nel folgorante Acciaio
Angelici Augurali Adoranti i miei occhi seguirti
Appassionatamente
Abolire Ah potere
Abolire distanza Astronomica che ci divide
Arrivare A te
Accostare pianissimo
Aleggiare nuvola soffio Alito
Allacciare Anche sì Anche Attimo solo
Ansia mia trepida Alla tua intrepida Audacia
Aderire Alata tua immagine
Addolcire tua forse-Amarezza
Aggirare A te intorno speranza chimera
Allucinazione
Affascinare tua strada distratta sulla mia Attesa
Arruffare i tuoi pensieri col mio riso monello
Addentare la frutta più acerba del tuo-mio desiderio
Accerchiare di spiralica melodia carezza l'Aurea tua velocità
Assalire il tuo Altissimo volo col mio Adolescente Amore
Assaporare con te l'Atroce profumo pericolo mortale
Avvinghiare tuo tenace sforzo di vittoria

[8] For illustrations, see Liede, *Dichtung als Spiel*, ii. 94–103.

Alla mia fragile nostalgia di felicità
Arroventare d'infernale fiamma i tuoi paradisiaci cieli
Ah Ah Allentare i tuoi nervi
Ammorbidire tua durissima volontà
Affralire Anima tua Argentea metallica in un bacio
Assurdo bacio
Assaporante Aromi Angelici Annunzio Araldo Assunzioni
Agonizzare insieme
Annegare
Azzurro tuo Acciaio nell'Acquamarina dei miei occhi[9]

Needless to say, no English translation is able to reflect the pattern of initial 'A's but words beginning with an 'A' in Italian have been given a capital letter here:

To shiver
Yearning
Yearning which Flares Burns Scorches
Yearning for Blue in the Dazzling sky of today
Yearning for Blue in the shining Steel
Angelic Augural Adoring my eyes follow you
Longingly
To Abolish All power
To Abolish Astronomical distance which divides us
To Arrive At you
To Approach very gently
To hover cloud puff Breath
To Connect Also yes Also only for a Moment
My trembling Yearning To your intrepid Boldness
To adhere to your lofty image
To Sweeten your perhaps-Bitterness
To go Around you hope chimera
Hallucination
To Charm your absent-minded way on my waiting
To Entangle your thoughts with my roguish smile
To Bite the greenest fruit of your-my desire
To surround with spiral melody caress your Golden speed
To assail your very High flight with my Adolescent Love
To Savour with you the Atrocious scent mortal danger
To clutch your tenacious effort for victory
To my fragile nostalgia for happiness
To make your heavenly sky red-hot with infernal flame
Ah Ah to Relax your nerves
To soften your hard will
To Weaken your Silver metal Soul in a kiss

[9] From the 'La preghiera alle macchine aeree' section of Maria Goretti's *La donna e il futurismo* (Verona, 1941), 111 f.

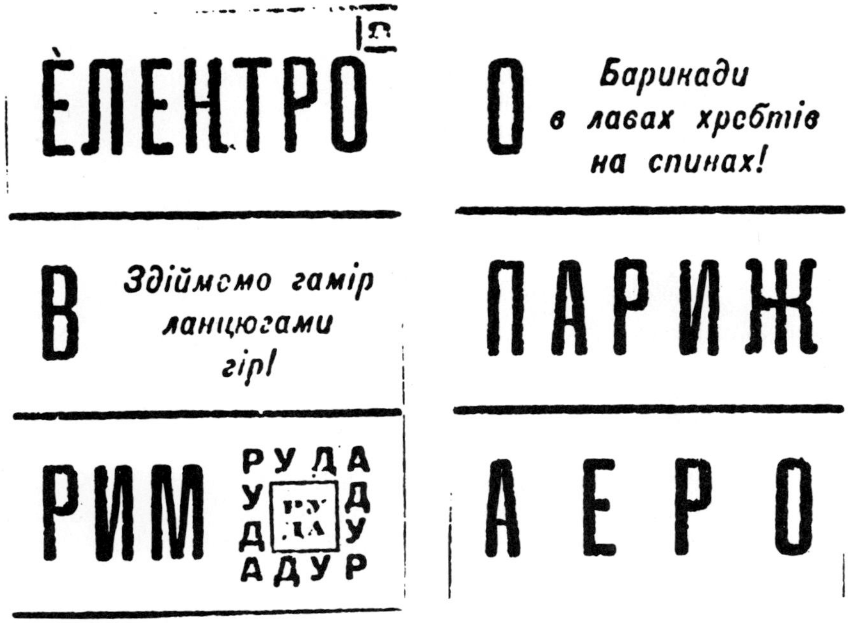

FIG. 4.7. Card Two of Mykhaylo Semenko, 'Kablepoema za okean', 1924

Absurd kiss
Savour Angelic Aromas Announcement Herald Assumptions
To Agonize together
To Drown
Your Azure Steel in the Aquamarine of my eyes

There are also a number of essentially tautogrammatic poems by Khlebnikov, usually those concerned with pagan deities, where similar games occur, as, for example, when the name of a character is repeatedly inserted into the text in anagrammatic form. In 'Perun' virtually every line contains a partial or complete anagram of the Slavic god's name. Similarly, the name of the Egyptian sun-god Ra becomes a key game-factor in 'Ra', although, as Vroon's analysis shows,[10] the situation is more complex than in the case of Perun, for *Ra* is also related to the name Razin and the district of that name between the Volga and the Don. Hence the result is not so much a playing on one single word but a whole complex of associations.

Acrostic patterns also figure in many of the cards of Semenko's 'Kablepoema za okean'. Thus, the sequence in Card Two (fig. 4.7) (Electro | In | Rome | Against | Paris | Aero) at the same time creates the word 'Europa' with its initial letters, when read vertically. And various other words ('machine', 'Asia', 'Africa',

[10] *Velimir Xlebnikov's Shorter Poems*, pp. 178 f.

'America') can be found in the cycle when the first letters of certain word-blocks are read vertically. Frequently, the effect is to join together various geographical names in the horizontal text in such a way as to produce the names of entire continents in the acrostic direction.

With Mayakovsky's 'Utro' we already encountered a tendency to play semantically with the inversions of words: thus the split word *rez-che* ('sharper') was reflected in the next two lines by the split word *che-rez* ('across'). Such local mirror effects can be found repeatedly in the poetry of both Mayakovsky and Khlebnikov, but the latter was also to go on to compose whole works on the palindromic principle, as if some new organizational structure had displaced that of conventional syntax. Khlebnikov's epic *Razin* and his shorter poem on the same principle 'Pereverten′' ('Turncoat') are bravura extensions of the single-line palindrome to the creation of entire works.

'I wrote "Pereverten′"', Khlebnikov recalls, 'without my conscious mind controlling the activity and it was only after having felt the lines . . . that I realized that they were like reflected rays of the future, cast by the subconscious "self" on to the sky of reason' (*SP* ii. 8 f.). In fact, the palindromic purity of the poem is so accomplished that it is difficult to credit that the work could be anything but a highly conscious construction.

(кукси кум мук и скук)

Кони, топот, инок,
но не речь, а черен он.
Идем молод, долом меди
чин зван мечем навзничь.
Пал а норов худ и дух ворона лап.
А что? я лов? Воля отча!
Яд, яд, дядя!
Иди, иди!
Мороз в узел, лезу взором.
Сóлов зов, воз волос.
Колесо. Жалко поклаж. Оселок.
Сани плот и воз, зов и толп и нас.
Горд дох, ход дрог.
И лежу. Ужели?
Зол гол лог лоз.
И к вам и трем с смерти мавки.

(*SP* ii. 43)

As the following translation of the first few nonsense lines shows, Khlebnikov's concern is with the reversal of words as sound-shapes alone, and not with any possible antithesis of meaning along a central axis, of the kind that often occurs in palindromes (e.g. T. Eliot/toilet or the Black Mass's God/dog).

Horses, cloppeting, monk
but not speech, he is black.
We go young through the valley of copper
Rank called from behind with a sword
Hunger long as the sword
He fell weak from wounds and breath of crow's feet
But what? Me a catch? The will paternal!
Poison, poison uncle!
Go, go! . . .

Krystyna Pomorska designates the central technique of 'Pereverten'' 'reverse homonymy', and she explicitly dissociates the effect from the earlier mannerist conceit: 'With all its similarity to the Baroque . . . palindrome . . . it is new here by virtue of its serious usage, whereas it formerly played the role of a humorous puzzle: when read backwards, the meaning changed from a serious message to a joke or to the negation of the former text.'[11] Various points can be made about this. First, in the Baroque context, a conceit was seldom merely a humorous game: the play between opposites was more often than not a matter of metaphysics, rather than mere *jeu d'esprit*. (This is most readily visible in those Baroque pictures of the man of the world which, when inverted, turn into a *momento mori*.[12]) The same conceit of inversion operates in the Baroque palindrome, but, as in most Baroque poetry, the antithetical has metaphysical significance. As if to emphasize both continuity and discontinuity within this tradition, the very title that Khlebnikov gives to his shorter work in this style, *Pereverten'*, uses the 'turncoat' image as a metaphor for its palindromic structure. He may deliberately avoid the standard Russian term (*palindrom*), but his poem represents a continuation of the genre, though with different functions, rather than a complete innovation. Nevertheless, there is one respect in which Khlebnikov's technique does need to be distinguished from a large portion of the palindromic poetry of the past: it uses the device to concentrate much more specifically on the materiality of the medium and to generate nonsense effects more than was usually the case with earlier palindromes.

There are various types of palindrome. A typological division can be made between *reciprocal* palindromes, i.e. those which produce the same utterance in either direction (single words like 'deified' or sentences like 'Madam, I'm Adam') and *non-reciprocal* ones. It is the second kind—the *non-reciprocal* palindrome—which is most frequently Khlebnikov's concern, no doubt because he is using the device as a strategy for unearthing new and unexpected relationships between words, rather than as a cerebral vehicle for wit. In addition, there

[11] *Russian Formalist Theory and its Poetic Ambiance*, p. 100.

[12] See, for instance, Mattaeus Merian's *Todten-Tantz* engraving, used as a frontispiece to Andreas Gryphius, *Cardenio und Celinde*, ed. Hugh Powell (Leicester, 1967). When inverted, the image of a nobleman in full regalia turns into a skull.

are what might be called *pure* and *impure* palindromes.[13] In the former, the exact mirror effect is achieved by reversing an utterance; in the second, one has a rought (acoustic) approximation, but not total fidelity to the original. In *Razin* Khlebnikov resorts to both types of palindrome, but he is extraordinarily able to maintain a largely 'pure' approach—which is particularly surprising when one considers how permissive his method of creating a suggestion of 'internal declension' was in a poem like 'Ya videl'. Admittedly, there are one or two minor respects in which his handling of the form might be considered less than rigorous. Not only is the Russian soft-sign discounted, but other small distinctions, e.g. between hard and soft 'e', are ignored, in the interests of an overall semblance of symmetry.

Peter Urban has argued that the challenge of the palindrome is different in various languages, and that Russian actually favours the form (by comparison with German):

> the palindrome is facilitated by the fact that in Russian there are neither auxiliary verbs nor articles (things which in German tend to restrict the palindrome to a small range of possibilities), by the fact that Russian is a synthetic, inflected language and finally by the fact that Russian possesses a large number of monosyllabic words, which Khlebnikov makes use of, together with numerous archaisms, regional expressions, uncommon and recherché words.[14]

One way of looking at the permutational poem by Apollinaire quoted earlier in this chapter would be to consider it as a form of lipogram, and a fairly radical one at that. Whereas most lipograms involve the removal of a limited number of letters from the available alphabet, one would have to assume that Apollinaire had gone so far as to remove twenty-one, leaving himself only five left with which to work.

The lipogram proper was exploited by a number of Futurists, for the particular reason that the removal of certain parts of the alphabet drew attention to the materiality of language. In *Dokhlaya luna* (*Croaked Moon*) David Burliuk published a number of poems deliberately excluding certain sounds of the Russian alphabet ('Bez *R*' ('Without R') and 'Bez *R* i *S*' ('Without R and S') even signal this in their titles). But the most thoroughgoing lipogrammatical exercise of the time is to be found in the *Sturmkreis*, in the work of Otto Nebel.

In his foreword to *Unfeig* (*Uncowardly*) Otto Nebel makes it clear that he wishes these permutational exercises to be received in the context of a Futurist displacement of grammar. Having specified that the work is a piece of

[13] Peter Urban distinguishes between 'completely arithmetical' and merely 'phonetically' symmetrical palindromes (*Velimir Chlebnikov: Werke*, i. 424).

[14] Ibid. 423.

absolute poetry . . . a figure using nine runes

U G T

E F N

I R Z

Nebel adds that 'these nine abstract poetic elements are, taken in isolation, devoid of any aesthetic meaning. They became constructive means and receive their life in the work both through their optical and their acoustic arrangement.' He refers to the thousands of permutations in which these 'runes' occur, explaining that these various 'sequences' are created 'without reference to grammatical or other intellectual concerns, inasmuch as they are at odds with or conflict with the nature of creativity and the logic of the work of art'.[15] Permutation, in other words, is as much a response to the irrelevance of grammar as telegraphic sentence-reduction or experimental non-linear layouts were at this time. Moreover, in the equation of grammar with 'intellectual concerns', one encounters the same anti-intellectualism that is the hallmark of Futurist linguistic reform and which figured in Khlebnikov's account of his use of the palindrome.

There are some sequences in the poet's parallel work *Zuginsfeld* (*Foray into the Field*) where Nebel actually bases his lipogrammatic word-games on the language of grammatical analysis, as in the following poem:

Tatform	Active
Leideform	Passive
Der Aktive	The active person
Der Passive	The passive person
Keine Verwechslung!	Don't confuse them!
Reinliche Trennung	Keep them tidily separate
(Sieh Reinlichkeit)	(Cf. 'tidiness')
Aktiva-Passiva	Actives-Passives
Tatform leidet nicht	Active doesn't suffer
Leideform tut alles	Passive does everything
Tatform tut nichts	Active does nothing
Das macht nichts	Doesn't matter
Sie tut nur so	It only pretends to
Sie tut sich gütlich	It makes free
Du meine Güte	Well, I never
Es tut sich was	It does something to itself
Er tut sich was zugute	He does something to his advantage
Zuguterletzt: Der Gute	Finally, the fellow
Mein Bester	My dear fellow
Du Reizender	You charming chap
Asbest	Asbestos

[15] *Der Sturm* (1924), 130f.

Herzallerliebster	Most liked of all
Nein, wie lieb	No, how nice
Und so liebenswürdig	And so kind
Und so weiter	And so on
Der Erstebeste	The first to hand
Wer ist der Beste	Who is the best
Wer zuletzt lacht	He who laughs last
Der Erste ist der Beste	The first is the best
Nein, der Letzte	No, the last
Am besten ist nichts zu lachen . . .[16]	The best thing is not to laugh . . .

Many sequences, if not automatically leading to neologisms, nevertheless play with the patterns of existent word-creation so as to highlight the 'flexibility' of the German language. For example:

Es kann etwas vorfallen	Something can happen
Bei Nacht	At night
Ein Vorfall	An incident
Ein Zufall	A chance event
Ein Einfall	A brainwave
Ein Durchfall	A calamity
Ein Ausbruch	A breakout
ein Durchbruch	A breakthrough
Bruchstrich . . .[17]	

The final word, a mathematical term for the dividing line in a fraction, here not only becomes self-referential, in the sense that it draws a line across one section of the poem and allows the poet to start afresh on a new sequence ('Ich stehe auf . . .'); it also establishes a new physical relationship with the word for 'breaking'—there in both the English 'fraction' from *frangere* and in the German for a fraction (*Bruch/brechen*)—thereby lending a physicality to the earlier abstract words and emphasizing their dependence on what was once very dramatic imagery.

Nebel's focus on language soon leads to a contrast between the word-conscious creative poet and the (passéist) language-abusing world around him:

Ver und Zer und Vers und Kitsch
Tinten-Stifte
Federvieh
Wer nichts hört, verschreibt sich was
Verschreibungen
Ver-Rechnungen
Ver-Zierungen

[16] *Das dichterische Werk*, i. *Zuginsfeld, Unfeig, Das Rad der Titanen*, ed. René Radrizzani (Munich, 1979), 27 f.

[17] Ibid. 73.

Ver-Zerrungen
Ver-Sammlungen
Ver-Nichtungen
Verdichtungen
  Dichter entstehen nicht
  Dichter entlaufen[18]

Frequently the effect of such associative word-plays is satirical.

Ich hüte mich
  Der hohe Hut
  Schön ist ein Zylinder
Die verfluchten Zivilisten
Die verfluchten Zivil-Listen
Die hohe Hut
Obhut
Kopfbedeckung
Fliegende Hüter
Meine Flieger
Flugstaffel
Stoffel
Staffelstab
Staffelei
AFLEI
(Artilleriefliegerei)
IFLEI
(Infantriefliegerei)
  Das neue Deutsch
  Ei ei[19]

Or again, an anti-war sequence:

Keile kriegen
Kriege keilen
Husch
Verpfuscht
Versiegen Leben
Kinder, Kinder
Vater Sieg ist Völkermord
Strömt herbei, ihr Völkerscharen
Und
Zermalmt die MOLOCHEI
  Aber vorsichtig[20]

The sheer untranslatability of many of Nebel's effects is a sign of how related many of his word-games are to the German language and once more just how

[18] *Das dichterische Werk*, i. (Munich, 1979), 87.
[19] Ibid. 73 f.
[20] Ibid. 155.

different the conditions were under which the various Futurists from different countries approached the task of language-revitalization.

Nebel himself did translate some of his work, but into visual and acoustic forms, not into some other language. In his final years he spent some time putting the early poems on to tape and bringing out the sound-patterning that the first *Unfeig* introduction had referred to. And in a posthumously published piece entitled 'Schauen und Lauschen' ('Looking and Listening'), he describes how he also gradually developed his lipogrammatical 'Nine-Rune-Fugue' *Unfeig* from its musical–fugal origins in the direction of a visual permutational pattern of coloured dashes and signs laid out in positions previously occupied by the words of the poem. The result he calls a 'gigantic mosaic', 'a panorama on four long word-banners'.[21] In other words, like many of the *zaum'* poets and acoustic patterners, Nebel tried to take his concern with the materiality of language to the ultimate limit: that of pure abstraction.

'The Futurist movement has undoubtedly dramatized the need for an adequate system of scientific poetics,' Victor Erlich observed in the 1960s. Indeed, he claimed that 'the movement was to become one of the main factors behind the emergence of Russian Formalism'.[22] One *zaum'* poet in particular, 'Alyagrov' (i.e. Roman Jakobson), was not only to become, with his *Noveyshaya russkaya poeziya*, one of the most discerning analysts of his fellow Futurists' linguistic literature, but he was later, in America, to form the bridge leading to the impressive scholarly analysis of Russian Futurist poetry by such commentators as Vahan Barooshian, Krystyna Pomorska, and Ronald Vroon. No such claims can be made for the impact of the linguistic literature of Italian Futurism. Not only did it fail to inspire any detailed analysis at the time, even in the intervening years it has suffered from relative neglect. And yet, as I have tried to show, a number of the experiments being carried out by the Russian Futurists did have their counterparts, albeit comparatively modest ones, in the work of some of the Italian Futurists.

[21] *UNFEIG: Eine Neun-Runen-Fuge: Zur Unzeit gegeigt*, ed. René Radrizzani (Zurich, 1960), n.p. A facsimile of these 'Schriftfahnen' is included as a fold-out at the end of Radrizzani's edition.

[22] *Russian Formalism*, p. 49.

5.
Futurist Primitivisms

'I CRAVED for savagery. . . . I should have liked to have gone naked with Picts and Celts, painted in woad or clothed in pelt and rags, shameless around fires or in dark caves.' 'In the prose countries, something was felt to have gone out of life, hence the attraction of the primitive.'[1] The attraction which Stephen Spender voices in these two contexts, one autobiographical, the other academic, was not just felt by the inter-war generation in the 'prose countries' of northern Europe. During the first few decades of the twentieth century the lure of the primitive was, in fact, palpable almost everywhere in avant-garde circles: in the Mediterranean regions as much as in London, Paris, or Berlin, not to mention in a number of Slavic countries. Indeed, it was one of the hallmarks of the avant-garde in both Europe and the New World. The atavistic excitement captured in the German Expressionist Wilhelm Klemm's words *Du großartiges Wort: primitiv! primitiv! | Ich könnte dich lallen bis zur Benommenheit* ('Primitive! Primitive! O you wonderful word. I could chant you until I fell into a trance'[2]) might have been echoed, with a variety of nuances and associations, by a whole host of artists and writers during the early decades of this century. Pronounced primitivist characteristics are to be found in one form or another in the work of some of the most significant figures of the period: most famously, in Picasso's *Les Demoiselles d'Avignon*, with its bold references to the African mask, and in the large-scale Cubist appropriation of *art nègre* that followed in its wake; but also in the archaic woodcuts and cultivatedly naïve canvasses of the artists of *Die Brücke* and *Der Blaue Reiter*; in the concern for ethnic forms that shaped the statues and woodcarvings of Gaudier-Brzeska, Epstein, Archipenko, and Brancusi; and in what might be thought of as the 'tribal elements' of Zurich Dada performance. Around this time, a series of influential treatises, including Carl Einstein's *Negerplastik* (*Negro Plastic Art* (1915)), Marius de Zayas's *On African Negro Art and its Influence on Modern Art* (1916), Alexander Shevchenko's *Neo-Primitivizm* (*Neo-Primitivism* (1913)), and Vladimir Markov's *Iskusstvo negrov* (*The Art of Negroes* (1919)), also played their part in sensitizing contemporary awareness to the aesthetic value of ethnic

[1] These quotations are taken, respectively, from Spender's *World within World* (London, 1953), 2, and *Journals 1939–1983*, ed. John Goldsmith (London, 1985), 136. On the internalization of the primitive, as part of a modern sensibility, instead of its projection on to some exotic culture, see Walter J. Ong, *The Barbarian Within* (New York, 1962).

[2] *Die Satanspuppe* (1922; repr. Nendeln, 1973), 53.

forms of art, thus fostering a receptivity to the new creative interaction between the artefacts of remote cultures and their modernist counterparts. In the same spirit, the *Blaue Reiter* almanac published pieces on 'The "Savages" of Germany' by Franz Marc and 'The "Savages" of Russia' by David Burliuk. Using the word with its ethnic connotation this time, Herwarth Walden was to declare in the 1920s that 'it's the savages who have the best art'.[3] Nevertheless, this large-scale cult of the savage and the primitive is by no means reducible to the negrification of European early modernism. When Oto Bihalji-Merin talks, in *Modern Primitives*, of a 'burst of archaic primitiveness into modern art', welcoming what he calls this 'decisive event of our time' as 'a reaction against technical virtuosity, disintegration, and abstraction', he is thinking not of any impulse from African ethnic art (to which such arguments also apply), but of the 'conscious *naïveté*' of a whole host of 'primitive' painters from Henri Rousseau and Grandma Moses to the Jugoslavian 'peasant naïves' of more recent years.[4] Yet even the culturally less eclectic primitivism of Bihalji-Merin's twentieth-century *primitifs* represents but one further strain within a much wider tendency—a tendency which also includes the way in which, in Germany, Wilhelm Worringer lent decisive theoretical support to the exploration of new Gothic forms of primitivism; and which embraces such things as the role played in Russian art by childlike folk elements in the paintings of Goncharova and Larionov, and the way in which the Hylaea Group of Black Sea Futurists, viewed by Markov as the virtual personification of primitivism,[5] derived much of their inspiration, not from the Congo or Polynesia, but from the Slavic and Siberian folk art of earlier centuries.

Kornei Chukovsky's *Futuristy* devotes considerable space to an amusingly satirical survey of the primitivist trend within early Russian Futurism. And, while his onslaught is directed above all against the Ego-Futurists, in particular Igor Severyanin, Chukovsky manages to direct his fire at a generous sweep of further targets *en route*. In the period just before the First World War, according to Chukovsky, 'there began to emerge in both literature and art the cult of the caveman, the fashion of the primitive and the savage. . . . Poets vied with one another with all their might to see who could bellow in the most bestial

[3] 'Zur Kunst der Neger und Südsee-Insulaner', *Der Sturm*, 17 (1926), 18. Wassily Kandinsky, one of the major influences on early *Sturmkreis* aesthetic theory, also writes of 'our sympathy for, our understanding of, and our kinship with the primitives' (*Über das Geistige in der Kunst*[10] (Bern, 1973), 21). See also Emil Filla's Czech essay 'O ctnosti novoprimitivismu', *Volné směry*, 9 (1913), 18–23.

[4] *Modern Primitives: Naïve Painting from the Late Seventeenth Century until the Present Day* (London, 1971), 276.

[5] See Vladimir Markov's chapter on the Hylaea Group in *Russian Futurism: A History* (Berkeley, 1968), 29–60, and J. E. Bowlt, 'Neo-primitivism and Russian Painting', in his *Russian Art, 1875–1975: A Collection of Essays* (New York, 1976), 94–111. Nils Åke Nilsson has documented connections between the primitivist concerns of the Hylaea Futurists and a whole range of further forms of (neo-)primitivism in the first decade of this century: 'Futurism, Primitivism and the Russian Avant-garde', *Russian Literature*, 8 (1980), 469–82.

manner.' That the Futurists had no monopoly on such concerns becomes clear when Chukovsky begins by citing Gumilov's remark that 'as Adamists, we turn . . . to the beasts of the forests. . . . Let us therefore reject the couches of a thousand years of culture.'[6] Further examples of a back-to-the-primitive urge are diagnosed: in the *Blaue Reiter* almanac (with its interest in Malaysian and New Caledonian ethnic art), in Mussorgsky's 'adorable barbarism', in the work of Khlebnikov (here quoted as declaring 'I must find shelter with the savages of Siberia'), Kamensky ('I have learnt wisdom from the sun-coloured savages'), in 'anti-intellectualist Bergsonism', and, of course, in numerous Futurist writings by Severyanin himself, duly censured for daring to claim 'I am inseparable from the savages. . . . My spirit aspires to the Primitive.' As further grist to his mill, Chukovsky offers a passage from the magazine *Soyuz molodezhi* (*Union of Youth*): 'It is time for your Apollo to die, your Apollo is rachitic and gouty. We'll crack his skull. We offer you another kind of Apollo: a black and twisted monster.'

For Chukovsky, 'the taste for the primitive' is not only present in Futurism's interest in the 'black and twisted monsters' of ethnic cultures; it has also awakened 'a particular predilection for children, for the soul of children'. Not only are their poems printed in *Sadok sudey* (*A Trap for Judges*), but an enthusiasm for children's drawings leaves its mark on 'A Slap in the Face of Public Taste' ('Poshchechina obshchestvennomu vkusu'), the *Blaue Reiter* almanac, and the naïve style of many Russian Futurist artists.[7] Chukovsky's *Futuristy* unfolds this panorama of current primitivist tendencies primarily in order to accuse Severyanin of characteristically climbing on to a further contemporary artistic band-waggon. But the wealth of illustrations of primitivism's fashionability that Chukovsky is able to amass by 1922[8] to support his picture of Severyanin the opportunist indicates just how widespread, and independent of Cubist *art nègre*, the tendency was at the time in Russia.

What still does not figure in Chukovsky's rich catalogue is any noticeable awareness of the gestures of iconoclasm and cultural anarchy with which the Russian Futurists were also associated. 'Throw Pushkin, Dostoevsky, Tolstoy, *et al.*, overboard from the Ship of Modernity', may be little more than promotional rhetoric in the context of 'A Slap in the Face of Public Taste' (*Manifesty*, p. 50). But then there is the sounding in Khlebnikov's 'Trumpet of Martians' ('Truba marsian') of 'the words of the new hostility' (*Manifesty*, p. 161). Or, the

[6] *Futuristy. Igor' Severyanin.—Al. Kruchenykh.—Vas. Kamensky.—Vl. Mayakovsky* (Peterburg, 1922), 10f.

[7] Ibid. 15.

[8] Chukovsky's *Futuristy* of 1922 is an extended version of two essays first published in 1914: 'Ego-futuristy i kubo-futuristy', *Shipovnik*, 22 (also in *Litsa i maski* (St Petersburg, 1914)). While the earlier essays have been reprinted ('Russian Titles for the Specialist Series', No. 83; Letchworth, 1976), the fuller version of 1922 has yet to be reissued. This is a pity, since the 1914 version says relatively little about the primitivist tendencies of the period.

same poet's question, 'Where are the great destroyers of books?' and his confession of longing 'for a great bonfire of books' (*SP* v. 174, 183). (In his *Vnuchka Malushi* (*Malusha's Granddaughter*) that conflagration is duly enacted.) Also, in Mayakovsky's 'A Drop of Tar' ('Kaplya degtya'), there are equally disturbing references to anarchy and destruction and the concluding thought that Futurism has Russia 'by the throat' (*Vzyal. Baraban Futuristov*, 1915). Such imagery is as much a part of the new primitivism as the less disturbing forms of ethnic eclecticism that Chukovsky diagnoses.

Gga (*Honk*), the first almanac of the Polish Futurists, contains a manifesto by Anatol Stern and Aleksander Wat entitled 'The Primitivists to the People of the World and Poland' ('Prymitywiści do narodów świata i Polski' (1920)), as if to be a Futurist was synonymous with being a primitivist. In western European literature, too, primitivism of one kind or another had by the second decade of the century become a major phenomenon, in both Futurism itself and in various related avant-gardes. It figured in the works of such radically different and differently radical writers as Apollinaire,[9] Wyndham Lewis, the Zurich Dadaists,[10] the *Sturmkreis* poets and many of the leading playwrights of German Expressionism.[11] 'I am the master of all that is half-uttered and imperfectly heard', Wyndham Lewis declares in his poem 'One-way Song', 'Return with me where I am crying out with the gorilla and the bird'. On the whole, though, modern primitivism has tended to be distinctly un-Rousseauistic in quality. Robert Goldwater's contention that 'where the nineteenth century thought of the primitive ... as calm and reasonable, the twentieth sees it as violent and overwhelming'[12] is as applicable to the literature of the period as he has shown it to be to the visual arts.

In its more productive manifestations, primitivism was a form of cultural revitalization, part of a search for less intellectualized, more vibrant sources of artistic energy. In both 'The Plastic Foundation of Futurist Painting and Sculpture' ('Fondamento plastico della pittura e scultura futuriste') and in *Pittura scultura futuriste* (*Futurist Painting and Sculpture*), Boccioni declares: 'we Italians need the barbaric to revitalize ourselves. Our race has ... always renewed itself by barbaric contacts' (*Gli scritti*, pp. 40, 138). But this renewal through recourse to the primitive unfortunately came to the fore at a time

[9] The subject is explored in Katia Samaltanos, *Apollinaire: Catalyst for Primitivism, Picabia and Duchamps* (Studies in the Fine Arts: The Avant-Garde, 45; Ann Arbor, 1981). Being primarily concerned with Apollinaire's art-criticism, Samaltanos does not consider any parallel primitivism in his literary works.

[10] There is much material on this subject in Christopher Middleton, 'The Rise of Primitivism and its Relevance to the Poetry of Expressionism and Dada', in his *Bolshevism in Art and Other Expository Writings* (Manchester, 1978), 23–37.

[11] See Regina Baltz-Balzberg, *Primitivität der Moderne 1895–1925 am Beispiel des Theaters* (Königstein, 1983), and Bernard Myers, *Expressionism: A Generation in Revolt* (London, 1963).

[12] *Primitivism in Modern Art* (New York, 1938; rev. edn. 1967), 95.

when, to quote Christopher Middleton, 'reason seems to have been hungering, or sickening, for primitive unreason'; and it was for that reason perhaps inevitable that the ensuing archaizing tendencies should have remained 'both a vital and a deadly force in social and cultural life ever since primitivism's epiphany around 1900'.[13] And here lies one of the principal moral problems raised by the new phenomenon. For artistic renewal of this kind can exact a high price.

'It would be agreeable', Lázsló Moholy-Nagy once remarked,

> to believe that inventiveness is to be found only in the works of men who have a fundamental respect for the integrity of each man, who have a feeling of social responsibility. But Marinetti [and he is by no means alone in this among contemporary modernists] exemplifies the case that a misinterpretation, or a malicious interpretation, of the storm signs of an imminent social crisis can also instigate inventions in literary techniques[14]

and significant inventions at that, as we have seen in the preceding chapters.

Certainly, no other avant-garde movement in the early part of the century was characterized by such a chaotic and morally ambiguous amalgam of positive and negative aspects of primitivism. This was probably in no small measure due to Futurism's chronological position as first in a series of adversary responses to much that the nineteenth century represented. Although many other factors, including the current vogue for aggressively vitalist Sorel- and Nietzsche-derived philosophies, for gestures of cultural anarchy and political and artistic chauvinism, and not least, Marinetti's own belligerent and volatile personality, all played their part in nurturing the general climate of irrationalism which preceded the First World War, but did not come to an end when it did.

In his exploration of some of the aesthetically enriching aspects of Expressionist and Dada primitivism, Christopher Middleton rightly points out that 'there is [also] a "false primitivism" which arises when the primitive or the archaic exerts a spell that paralyses critical and moral judgement'.[15] In his investigation weighted towards the positive aspects of the new aesthetic primitivism in German literature, however, Middleton (as far as one can see) encounters little difficulty in holding the two kinds apart. When it comes to Futurism, any such polarization becomes less easy. For here, the cult of the archaic, the desperate hunt for new revitalizing possibilities—the 're-barbarization of art', as it has even been called—were often to give rise to a

[13] *Bolshevism in Art*, p. 23. Although Futurist primitivism is primarily associated with the 'heroic' early period of the movement's history, even the post-war Futurists frequently presented themselves as primitives. T. A. Bracci and Fillia, in *alfabeto spirituale* (Turin, 1925; repr. in Enrico Crispolti, *Il Secondo Futurismo. 5 pittori + 1 scultore. Torino, 1923–1938* (Turin, 1962), 272 f.), still regard their main goal as the creation of a 'primitive sensibility of lyricism'.

[14] *Vision in Motion*[6] (Chicago, 1961), 303.

[15] *Bolshevism in Art*, p. 32.

disconcertingly ambiguous fusion of the morally regressive and the experimental, the politically reactionary, and the aesthetically avant-garde. The result is a critical predicament that Frederic Jameson has addressed in a related context.

Exploring 'the affinities between protofascism and Western modernism' in the case of Wyndham Lewis, Jameson finds himself confronted by a dilemma similar to that created by many Futurist works. 'The most urgent and visceral issue for any reader', as he puts it, is 'why he or she should be expected to find aesthetic pleasure and satisfaction in a work whose impulses are often so ugly or ideologically so offensive.'[16] Or, put more bluntly, how can the critic reconcile the negative with the positive aspects of primitivism? How can the reactionary and atavistic element in many avant-garde writings be squared with other modernist features which make them of such aesthetic interest?

Jameson endeavours to justify book-length scholarly preoccupation with Lewis's writing, despite such qualms, because it '"produces" the ideological as an object for our aesthetic contemplation and our political judgment'.[17] On such charitable (and presumably non-intentionalist) grounds—which appear even more contentious, when one recalls that, from his sub-title onwards, Jameson has approached Lewis as a casebook illustration of certain 'protofascist' tendencies in modernism—Jameson ventures the opinion that Lewis's works can only 'be even more painful for protofascism itself, which must thereby contemplate its own unlovely image and hear blurted out in public speech what even in private was never meant to be more than tacitly understood'.[18] Such a hypothesis has to become, for Jameson, the ideological justification for an uneasy attraction to 'Lewis's style, the only true English Futurism, an immense hangar in which we may still learn to tap the almost extinct resources of language production'.[19]

This reader, for one, remains unconvinced by such arguments, even in the case of Wyndham Lewis's work; and they certainly do not transfer unproblematically to Futurist primitivism in its even more abrasively atavistic forms. Yet, although the dilemma that Jameson pinpoints here is occasioned by so many works of Futurist literature, it tends to be passed over in diplomatic silence by all but the movement's most polemical detractors. Such reticence is in fact part and parcel of a tendency in recent years to whitewash Futurism—to market it, in homogenized form, as no more than one further example of the avant-garde,[20] rather than at the same time one of extreme modernism's more

[16] *Fables of Aggression: Wyndham Lewis, the Modernist as Fascist* (Berkeley, 1979), 20.

[17] Ibid. 22. [18] Ibid. 23 [19] Ibid. 86.

[20] In general, it has been remarked, 'the neo-primitive current . . . within the avant-garde . . . remains the most unpopular' (Renato Poggioli, *The Theory of the Avant-Garde* (Cambridge, Mass., 1968), 49), which is as true of academe as of the general public's response. Fortunato Bellonzi's complaint that 'there is a phase of Marinetti's writings . . . for which he has yet to receive the recognition due to him: a certain primitivism' is as true today as when it was made in Feb. 1954 in *La fiera letteraria*

insidious manifestations. This policy is an unfortunate development, for Futurism would not have been Futurism without its primitive component; and that component, as we shall see, had its good and its bad sides. Hence, a concern with the negative aspects can be as instructive as a focusing only on the positive features. What remains clear is that one cannot simply use the innovative aspects of Futurist primitivism as a justification, or even a whitewash, for its reactionary components.

In her study of Apollinaire and the rise of primitivism, Katia Samaltanos interprets early twentieth-century primitivism as a phenomenon which 'overreaches and subsumes all other "isms" of the period'.[21] Yet it could be argued that the primitive is essentially a strain—or a variety of strains—running through a number of avant-garde movements, not some single umbrella development under which various other 'isms' can be wholly subsumed. How these strains manifest themselves, and where and how the specifically Futurist forms of primitivism differ from their main modernist analogues are the particular questions which require attention, if such a central but complex component of literary Futurism is to receive its due.

One thing that is undeniable, and furnishes us with a convenient starting-point, is that a concept of 'primitivism' was from the outset part of the Italian Futurists' own self-image. Carlo Carrà may have grumbled at the time that blanket-words like 'primitive' were 'too vague, too indeterminate and elastic' to be of much assistance in an aesthetic context,[22] but this was a feature which actually contributed to the term's appeal in the early proselytizing days. For primitivism was—like the cult of the telegraphic—in the air at the time. By 1914, indeed, 'there were already signs of primitivism becoming a cliché'.[23] Thus, by latching on to a watchword of the moment, the early Futurists were once more able to align themselves with a fashionable and acknowledgedly avant-garde cause, while noisily making much of the fact that their own particular brand of primitivism was in no way to be confused with that espoused by any other coterie. The attractions and rewards of such a promotional strategy may well account for the Italian Futurists' eagerness in early manifestos to

(although I would prefer to view it as a continuing element, rather than a 'phase'). Moreover, the point applies not only to Marinetti's own brand but to most other Italian forms of Futurist primitivism, whereas their Russian counterparts, who were seldom as reactionary or as aggressive, have enjoyed more attention in this respect. Even if the institutionalized control over Italian Futurist scholarship—about which Christiana J. Taylor has some pungent things to say in her *Futurism: Politics, Painting and Performance* (Studies in the Fine Arts: The Avant-Garde, 8; Ann Abor, 1979)—is no longer a disincentive, the movement's apologists have tended to shy away from the uncomfortably regressive aspects of Futurism.

21 *Apollinaire: Catalyst for Primitivism*, p. xiv.

22 *Guerrapittura: Futurismo politico, dinamismo plastico, 12 disegni guerreschi, parole in libertà* (Milan, 1915), 71.

23 *Bolshevism in Art*, p. 30.

identify themselves as 'primitives': the 'primitives of a new sensitiveness, multiplied hundredfold', as the Italian painters proudly announced at the Sackville Gallery, London, in March 1912.[24] In the Bernheim-June catalogue of the previous month, they had styled themselves equally programmatically as 'the *primitives* of a completely renewed sensibility', and they had specifically drawn French attention to what they termed 'the brutally realist aspects of our primitivism'.[25]

Given that these expository remarks figure in statements made for foreign consumption, in contexts where the Italians found themselves obliged to explain their new style to London and Paris audiences, it is not inconceivable that they had merely resorted to the term as a convenience-label, one guaranteed to situate and delineate their interests within the more familiar topography of the European avant-garde landscape. (This was surely one of the reasons, in addition to any polemical impact, why Apollinaire chose to define Futurism in 1912 as 'an Italian imitation of two schools in French painting which have followed one another in the last few years: Fauvism and Cubism'.[26]) Yet one finds the Italians elsewhere opting for the term 'primitive' when this cannot be accounted for by any such international, mediating perspective. In a private letter to Vico Baer of 1 July 1913, for example, Boccioni describes what he calls his Futurist 'primitivism' as 'the extreme point of a form of complexity' (*Gli scritti*, p. 370). And in *Pittura scultura futuriste* he repeatedly puts forward the view that, in contrast to the retrospective primitivism of the Cubists, the Futurists are to be seen as '*the primitives of a new sensibility that has been completely transformed*' (e.g. *Gli scritti*, p. 132).

If such manifesto generalities have an on the whole innocuous ring about them, it would nevertheless not be difficult to put together a well-stocked chamber of horrors relating to Italian Futurist primitivism, a museum which would serve as a reminder that there is at the same time more going on here than a healthy vitalist renewal of tired artistic traditions. The most widely remembered article of the Futurist manifesto-programme that has a bearing on

[24] 'Futurist Painting: Technical Manifesto', in *Exhibition of Works by the Italian Futurist Painters*, quoted from *Futurist Manifestos*, ed. Umbro Apollonio (London, 1973), 29. This phrase does not occur in the Italian original (*Archivi*, i. 67); there the reference is simply to 'our acute sensibility'.

[25] 'Les Exposants au public' (*Archivi*, i. 107 f.).

[26] 'Les Futuristes', in *L'Intermédiaire des chercheurs et des curieux* of 10 Oct. 1912, repr. in *Chroniques d'art (1902–1918)*, ed. L. C. Breunig (Paris, 1960), 262 f. This is a deliberate echo of the claim, in 'Les Exposants au public', that the Futurists had 'surpassed' the advances of both Cubism and Fauvism (*Archivi*, i. 118). In *Le Petit Bleu* of 9 Feb. 1912 Apollinaire even establishes a Fauvist pedigree for the Italian Futurists, with Severini's paintings being 'too influenced . . . by Van Dongen' and Carrà being dismissed as 'a sort of more vulgar version of Rouault' (*Chroniques d'art*, p. 213). Carrieri refers to the painters loosely as 'the "Fauves" of Motion and Velocity' (*Futurism* (Milan, 1963), 40 f.) and Marianne Martin cites the 'Fauve tendencies of the Futurists' (*Futurist Art and Theory, 1909–1915* (Oxford 1968), 41 f., 54). In the light of these remarks, a detailed comparison of the two movements would seem to be a desideratum.

the cult of the primitive is a declaration of open aggression. 'We will destroy the museums, the libraries, the academies of every kind', Point Ten of the 'Foundation and Manifesto of Futurism' ('Fondazione e Manifesto del Futurismo' (1909)) begins (*TIF* 10). And one hears in another early manifesto ('Against Past-loving Venice' ('Contro Venezia passatista', 1910)), the equally philistine and ignominious: 'Let us hasten to fill in Venice's smelly little canals with the shards from its leprous, crumbling palaces. Let us burn the gondolas' (*TIF* 30). 'Let's bomb the Academies, the Last Stronghold of Pacifism' ('Bombardiamo le accademie, ultimo residuo pacifista'), the title of a later manifesto by Prampolini parrots (*Archivi*, i. 30; written in 1913, the piece was not published until 1 February 1918, in *Il Fronte Interno*). In 'Let's Murder the Moonshine!' ('Uccidiamo il Chiaro di Luna!') of the previous year, a dreaming Enrico Cavacchioli is heard to mutter: 'I want expensive playthings to smash . . . cities to raze to the ground, human anthills to kick aside' (*TIF* 17). The same Cavacchioli published 'Inno alla crudeltà' ('Hymn to Cruelty') in 1914, and his poem 'Rivoluzione' ('Revolution') includes the exhortation to kill all useless people ('Uccidiamo gli uomini inutili!'). 'We shall have to hang or shoot anyone who dissociates himself from our idea of a great Futurist Italy,' Boccioni, usually rated the voice of relative reason among the hot-headed Futurists, proclaims in *Pittura scultura futuriste* (*Gli scritti*, p. 81). In the early manifestos, at least, the Futurist ethos would at times seem to have amounted to little more than a series of formulae for cultural vandalism and injunctions to large-scale aggression against the passéists and their world. But the prevalence of such exhortations in the early manifestos suggests that this was, in some measure at least, an attention-seeking ploy. (Even subsequent paragraphs of the first manifesto make it clear that what is really at stake is a rhetoric to establish attitudes to the past, not some call for direct action. The fantastic side of such rhetoric is well to the fore in Marinetti's attacks on the fictitious cities of Paralisi and Podagra, Paralysis and Gout, which play as large a role in his invective as passéist Rome and Madrid, against which he also delivered onslaughts.) Despite fantasies about destroying Venice, murdering the 'moonlight' and bombing the academies, probably the closest the Futurists come to actual iconoclasm is during the scene in Marinetti's *Runio Clacla* when various paintings are destroyed. In fact, if a more deep-seated primitivism finds its way into early literary Futurism, then it is not primarily in the form of anti-passéist gestures of this kind. It assumes more reactionary guises than aggressive cultural posturings, 'slaps in the face of public taste', attacks on sentimental moonlight, jejeune Venice, and paralytic academies.

On the landscape of early Futurist literature, a much more insidious primitivism was to rear its ugly head very early on—insidious not least because it was feeding off and exacerbating a climate of violence which was about to find a ready political outlet. Marinetti's callous catch-phrase 'war, the world's only hygiene', merits the unchallenged place of honour in any such chamber of primitivist horrors. But room would also have to be reserved for the section 'The Rape of the Negresses' and many of the gratuitous slaughter-scenes in *Mafarka le futuriste*. And one would not have to search as far as Futurism's later love-affair with Mussolini's Fascism to find embarrassing displays of regression and mindless calls to violence. The 1912 anthology *I poeti futuristi* already comes up with a wealth of embarrassing material. Enrico Cardile's 'Oda alla violenza' ('Ode to Violence'), praising violence as the 'ultimate force, the ultimate muse, the ultimate fury' and exhorting its readers to 'make of their entire lives a battle,'[27] or Paolo Buzzi's 'Inno alla guerra' ('Hymn to War'), with the lines 'I want to fight beneath a cataract of bullets. . . . I want to destroy the enemy in their millions', come readily to mind.[28] Or one thinks of Buzzi's 'Canto dei reclusi' ('The Inmates' Song'), with its image of lunatics (the Futurists?) waiting to rise up in revolt against society. 'Aggressivo' is an epithet repeatedly attached to 'Futurismo' in the theoretical writings of the period (e.g. *Archivi*, i. 349, 351), as if the two ideas were naturally connected, like 'Venice' and 'passéist'. But if Futurist primitivism had simply been a matter of belligerence, of the regressive cult of violence and the threat of cultural anarchy, all of which it happily embraced, the literature displaying such sentiments would amount to little more than one further lamentable instance of a widespread irrational tendency in modern European culture just before the outbreak of the First World War, when the influence of Sorel's *Réflexions sur la Violence*, Stirner, and Zarathustra was strong (though Jeffrey Herf's *Reactionary Modernism* (Cambridge, 1984) shows that these tendencies were to be felt just as powerfully in the decades following the war). But, in fact, there is more to the phenomenon than this, and it is the particularly Futurist turn to the idea that rescues it from being synonymous with crude regression. Boccioni, it should be noted, talks more particularly of the primitivism of a *new* sensibility, and it is in the underlying paradox to this phrase that one must look to find something more distinctive in the Futurist cult of the primitive.

Having discerned three main tendencies in Russian Futurism ('urbanism', 'primitivism', and 'anarchy'), Kornei Chukovsky remarks: 'the return to the simplicity of the Stone Age might at first sight appear to contradict the first tendency' and strike many as 'incompatible with it'. But, he argues, 'the contradiction between them is illusory—they really complement one another and the one tendency would be impossible without the other'. The main inference which Chukovsky draws from this—'the more Man progresses towards the skyscraper, the more he will dream passionately of the cave'[29]—pinpoints a

[27] *I poeti futuristi*, ed. F. T. Marinetti (Milan, 1912), 179 f. See also 'La barbarie', in Luciano Folgore's *Il canto dei motori* (Milan, 1912), 151, and Enrico Cavacchioli's reference to cruelty as 'dea madre' in *Cavalcando il sole* (Milan, 1914), 125.

[28] *I poeti futuristi*, pp. 27–38, at p. 31.

[29] *Futuristy*, p. 37.

connection between the primitive and the modern which is at its seemingly most contradictory, not among the Russian Ego-Futurists he is discussing, but in the Italian Futurist notion of the 'primitivism of a new sensibility'.

In 'Destruction of Syntax' ('Distruzione della sintassi—Immaginazione senza fili—Parole in libertà' (1913)) Marinetti defines this new sensibility in terms of those features of modern life that have allegedly transformed man and in doing so changed his psyche as well as his mode of existence. Among the 'modifiers' enumerated, the following have a special bearing on Futurism's much-trumpeted primitivism of a renewed sensibility:

> 1. The acceleration of life, which today has a rapid rhythm. The physical, intellectual and sentimental balancing on the tightrope of speed which is stretched between contradictory magnetisms. Multiple and simultaneous awareness in a single individual. . . .
>
> 12. Man multiplied by the machine. A new mechanical sense, the fusion of instinct with the efficiency of the machine. . . .
>
> 14. The new tourist sensibility of ocean liners and grand hotels (annual synthesis of diverse races). The passion for the city. The negation of distances and nostalgic solitudes. . . .
>
> 15. The earth made smaller by speed. New sense of the world. That is to say: men will in turn gain the sense of their homes, of the areas where they live, the sense of the city, the sense of their geographical region and the sense of their continent. Today man possesses the sense of the whole world; he has little need to know what his ancestors did, but needs assiduously to find out what his contemporaries are doing in all parts of the world. Consequently, it is necessary for the individual to communicate with all the peoples of the earth. . . . There is a huge increase in the sense of humanity and an urgent need to establish at all times our relations with entire mankind. (*TIF* 58–60)

This smacks of the kind of cosmopolitanism we found associated with man's telegraphic capabilities. For this new sensibility, journalism would appear to have replaced history, globalism taken over from regional allegiances, and the future displaced the past—and sometimes even the present. The historical shift in emphasis which Michael Bell traces between a 'primitivism of conventional location, such as the rural retreat or the Pacific paradise' and 'primitivism as a mode of sensibility'[30] is to a large extent already discernible here, even if the paradigm-change is far from absolute. Certainly, much of the stress in the Futurist context is on a new modernist primitivism *qua* state of mind, not on the geographical exoticism of a Tahiti or the Gabon; yet one also finds that the new-felt globalism discussed by Marinetti still privileges certain technically more progressive locations. Despite the displacement of any cult of distant

[30] *Primitivism* (London, 1972), 58 f. In 'Il significato del futurismo' (*L'Avvenire d'Italia* (Bologna, 1910), 3), P. Arcari already makes the point that the Futurists were replacing history with geography. This was the beginning of what Marshall McLuhan was later to call 'the age of implosion . . . an implosion in which everybody is involved with everybody' (*Counterblast* (London, 1970), 35).

exotic settings,[31] some places still remain qualitatively more 'global', more cosmopolitan (more Futurist, one might even say) than others, and therefore reflect the renewed sensibility more fully. A predilection for characteristically Futurist locations often plays a major role in evocations of the new synthesis of the primitive and the hyper-modern. For this reason, Marinetti can tell the English, 'Why, London itself is a Futurist city!' and is able to praise 'its brilliant hued motor buses, enormous glaring posters, coloured electric lights that flash advertisements in the night' and can enthuse about the London Underground, where, the poet confides, 'I got what I wanted—not enjoyment, but a totally new idea of motion, of speed'.[32] Hence, too, even within this allegedly Futurist city, the Strand is singled out by both Marinetti and Nevinson as one of the most Futurist of streets.[33] For Vasari, Marinetti, and Döblin, parts of Berlin have a similarly Futurist aura, as does Mannheim for Buzzi. If the bustling street of the modern city is one paradigmatically Futurist location, then another is the busy seaport so often depicted in the paintings of Felix Delmarle and Nevinson and in the poetry of Pennone and Buzzi (for instance, the evocations of Kiel and Hamburg in Buzzi's *Aeroplani*). Marinetti himself came relatively late to this topos, with his *poema non umano dei tecnicismi* (*the non-human poem of technicisms*) (1940), but his picture of the port of Genoa there rivals the paintings of Delmarle in vitality.

In 'The Achievement of Futurism', Vladimir Mayakovsky refers to the 'new urban elements which the poets of the past did not know'. His catalogue is very similar to the items that Marinetti and Boccioni were referring to in their writings about the new sensibility. 'Telephones, aeroplanes, express-trains, lifts, rotating machines, pavements, factory chimneys, soot and smoke—these are the elements of beauty in the new urban landscape.'[34] Many of these same items recur, now assembled within the framework of a quintessentially Futurist journey, in Mayakovsky's poem 'Puteshestvie s mamoi' ('A Tour with Mama'):

> Вселенная вся семьею засеяна.
> Смотрите,
>   мачт корабельных щетина —
> в Германию врезался Одера клин.
> Слезайте, мама,
>   уже мы в Штеттуне.

[31] For Katia Samaltanos (*Apollinaire: Catalyst for Primitivism*, p. 4), the Cubist concern with African art heralds a general turning away from the dominant paradigm of nineteenth-century orientalism. In this respect, it could be argued, one new location is merely being privileged in lieu of another, which rarely happens with Futurist primitivism.

[32] *Evening News*, 4 Mar. 1912.

[33] No doubt Nevinson's painting *The Strand* (*c.*1914) was influenced by the flattering remarks Marinetti made to the press about the thoroughfare during his London visit.

[34] 'Dostizheniya Futurizma' (*PSS* i. 453 f.).

Сейчас,
мама,
несемся в Берлин.
Сейчас летите, мотором урча, вы:
Париж,
Америка,
Бруклинский мост,
Сахара,
и здесь
с негритоской курчавой
лакает семейкой чай негритос.[35]

Through the whole universe my family is sown.
Look,
the masts of ships bristle on edge—
into Germany the Oder has thrust its wedge.
Alright, Mama,
we're already in Stettin.
Now,
Mama,
we're driving to Berlin.
Now you're flying with motor dinning:
Paris,
America,
Brooklyn Bridge,
Sahara,
and here
a curly-headed picaninny
is guzzling her negro-mammy's tea.

A poem of this kind presents the tokens of the new sensibility in terms of an epic, world-wide perspective. But where Marinetti, Boccioni, and Malevich present them as items of a new external *beauty*, Mayakovsky is more concerned with the subjective excitement that they generate in him.

Other works by Mayakovsky focus on one specific location as the epitome of the new reality, just as Marinetti does in the case of London and Buzzi in the case of certain German ports. For Mayakovsky, during his 1925 visit, New York was one such place.

A man gets dressed by electric light, in the streets electric lights, houses with electric lights, evenly cut by windows like the pattern of a publicity poster. Houses spread endlessly, traffic lights flash with coloured signals, movements double, triple, multiply on the asphalt, on the rain-licked mirror. In the narrow groves of buildings some adventurous wind roars, tears, rattles the signboards, tries to sweep people off their feet

[35] From *Pro eto* (*PSS* iv. 158).

and escapes unpunished, unarrested by anyone, down the many miles of dozens of avenues cutting through Manhattan Island—from the ocean to the Hudson. . . .

Down below, a mass of human flesh is flowing: first the black mass of negroes, who do the hardest and most unpleasant work. Then—at about seven—a sea of whites. They go in one direction, hundreds of thousands of them, to their places of work. . . .

The crowd flows, filling the entrance holes to the subway and the covered passages of the elevated railway, going up in the air on two levels and three parallel tracks in fast trains, seldom stopping and local trains stopping every five blocks. . . . Lights go on all along the entire twenty-five mile long Broadway . . . this is, the Americans say, the Great White Way. It is really white and one really has the impression that it is brighter there at night than in the daytime, because there is light everywhere in the daytime, but this White Way, contrasted against the night, is bright as day. The street-lamps, the dazzling lights of advertisements, the glow of shop-windows and windows of the never-closing stores, the lights of illuminating huge posters, lights from the open doors of cinemas and theatres, the speeding lights of automobiles and trolley cars, the lights of the subway trains glittering under one's feet through the glass pavements, the lights of inscriptions in the sky.

Brightness, brightness, brightness. . . .

There is light in the restaurants and in the theatre district. . . . In the daytime, the port is a most interesting quarter. There is a constant roar here—work, shots, shouts. The ground is shaken by cranes which unload ships, and are almost capable of lifting up a whole house at once.[36]

More than a decade before Mayakovsky's America visit, Sant'Elia had given theoretical expression to this urbanist element of the new primitive sensibility in similar terms: 'We are the men of the grand hotels, the railway stations, the immense streets, colossal ports, covered markets, luminous arcades and beneficial demolitions.'[37] As Sant'Elia's architectural plans make clear, he was reacting less to a modernism that he saw already in evidence around him than to a future of the mind that he hoped his blueprints would one day help create.

Kasimir Malevich, also at one stage a self-declared Futurist,[38] paints an

[36] 'My Discovery of America' ('Moe otkrytie Ameriki' (*PSS* vii. 300–3)), quoted from Wiktor Woroszylski, *The Life of Mayakovsky* (London, 1972), 370–2. In the context of Italian Futurism, two poems which might have been similarly entitled 'My Discovery of Germany' are Buzzi's 'La notte di Francoforte sul Meno' (*Versi liberi* (Milan, 1913), 150–5), hymned as the quintessential 'Moderna città' (p. 153), and 'Al porto d'Amburgo' (*Versi liberi*, pp. 244–53), Hamburg being seen as the embodiment of 'l'anima transoceanica' (p. 248).

[37] 'L'architettura futurista' (*Archivi*, i. 82 f.).

[38] According to Malevich, 'whoever has not trod the path of Futurism as the exponent of modern life is condemned to crawl for ever among the ancient graves and feed on the crusts of the past'. Nevertheless, Malevich does go on to announce that: 'We have abandoned Futurism; and we, the most daring, *spat on the altar of its art*' ('From Cubism and Futurism to Suprematism', in his *Essays on Art: 1915–1933*, i, ed. Troels Andersen (London, 1967), 27).

equally enthusiastic imaginative picture of the throbbing Futurist metropolis:

> The new life of iron and the machine, the roar of automobiles, the glitter of electric lights, the whirring of propellers, have awoken the soul which was stifling in the catacombs of ancient reason and has emerged on roads woven between earth and the sky.[39]

> We town-dwellers do not know the forests, the fields, the flowers—only the tunnels of the streets with their noises, flashes, their perpetual circulation.[40]

The 'awakening of the soul' which Malevich refers to is seen by Marinetti as the formation of a new kind of (Futurist) man with an enlarged global awareness (a prefiguration of McLuhan's 'global village' idea):

> Those who today use the telegraph, the telephone, and the gramophone, the train, the bicycle, the motorbike, the car, the ocean liner, the dirigible, the aeroplane, the cinema, the great daily newspaper (synthesis of a day in the life of the world) do not think that these diverse forms of communication, transportation and information exercise a decisive influence on their psyches.

But in fact, according to Marinetti, modern man is being radically transformed by his inventions.

> An ordinary person can within the space of a day travel by train from a dead little town with deserted squares, where the sun, the dust, and the wind amuse themselves in silence, to a great capital bristling with lights . . . The inhabitant of a mountain village is able, by reading a newspaper, to tremble each day with anxiety at the uprising in China, the London suffragettes and those in New York, Dr Carrel and the heroic dog-sleds of the Polar explorers. (*TIF* 57 f.)

Italian Futurism's 'primitives of a new sensibility' are those who are revitalized by such a transformation of their horizons. 'Primitives', because the effect is to invigorate them, to return them to the essential and the intuitive machine-reality, uncluttered by the passéist intellect. From Boccioni's initial coining of the phrase 'primitives of a new sensibility' onwards, the idea echoes through Italian Futurism. Although primarily associated with the early years of the movement, it is still present in Fillia's call for a 'new primitivism' without Romanticism (in 1934) and Marinetti's and Tullio Crali's talk of the 'new primitive man' in their 'Manifesto of Musical Words' ('Manifesto dei parole musicali') of 1940.[41]

Throughout the history of the movement, the Italian Futurists were intent on recapturing something they felt had been lost as a result of the intellectualization of modern life. (There is a parallel here to the attack on the Socratic in

[39] *Essays on Art*, i. 29.

[40] Letter to *Trudovaya Gazeta* (26 Jan. 1914), quoted in *Essays on Art*, i. 242.

[41] For details, see Noëmi Blumenkranz-Onimus, *La Poésie futuriste italienne: Essai d'analyse esthétique* (Paris, 1984), 198 f.

Nietzsche's *Birth of Tragedy*.) And in what they saw as the new exciting possibilities of technology and war, they assumed that this lost quasi-Dionysian vitality was to be rediscovered. Certainly, theirs was to be a different kind of primitivism from their predecessors'. 'In our age', according to Malevich, 'there can be no such thing as the primitive in art, for we have passed through the primitiveness of the prehistoric drawing, and the new approach is an apparent primitivism—moving towards a counter-primitivism.'[42] The machine, war, and the city were for this 'counter-primitivism' what Tahiti was for a Gauguin. Referring in *Pittura scultura futuriste* to the connections between primitivism and artistic renewal, which is but a particular instance of a requisite renewal of modern man in general, Boccioni once more enumerates the fetich-objects of the new sensibility: the 'gramophone, cinematography, luminous advertisements, mechanical architecture, skyscrapers, the Dreadnought and the ocean liners', all of which he describes as being specifically 'anti-artistic' (*Gli scritti*, p. 137). The sub-text of this is that Futurism's new sensibility is more than just the mere *product* of such modern phenomena; for—since they are largely industrial artefacts—any truly modern primitivism will arrive at an artistic renewal only through interaction with the (anti-artistic) real world, not from feeding parasitically off the aesthetic styles of other periods and cultures in the Cubist manner. (Boccioni refers dismissively to the 'idols and fetich-objects from Central Africa' scattered around the studios of Montmartre and Montparnasse at the time (*Gli scritti*, p. 138)). In one of their more obviously Futurism-inspired statements, the English Vorticists describe themselves in *Blast No. 1* as 'Primitive Mercenaries in the Modern World', arguing that 'the Art-instinct is permanently primitive' and that 'modern life serves modern man as Nature did more technically primitive man'.[43]

The Italian Futurists were certainly adamant that their brand of primitivism must have little to do with the eclectic, elegiac kind that they saw proliferating in various modernist quarters at the time. In 'Modern Life and Popular Art' ('Vita moderna e arte popolare' (*Lacerba* (1 June 1914), 167)), Carrà warns against the wrong kind of primitivism—which he associates with the 'ready-made inspiration' to be derived from plagiarized archaic motifs. Often it is when attempting to distinguish their own mode of primitivism from related contemporary trends ('false primitivisms', as the orthodoxy has it) that the artists come closest to delineating the associations the concept held for them. In *Pittura scultura futuriste*, for example, Boccioni sums up the Italians' reaction to Cubist primitivism, which was the dominant counter-mode of the time and, for that reason, the Futurists' particular *bête noire*: 'Other artists have sought the authentic in the artistic manifestations of primitive cultures and in those of Central Africa. That was partly the right thing to do, but in doing so they placed

[42] Malevich, 'On New Systems in Art', *Essays on Art*, i. 90.

[43] *Blast No. 1* (London, 1914), 32 f.

themselves too much in the realm of culture'—by which Boccioni means that such art becomes too museum-orientated, too 'academic' in its inspiration, and hence out of touch with the very realities of the twentieth century which it should learn to emulate. Which is why he goes on to boast that

> We Futurists present for the first time the example of an enthusiastic 'yes' on the part of Man to the form of civilization which has come into being before our very eyes. There is a profound, primordial significance in the completely new imitation of modern dances, in clothing, in the colours and architecture of a chanteuse. . . . We admire the waiter, the bon viveur, who are geometricized through the black and white of their attire; the light cascades of a cocotte between the lights and glitter of crystal; the serious severity of the surgeon's white tunic, the incisive and calculated hardness of a train-driver, of the pilot and the driver of a 200 HP car. . . . It is in this sense that we have called ourselves primitives. (*Gli scritti*, p. 133)

Even if his paean to a hyper-modern technological primitivism cannot avoid occasional lapses into modish bohemianism, Boccioni is unequivocal enough in his explanation of how little this has to do with recent developments in French Cubism:

> The study and therefore the influence of the archaic art of antiquity, of the Negroes, of wood carvings, of Byzantine art, etc., has saturated the paintings of our young friends in France with archaism, which is another evil brought about by their obsession with the past, another 'cultural' phenomenon like the influence of the Graeco-Roman world. Even if these influences from rudimentary arts are accepted for what they have that is new, and even if they have helped to free us from classicism, they remain harmful to the development of a completely modern plastic outlook.

Although 'we call ourselves primitives', Boccioni adds, 'we Futurists do not feel the need to plunder antiquity or Negro art for inspiration'. On the contrary, he makes it clear that no 'Futurist, whether painter or sculptor, is affected by the kind of archaism which brings with it a hieratic immobility and the kind of antique solemnity which is repellent'. Yet he can still with justice claim that 'there is a *barbaric* element in modern life in which we find inspiration'. In his use of the phrase 'higher barbarism' (*Archivi*, i. 193), Boccioni betrays his anxiety to dissociate the Futurist variety from any more archaic primitivism. As Carrieri points out, 'Boccioni never misses a single opportunity of shooting his arrows at the neo-primitives, whatever may have been their origins—African or Northern.' He showed 'no sympathy for negro art or for archaic art in general and he was unable to explain to his own satisfaction why African sculpture had had so much influence on the artists he admired. . . . By nature, temperament and education, Boccioni was opposed to all primitive and archaic phenomena.'[44] In any case, since Boccioni went so far as to single out the

[44] *Futurism*, pp. 69 f.

Italians as 'the sole primitives of a new, completely transformed sensibility' (*Gli scritti*, p. 132), this gave him a theoretical platform from which to attack foreign artists unable to share Italy's privileged position. 'Archipenko's sculpture has lapsed into the archaic and the barbaric,' he complains in a letter to Vico Baer. 'This is a mistaken solution. Our own primitivism should have nothing to do with antiquity' (*Gli scritti*, p. 370)—a critical point he drives home with even greater vengeance in another context:

My charming friend Archipenko has introduced into sculpture a strong destructive contribution. [Coming from a Futurist, this is high praise indeed!] But his most interesting work, when it is free of Cubism, precipitates into Negro or oriental sculpture; if it moves out of Cubist realism, it falls into Byzantine insipidness, into archaic, skeletal, rigid and trivial silhouettes which are so characteristic of all primitive art of all epochs. It would seem that a pure Russian like Archipenko, as a representative of a young race—as he is wont to call it—should be free of any contacts with the past. But there is in him, instead, an absolute inability to transcend culture which is based, it is agreed, on the primitive, but which, if it is not transcended as an inspiration, remains extraneous to the modern sensibility. (*Gli scritti*, p. 186)

The final remark is important: at the same time as reaping praise for trying to revitalize modern art, Archipenko is ritually taken to task for his failure to respond adequately to the modern world.

As the originator of the phrase 'primitives of a new sensibility' and the leading theoretician among the Futurist painters, Boccioni invariably remained in the vanguard of the vociferous onslaught on false primitivisms. But Marinetti and his literary cohorts can be found not far behind, engaging in similar skirmishes in other quarters. The manifesto 'Vital English Art' (1914), signed jointly by Marinetti and Nevinson, spares no punches when it comes to 'that most grave of all maladies—passéism', examples of which include 'the sickly revivals of mediaevalism, the Garden Cities with their curfews and artificial battlements, the Maypolo Morris dancers . . . Pre-Raphaelites, Neo-primitives and Paris'. Tailoring their argument to what was essentially an English Vorticist context, Marinetti and Nevinson continue with the jibe that all these 'pioneers suffering from arrested development' signally failed to acknowledge the error of emulating primitive art from the past: 'if it was once necessary in painting and sculpture to have *naïveté*, deformation, and archaism, it is only because it was essential to break away violently from the academic and disgraceful before going further towards the plastic dynamism of painting'.[45] Just as Boccioni was prepared to concede that Cubism's turning to Negro art was 'partly a right thing to do' and Soffici was later (in *Cubismo e Futurismo*) to present Cubism as a necessary interim stage to Italian Futurism, so here Marinetti and Nevinson are fundamentally in agreement with the Vorticists that some kind of primitivist

[45] *Lacerba* (15 July 1914), 209.

renewal is called for; but when it comes to the question of the form that it should take, the two movements' spokesmen part company totally.

The Futurists' targets in the England of the period were identifiable enough, for there was a widespread parallel to the Cubist concern with ethnic art. Nevinson's autobiography records that Wyndham Lewis talked to him 'a great deal about the African mask' at this time.[46] And in an attempt to sum up the trend of the previous few years in a letter to the *Egoist* of 16 March 1914, Henri Gaudier-Brzeska declared that 'the modern sculptor . . . is continuing the tradition of the barbaric peoples of the earth (for whom we have sympathy and admiration)'. (On 28 November 1912 he had written to Sophie Brzeska of a visit to the British Museum where he 'took particular notice of all the primitive statues, negro, yellow, red and white races, Gothic and Greek'.[47]) His own *Red Stone Dancer* (1913) and 'hieratic' bust of Ezra Pound (1914), together with Jacob Epstein's *Rock Drill* (1913–15), epitomize the vogue for neo-primitivism in contemporary English visual art to which the Italians were taking such raucous exception—although there is arguably little in literary Vorticism to correspond to this return to the archaic in the plastic works.[48] Back in Italy itself, D'Annunzio had in any case already fought and won the battle against the recent vogue for 'Africanism' in literature.

Surprisingly, given the realities of the situation, Apollinaire himself showed an eagerness to deny any influence on Cubism from African Negro, Italian, or even fourteenth-century German 'primitives'; ironically, he even lampooned Marinetti for turning to *art nègre* in a way that no self-respecting Cubist would have dreamt of doing.[49] Such wilful misrepresentation was probably, as Harry E. Buckley surmises, the result of an 'anxiety to depict the [French] avant-garde as heroic explorers entering territory without any possible precedent'.[50] The desire actively to foster such an uncompromising self-image must also have played some part in the Italian Futurists' denial of any affinities between their own modernist primitivism and the kind of neo-primitivism they lamented in England and France. In reality one only has to examine certain works by Boccioni and Carrà (for example, the latter's 1913 *Ritratto di Boccioni* and 1916 *Ritratto di Rémy de Gourmont*, not to mention Depero's later stage-figures), to recognize that a marked ethnic primitivism *was* influencing their

[46] *Paint and Prejudice* (London, 1937), 43.

[47] Quoted by Richard Cork in *Vorticism and Abstract Art in the First Machine Age*, i. *Origins and Development* (London, 1976), 175.

[48] A discussion of the salient features of Vorticist primitivism can be found in Reed Way Dasenbrock's 'Using the Past: The Ideogrammatic Loop', in his *The Literary Vorticism of Ezra Pound and Wyndham Lewis: Towards the Condition of Painting* (Baltimore, 1985), 78–85.

[49] 'La Peinture nouvelle', *Les Soirées de Paris* (May 1912). Cf. the image of Marinetti in Apollinaire's *Le Poète assassiné* (Paris, 1916), 134: 'autour d'un feu de bivouac M.r.n.tt. s'exerçait à parler petit nègre.'

[50] *Guillaume Apollinaire as an Art Critic* (Studies in the Fine Arts: Criticism, 11; Ann Arbor, 1981), 74.

techniques, even if not as overtly as was the case with some of the Cubists' work.[51]

Another contemporary trend from which the Futurists were anxious to dissociate their own primitivism, yet one which strongly influenced it, was Nietzschean vitalism, polemically dismissed by them as a form of 'false' neo-primitivism. The charge of historicism levelled at Nietzsche in Marinetti's 'Against the Professors' ('Contro i professori' (1915)) is part and parcel of a general attack on the museal and culturally eclectic aspects of modern primitivism. Nietzsche is a 'professor', rather than someone whose primitivism belongs to the 'renewed sensibility'.

Nietzsche remains, for all his concern with the future, one of the most intransigent defenders of the greatness and beauty of antiquity. He is a passéist who wanders on the peaks of the mountains of Thessaly and whose feet are unfortunately fettered by long Greek texts. His Superman is the product of the Hellenic imagination, cobbled together from the three great putrefying corpses of Apollo, Mars, and Bacchus. We oppose to this Superman, born amidst the dust of libraries, the man multiplied by his own strength, the enemy of books. (*TIF* 263)[52]

This 'multiplied man' is a figure we shall shortly encounter in 'La Guerre électrique' ('The Electric War') and *Mafarka*, where we shall nevertheless find him to be closer to Zarathustra than Marinetti would have us believe.

While vigorously attacking retrospective ('cultural') neo-primitivism of various kinds, the Euro-centric Italian Futurists apparently remained unaware of any contemporary full-scale embodiments of the kind of modernist primitivism they espoused. As late as December 1919 Papini declared that Futurist ideas were a 'genuine product of Italy' and that such things would have been 'ridiculous' in Russia or America.[53] Yet the fact that a redoubtable American parallel did exist is well worth noting, for the *modernolatria* of Italian Futurism has often enough been dismissed as the yearning of an industrially backward country for a technology which it scarcely possessed.

[51] See the discussion of Boccioni's sculpture in John Golding's *Boccioni's 'Unique Forms of Continuity in Space'* (Newcastle upon Tyne, 1972) and of Carrà's work by Ezio Bassani in 'Italian Painting', in William Rubin (ed.), *'Primitivism' in 20th Century Art: Affinity of the Tribal and the Modern*, ii (New York, 1984), 405–14. Depero's puppet 'The Great Savage', designed for inclusion in his and Clavel's *Plastic Dances* (1918; ill. Kirby, *Futurist Performance* (New York, 1971), 108 f.) is an example of a work that is as archaically primitive as anything being attacked in the Cubist camp.

[52] In his introduction (*TIF*, p. ci), De Maria suggests that the 'professorial' charge levelled against Nietzsche in all probability derives from Sorel.

[53] *L'esperienza futurista* (Florence, 1919), 29. Papini is here recycling his own earlier remark that 'in Russia or in America Futurism would be ridiculous: it is a necessarily Italian product' ('Il significato del Futurismo', *Lacerba* (1 Feb. 1913), 25), though what Soffici primarily has in mind is the weight of the past that creates such a reaction, not the technology to which it turns as an antidote. Cf. Trotsky's remark: 'A phenomenon was observed which has been repeated in history more than once, namely, that the backward countries . . . reflected in their ideology the achievements of the advanced countries more brilliantly and strongly' (*Literature and Revolution* (Ann Arbor, 1971), 126 f.).

Dickran Tashjian's account of the 'Skyscraper Primitives', the circle of American writers grouped around Robert Coady's magazines the *Soil* and later *291*,[54] reveals a number of features which are markedly akin to Boccioni's 'primitivism of a renewed sensibility'. Historically, Tashjian depicts these writers as a coterie of modernists 'affected by Dada's fascination with the machine . . . who in turn celebrated American technology'. 'This new primitivism', he remarks, 'was paradoxically urban in locus and sensibility',[55] i.e. the same paradox that Chukovsky sees in Russian Futurist primitivism and which is the hallmark of the sensibility discussed in *Pittura scultura futuriste*.

In the first issue of the *Soil* Coady has an essay entitled 'American Art' which could, were it not for the local overtones of some of its illustrations, just as well have graced a Futurist declaration, for it practically amounts to a list of the 'modifiers' of the modern sensibility, in the Italian sense. Coady itemizes examples of what he provocatively calls 'American Art':

> The Panama Canal, the Sky-scraper and Colonial Architecture. The East River, the Battery and the 'Fish Theatre'. The Tug Boat and the Steam-shovel. The Steam Lighter. The Steel Plants, the Washing Plants and the Electrical Shops. The Bridges, the Docks, the Cutouts, the Viaducts, the 'Matt M. Shay', and the '3000'. Gary. The Polarine and the Portland Cement Works. Wright's and Curtiss's Aeroplanes and Aeronauts. The Sail Boats, the Ore Cars. Indian Beadwork, Sculpture and Decorations, Music and Dances. Jack Johnson, Charlie Chaplin, and 'Spike' in 'The Girl in the Game'. Annette Kellerman, 'Neptune's Daughter'. Bert Williams, Rag-time, the Buck and Wing and the Clog. Syncopation and the Cake-Walk. The Crazy Quilt and the Rag-mat. The minstrels. The Cigar-store Indians. The Hatters, the Shoe-makers, the Haberdashers and the Clothiers. The Window Dressers.[56]

Understandably there is, when viewed from a strictly Futurist point of view, a certain lack of rigour to this list. The East Coast 'primitives' frequently show a remarkable alacrity to abandon their skyscraper perches and go off West in their hunt for ethnic American artefacts and customs. Some of the items cited are touchingly folksy. Any pronounced sense of modern dynamism is also lacking, though the naming of specific people (and even buildings) as embodiments of the new spirit is an effective propaganda strategy which the Futurists them-

[54] *Skyscraper Primitives: Dada and the American Avant-Garde 1910–1925* (Middletown, Conn., 1975). The label 'skyscraper primitive' was first applied to Coady by Gorham Munson in an article of that title in the American little magazine *The Guardian* (Mar. 1925), 164–78.

[55] *Skyscraper Primitives*, p. x. Tashjian treats the Skyscraper Primitives in terms of a Dada-reception—doubtless influenced in this by the presence of such link-figures as Picabia and Cravan—without noting the equally striking points of contact with Futurism. There are various similarities between Picabia's interests and those of the Italian Futurists (see *FF* 539). In any case, Cravan is known to have moved in Orphist circles familiar with, and influenced by, Italian Futurism. See Enrico Crispolti, *Il mito della macchina e altri temi del futurismo* (Trapani, 1969), 16 f.; and, on Cravan's New York conference on Futurism and Dada, José Pierre, *Le Futurisme et le Dadaïsme* (Lausanne, 1971), 69.

[56] *The Soil* (1 Dec. 1916), 3.

selves never hit upon.[57] Moreover, unashamed patriotism evidently retains more attraction here than the lure of any cosmopolitan *modernolatria*. But the same number of the *Soil* also contains Arthur Cravan's poem 'Sifflet', which to some considerable extent does expand the *Soil* catalogue in a Futurist direction:

> The rhythm of the ocean cradles the transatlantics,
> And while the heroic express arriving at Havre
> Whistles into the air, where the gases dance like tops,
> The athletic sailors advance, like bears.
> New York! New York! I should like to inhabit you!
> I see there science married
> To industry,
> In an audacious modernity,
> And in the palaces,
> Globes,
> Dazzling to the retina
> By their ultra-violet rays;
> The American telephone,
> And the softness
> Of elevators. . . .[58]

'Audacious modernity' effectively sums up one of the salient features of much Futurist primitivism. Indeed, there is decidedly more dynamism to Cravan's poem than in Coady's list or, for that matter, in most of the poetry of *I poeti futuristi*. What is still missing, and what makes the American Skyscraper Primitives still substantially unlike their Italian counterparts, is the aggression and iconoclasm that is so much part of the Futurist sensibility.

'The Founding and Manifesto of Futurism' identifies this element very forcefully: 'We intend to sing the love of danger. . . . Except in struggle, there is no beauty. No work without an aggressive character can be a masterpiece. . . . We will glorify war—the world's only hygiene—militarism, patriotism, the destructive gesture of freedom-bringers, beautiful ideas worth dying for' (*TIF* 9f.). In 'Weights, Measures and the Prices of Artistic Genius' ('Pesi, misure e prezzi del genio artistico' (1914); *Archivi*, i. 43), Corradini and Settimelli describe 'the Futurist conception of art' as 'essentially modern . . . and brutal';

[57] The nearest the Futurists come to this technique is in the ritual enumeration of their own number and kindred spirits: e.g. the 'Synthetic Panorama of all the Inventors of Modern Art', signed by Marinetti, Alberto Sartoris, and Giuseppe Tarragni in 1938 (cited in Crispolti, *Il mito della macchina*, pp. 794 ff.), which reads like a *Who's Who* of the Futurists, their allies and mentors. Comparable rolls of honour can be found in Marinetti's introduction to many a 'Poesia' publication: e.g. the preface to *Gli indomabili* ('Free-word Style' ('Lo stile parolibero'), *TIF* 842). But such lists do little more than establish the credentials of the 'in'-group; they hardly have the representativeness of modernity that Coady's inventory displays.

[58] *The Soil* (1 Dec. 1916), 36. The original French version of 'Sifflet' had appeared in *Maintenant*, 1 (1912), 1.

the two go together. And, although Boccioni can write figuratively of the 'higher barbarism' of the modern primitive (*Gli scritti*, p. 89), when a Buzzi decides to describe himself as a 'poor son who adores barbarism'[59] he has something more dangerously atavistic in mind.

The cult of an aggressive, virile primitivism is much more part of the *Italian* Futurists' self-image than of any other related faction. Within the various complex sub-groupings of Russian Futurism, by contrast, a far more heterogeneous state of affairs obtained. More often than not the Russians were, as Chukovsky's picture suggests, resorting to the very kind of 'archaic' primitivism that Boccioni and Marinetti object to. (That the Italians did not as a consequence include the Russian Futurists in their rogues' gallery on this account was probably less a mark of tact or solidarity than an indication of how ill-informed they were of just what was going on in Russia.)

The subject of Alexander Shevchenko's *Neo-Primitivizm* (1913) would seem to be the very antithesis of the modernist primitivism advocated in *Pittura scultura futuriste* and in some of Malevich's early writings. The primitive, as presented here, does still retain a markedly Rousseauistic aura and is, for that reason, exalted *in contrast to* the hostile world of modern technology:

> WE WHO ADVOCATE NEOPRIMITIVISM AS THE ARTIST'S RELIGION SAY: . . . The Earth and Nature remain only a memory, like a fairytale about something beautiful and long past. The factory town rules over everything. The movement, the never-ending commotion, the obscure nightmares and visions of the town are continually replacing each other. In the light of the daytime sun darkened by houses, in the bright light of the electric suns of night, life presents itself to us as quite different, replete with different forms new to us.
>
> The world has been transformed into a single, monstrous, fantastic, perpetually moving machine, into a single huge, inanimate, automatic organism, into a single gigantic whole constructed with a strict correspondence and balance of parts. . . . We, like some kind of ideally manufactured mechanical man, have grown used to living—getting up, going to bed, eating and working according to the clock—and the sense of rhythm and mechanical harmony, reflected in the whole of our life, cannot but be reflected in our thinking and in our spiritual life: our art.

In the Italian context, or even in the case of Mayakovsky in America, such a diagnosis would have naturally led to the promotion of similarly mechanical modernist principles in the arts. In fact, the 'manufactured mechanical man' Shevchenko describes is to loom large in both the first and second phases of Italian Futurism. But for Shevchenko and his associates, the predicament is hardly something to be welcomed with modernist enthusiasm. Old allegiances die hard. 'We are striving to seek new paths for our art, but we do not reject the old completely, and of its previous forms we recognize above all—the

[59] *Aeroplani* (Milan, 1909), 103. No doubt, Carducci's *Odi barbare* helped to popularize this use of 'barbarism' with positive associations in Italy.

primitive, the magic fable and the old East.' The East, in fact, frequently plays a similar role in the primitivist renewal of Russian art and literature to the one that Africa plays in much of western Europe at this time. (Livshits's image of the 'one and a half-eyed archer', 'squinting at the West with only half an eye' and with the other one turned fully-open to the East, is an apt emblem when it comes to primitivist inspiration.) But the East was still only one source of the primitive for the Russian Futurists:

The simple, unsophisticated beauty of the *lubok*, the primitive, painted sign-board, the severity of the primitive, the mechanical precision of construction, nobility of style, and good colour brought together by the creative hand of the artist-ruler—that is our password and our slogan.

The aesthetic values Shevchenko attributes to the *lubok* are more Romantic in spirit than bearing much resemblance to the *modernolatria* central to much Italian Futurism.

Primitive art forms—icons, *lubki*, trays, signboards, fabrics of the East etc.—these are specimens of genuine value and painterly beauty. . . .

For the point of departure in our art we take the *lubok*, the primitive art form, the icon, since we find in them the most acute, most direct perception of life—and a purely painterly one, at that.

We, like the primitivists and like Eastern artists, consider the most valuable and most productive work to be that which is guided by impression.[60]

Shevchenko saw this synthesis of older traditions and modern life expressed in the very term 'neo-primitivism': 'The word . . . on the one hand testifies to our point of departure and on the other—with its prefix, *neo*—reminds us also of its involvement in the painterly traditions of our age.'[61]

As for the Hylaea Group, they would on the whole have been in agreement with these sentiments. Although seen as part of Cubo-Futurism, 'Hylaea did not mean Cubism at all', according to Markov, 'but primitivism.'[62] The Burliuks, Goncharova, Khlebnikov, and Larionov regarded themselves, like Shevchenko, not so much as out-and-out primitives, but as '*neo*-primitives'.[63] And Livshits's memoirs bear witness to the way in which such neo-primitivism was bound up with the cult of the primitive in the ethnological sense, and remained in that respect closer to the elegiac, retrospective primitivism of the Cubists and Vorticists than that of Marinetti's circle.

[60] *Neo-Primitivizm. Ego teoriya. Ego vozmozhnosti. Ego dostizheniya* (Moscow, 1913), quoted from 'Neo-Primitivism: Its Theory, its Potential, its Achievements', in John E. Bowlt (ed.), *Russian Art of the Avant-garde: Theory and Criticism 1902–1934* (New York, 1976), pp. 44 f. According to Juliette Stapanian (*Mayakovsky's Cubo-Futurist Vision* (Houston, Texas, 1986), 10), Marinetti did refer to the Russian Futurists as 'sauvages'; however, it is not recorded whether this was a term of abuse or praise.

[61] *Russian Art of the Avant-garde*, p. 47.

[62] *Russian Futurism*, p. 35.

[63] Benedikt Livshits, *Polutoraglazyy strelets* (Leningrad, 1933), quoted from John E. Bowlt's translation *The One and a Half-Eyed Archer* (Newtonville, Mass., 1977), 14.

Here is Livshits anecdotally reminiscing about the role of such primitivism within the Hylaea Group:

On the subject of primitives. Among the many inhabitants of Chernianka who used to come and stare at the 'boss's little ones' was a man who was much enticed by the Burliuks' painting and saw it as his own vocation. . . .

The Burliuks furnished him with a canvas, brushes and paints and made him into a second Rousseau, exhibiting his paintings alongside theirs. . . .

This penchant for the primitive, particularly for the domestic sign-painting of laundries, barber-shops and other provincial establishments and businesses, which exerted an influence on the work of Larionov, Goncharova and Chagall, prompted Burliuk to spend all his money on hand-painted sign-boards. . . . The rich collection which David put together probably perished in Pushkino where the Burliuks lived in 1914.

The love of popular art, the attraction towards the primitive in all its forms, towards the art of Polynesia and ancient Mexico were, in the Burliuks' case, not the caprice of jaded taste, not the epicureanism of people like Sergei Makovsky.

No, this passion ran deeper.[64]

Ilya Zdanevich's cult of the work of Niko Pirosmanishvili (1862–1918), a naïve painter from Georgia, fits very much into this general tendency. Camilla Gray adds substantially to the list of sources for this eclectic form of neo-primitivism, citing

the turning to national folk-art traditions for that directness and simplicity which Gauguin and Cézanne had taught them to appreciate; . . . embroidery from Siberia, traditional pastry forms and toys, . . . peasant woodcuts . . . were the sources from which Larionov and Goncharova drew their inspiration in this new primitive style. . . . Another national tradition which influenced this 'Primitivist' style was that of icon painting introduced by Goncharova.[65]

Now, it may seem a long way from pastry forms to forms of poetry, but these artistic preoccupations with the primitive do have their equivalents in Russian literary Futurism. Kruchenykh's suggestion that the Futurists should 'emphasize the importance in art of the abruptness of the dissonances and of the pure primordial courseness' ('New Ways of the Word' ('Novye puti slova', *Manifesty*, p. 70)) is thinking in this direction.

Khlebnikov's poem 'Perun', for instance, addressed to the ancient Slav god, might be thought of as an equivalent, going this time back to mythology rather than just folk-art roots. As the following extract shows, the primitivism lies not only thematically in the evocation of the god, but in the archaic sound-patterns of the chant:

[64] *The One and a Half-Eyed Archer*, p. 53.

[65] *The Russian Experiment in Art 1863–1922* (London, 1971), 97. On the importance of the *lubok*, see also Stapanian, *Mayakovsky's Cubo-Futurist Vision* (pp. 177 f.) and *The Lubok: Russian Folk Pictures*, introd. Alla Sytova (Leningrad, 1984).

Первый и первый и первый Перун.
Пещерный Перун. Перун пещерный.
Пением первый первых пещер.
Первою палкой пари
По пушистой поре [пороха],
Пернатым порогом пляшущей пены.
Палицей пологого пения
Порохи пламенем пой.

Primal and primal and primal Perun.
Primitive Perun. Perun of the pits.
Primal by praising in primal pits.
Plunge like a primitive pike
Past the puberulent period of powder,
Plumed like the rapids of prancing spray
Pounding your club of praise,
Puff in the powdery blaze.[66]

The ethnic primitive was not only more important than the modern primitive, in such contexts; it was also viewed as an antidote to the modern technological world. Among the Polish Futurists, for instance, there was even a move to reject the manifestations of the new machine-world in the name of a more primeval form of primitivism. Anatol Stern's 'Czcicielowi maszyn' ('To the Worshipper of Machines') (from *Anielski cham* (Warsaw, 1924)) reads like a deliberate rejection of Italian Futurist *modernolatria*, as well as its equivalent in the poetry of Czyżewski and Jasieński:

powiadasz: spójrz-no uważnie ta fabryka to dom
w którym wykuwają JUTRO nieśmiertelne maszyny
powiadasz: i ty maszyną jesteś maszyną tylko — chociaż
pod wielu względami ustępujesz tym żelaznym aniołom
radzisz mi się przyjrzeć jak hangaru podnosi się ramię
a transmisyjne szerokich marzeń pasy
szumią w tej jasnej hali nie gorzej niż w głowie

tak tylkoże w pewnej chwili ujrzysz wyraznie roznicę
ujrzysz drobną różnicę pomiędzy maszyną a mną
w tej chwili
gdy leżąc na wznak
z trudem wykrztuszę z siebie wielką flaszę
pełną mistycznej — tak! — mistycznej pełną męki flaszę
i skrawków liliowych spojrzeń
i kantorów domu handlowego „SMITH & C-NY"
i słów ze stłumionym R
i zjedzonych przez mole akwizytorów

[66] *SP* iii. 12–15. For this translation I am indebted to Gary Kern's inventive version in *The Ardis Anthology of Russian Futurism*, ed. Ellendea Proffer and Carl R. Proffer (Ann Arbor, 1980), 23.

i ukamienowanych szpulkami bawełny wyniszczonych dzieci
i niewiadomo czemu

i niewiadomo czemu —
(o czarnoziemie twórczego nieporządku!
klasycyżmie podświadomego!!)
— niemych krzyków
tego mostu z rozpaczliwie
załamanymi rękoma
który biegnie w znienawidzoną
— znienawidzoną? —
pełną jaskółczego świergotu i lotu złotych krzyżów i chmur
nieskończoność

tego nie dojrzysz
w najnowocześniejszej
amerykańskiej maszynie

you say: look carefully this factory is the place
where *tomorrow* will be cast by immortal machines
you say: you too are a mere machine—although
in many ways you are inferior to these iron angels
you advise me to observe the hangar's shoulder go upwards
and the transmission belts of wide dreams
are singing in this bright hall not unlike in my head

yes only at a certain moment will you perceive the difference
you'll see a slight difference between the machine and me
at the moment
when lying on my back
with difficulty I will cough out a huge bottle
full of mystical—yes!—full of mystical sufferings
and shreds of lilac looks
and counters of the store 'SMITH & CO.'
and words with a muffled R
and moth-eaten salesmen
and exhausted children stoned by spools of cotton
and you don't know why

and you don't know why—
(o fertile soil of creative disorder!
o classicism of the subconscious!!)
—of muted screams
this bridge of desperately
wrung hands
which leads into the detested
—detested?—
full of swallows' twitter and the flight of gold crosses and clouds
infinity

this you won't see
in the most modern
american machine

In their introductory primitivist manifesto *Gga*, Wat and Stern declare: '*we destroy the city*, all mechanism—aeroplanes, trams, inventions, the telephone. In their place, primitive means of communication.' As Bogdana Carpenter explains, what this new primitivism of communication signifies can most easily be appreciated in the title *Gga* itself: 'an onomatopoeic sound in Polish that imitates the honk of a goose.... The sound *gga* suitably sets the tone... the airplane or automobile of Italian futurism is replaced in its Polish version by an animal, and technology is replaced by jocular primitivism.'[67]

Something else that Chukovsky's survey of Russian primitivism fails to bring out is the fact that the primitive could be a matter of technique as well as subject-matter. A whole variety of features—the use of Old Church Slavonic letters, deliberately crude handwritten words, books covered in sackcloth or printed on wallpaper, primitivist illustrations in the manner of either folk-art or children's drawings, naïve drawings, and woodcuts—enriched the repertoire of Futurist book-design and layout, as Compton and Janecek have amply demonstrated. Nikolai Burliuk's remark that 'only barbarism can save art' (*Manifesty*, p. 79) certainly makes some sense in the context of Futurism's turning to primitivist techniques in both painting and literature (even if the process of reinvigoration may seem less self-evident when it comes to Chukovsky's primitivism of content). But, unlike the Rousseauists, to whom they reveal certain Romantic affinities, the Hylaea artists are not concerned with the cultivation of a specifically primitive sensibility or the championing of some way of life. Their attention is directed solely at the renewal of art and literature.[68]

On the strength of the differences so far observed, mainly in Futurist theorizing about the subject, it would appear possibile to erect a series of contrasts between the Italians' primitivism and that of their Russian counterparts, as well as that of the Cubists and Vorticists.[69] The one, let us call it 'neo-primitivism', is

[67] *The Poetic Avant-garde in Poland, 1918–1939* (Publications on Russia and Eastern Europe of the School of International Studies, University of Washington, 11; Seattle, 1983), 7.

[68] A certain Arthur Hight wrote in *New Age* (8 Jan. 1914, 319) that a would-be primitive like T. E. Hulme really 'ought, were he consistent, to be squatting naked on Easter Island surrounded by the prehistoric Art he admires, and dieting himself on roots and toadstools after the manner of savages'. Yet this is to confuse aesthetic primitivism, as a form of artistic renewal and that alone, with primitivism as a way of life. Aesthetic primitivism was not in most cases intended to be livable. On the other hand, as a total sensibility, that of the Italian Futurists was. For the background to the debate surrounding Hulme's influential ideas on the subject, see Michael H. Levenson, 'In Search of a Primary Pigment', *A Genealogy of Modernism: A Study of English Literary Doctrine 1908–1922* (Cambridge, 1984), 120–36.

[69] Cf. also the primitivist context of Pound's Fenollosa-reception. Having stated that 'All nouns come from verbs', Pound continues: 'To primitive man, a thing only IS what it *does*. That is Fenollosa,

retrospective, archaizing, 'sentimental' (in Schiller's sense), and to some considerable degree feeds eclectically off other, usually older cultures. When applied to the avant-garde of the early part of this century, this is often the main, if not the only, connotation of 'primitivism' within art-history—which is why Futurism is either skimpily treated or scarcely mentioned at all in the classic studies by Bihalji-Merin, Goldwater, Rubin, and Laude.[70] But there is the other form: a modernist primitivism, one which is not backward-looking to the same degree (if at all), but seeks to create a new sensibility appropriate to its own culture, especially those elements of the modern world which point towards the future.

One only has to look at that early classic of Futurism, Marinetti's *Mafarka le futuriste*, however, to realize the extent to which any typological contrast between, say, archaic Cubist neo-primitivism and Boccioni's 'primitivism of a renewed sensibility' risks over-neat polarization. In fact, with the exception of the episode near the end when the hero Mafarka-el Bar constructs an artificial machine-son, the movement's notorious 'African novel' is often as regressively archaic in its primitivism as anything Boccioni ever chose to condemn in his attacks on Cubism.

The first section, provocatively entitled 'The Rape of the Negresses', was no doubt one of the parts that gave rise to the lawsuit against the novel's author on grounds of indecency.[71] In addition to the sensationally flaunted mass-rape scene, *Mafarka* indulges in almost every conceivable form of calculated excess. Despite occasional allusions to modern weaponry, the setting is on the whole a timeless primitive desert world where Arab fights black and where the scimitar, the cavalry charge, and the barbarities of hand-to-hand combat loom larger than the machinery of modern warfare.[72] Both the rape of the negresses and the

but I think the theory is a very good one for poets to go by' (Ezra Pound, *Selected Letters*, ed. D. D. Paige (New York, 1971), 82).

[70] It is symptomatic of the general narrowing of the concept within contemporary academia that, in his introduction to *'Primitivism' in 20th Century Art*, William Rubin feels able to define the term solely as 'the interest of modern artists in tribal art and culture' (i. 1). In Paris, he points out, 'the term "art nègre" began to be used interchangeably with "primitive art"' (i. 3). But this did not happen only in Paris: the same limiting equation is much in evidence among the Dadaists. Richard Huelsenbeck's pairing of 'the primitive' with 'das Hottentottische' and his remarks about 'the primitiveness of which our age seemed to be giving signs, in its predilection for Negro sculpture, Negro literature and music' (*Dada siegt* (Berlin, 1920), 12 f.) is one such example. Middleton (*Bolshevism in Art*, p. 288) cites this and a letter from Hugo Ball in which primitivism is seen as 'absolute Negerei' (*Briefe 1911–1927* (Einsiedeln, 1957), 66). On the more circumscribed subject of the impact of *art nègre*, the classic study remains Jean Laude, *La Peinture française (1905–14) et 'l'Art nègre'* (2 vols.; Paris, 1968).

[71] Details of the case are recorded in 'Processo di Mafarka il futurista' (*TIF* 511 f.). The official charge was that the work was 'an outrage to public decency'. No doubt because its offence was not against standards of *sexual* morality, 'La Guerre électrique' was never subject to any such litigation.

[72] This is to some considerable degree true of the writings in which Marinetti celebrates the Libyan and Balkan campaigns. In its image of battle, *Zang tumb tumb* is qualitatively less different from

summary way in which the Arab leader metes out justice to his captured black adversaries might, under other circumstances, have led one to suspect the presence of a particularly anti-black, colonialist racism, but in fact Mafarka is far from prejudiced in his cruelties: it is obviously immaterial to him what colour the victims of his outbursts happen to be.

Despite the foreword's proclamation that the work represents the Futurist glorification of war, *Mafarka* begins *after* a battle is over. Again, as with the collage *Après la Marne*, we witness the absolute power of the victorious general, not the relative haphazardness of a chaotic battle-situation. Instead of any orgy of indiscriminate slaughter in the heat of the moment, we encounter the determined callousness of a man who sees battle as a heightened experience and who is programmatically deaf to any humanitarian pleas. 'What, you come to beg me to stop the war in order to avoid starving the people! You speak of the deprivation brought about by the campaigns! I know all about that! And I don't care a damn!'[73]

For the soldier Mafarka, war has its *raison d'être* as the supreme form of intensified experience, not as a means to some strategic end. 'Yes, it is true, I have singularly sensitive nostrils this morning, thanks to all the perfumes of the corpses with which God has blessed me. . . . I only see stretched out before me the immense geometric table of death' (pp. 12, 14). This is a rather different phenomenon from the deadpan claim of 'The Technical Manifesto of Futurist Painting' ('La Pittura Futurista. Manifesto tecnico') that 'the suffering of a man is of the same interest as the suffering of the electric light' (*Archivi*, i. 66). For, whereas the manifesto statement is preceded by the declaration that the Futurists' 'renovated consciousness does not permit [them] to look upon Man as the centre of universal life', the perspective in *Mafarka* is still just that: the emphasis on heightened individual sensitivity and on perverted self-gratification makes Mafarka into a kind of Des Esseintes of the battlefield.

Just as the 'Foundation and Manifesto of Futurism' presents a picture of a Marinetti reborn after his perilous fall into a ditch, so here the post-battle scene introduces a renewed Mafarka, made more aware of his potential by the recent intensified experience of combat:

This morning's battle has just piled up unforeseen amounts of manure. Won't all the heaps of black bodies, shining, steaming, and almost turning to liquid on the dusty

Mafarka than one might have expected it to be from one's sense of the impending First World War as the first major mechanized conflict. With the infamous exception of the later Abyssinian War, Marinetti did not see active service on fronts where modern warfare had come into its own, and it no doubt suited his purposes to be involved in more atavistic forms of combat.

[73] *Mafarka le futuriste: Roman africain* (Paris, 1909), 12. References in parenthesis after quotations will be to this French original, in my translation. The Italian version, *Mafarka il futurista: Romanzo*, translated by Decio Cinti, did not appear until later the same year in Milan.

green of the plains soon change into pools of liquid dung with rich ebony reflecting surfaces to gladden the mean eyes of the Sun who owns this land? (p. 13)

There are touches of the *roman charogne* to such passages, but the author of *Mafarka* clearly owes as much to a popularized version of the Nietzschean Superman as he does to Symbolism and *fin-de-siècle* writing.

Frequently in the early part of *Mafarka* the hero can be observed reviewing his sense-impressions after battle, just as he might elsewhere marshall his troops or his captured negresses. If one such section is predominantly acoustic—registering such sounds as 'the whinnying of the horses, the sound of cows, the rattle of chains, the cry of the Negroes beneath the nails of the whiplashes' (p. 4)—another will consist of a mixture of olfactory impressions and general synaesthesia: 'a torrid naked wind, with its melodious body all dripping with salty seawater like that of a diver, an acrobatic wind that jumps as it passes above the citadel and throws at Mafarka's feet violet aromas, acrid smells, and the red shouts of sailors' (p. 11). Mafarka's hypersensitivity colours the detailed descriptive passages in much of the novel, but it is only one, and by no means the most regressive, of the novel's forms of primitivism. That without doubt comes with some of the early scenes of violence.

The notorious 'Rape of the Negresses' episode has been prepared for by Marinetti's foreword:

I said to them: 'Despise women!' . . . it is not the animal value of Woman that I am talking about, but their sentimental significance.

I want to combat the gluttony of the heart, the abandon of half-parted lips that drink in crepuscular nostalgia. . . . I want to defeat the tyranny of love, the obsession with the one and only woman, the grand romantic moonlight that bathes the façade of the Brothel. (pp. xiii–ix)

A scene of mechanical loveless rape might, in theory at least, have represented an adequately Futurist response to the despised Romantic obsession with Woman in the singular. In setting about the women prisoners, Mafarka's troops could easily have been functioning as faithful warriors, extensions of their leader's—and their creator's—wishes. Yet throughout the real object of attack is not so much the negresses in the novel as the passéist romantics in the world outside it (for their sensibilities are also being effectively violated in this chapter). But we are presented with a substantially more complicated scenario. For a start, Mafarka has not stage-managed this orgy of group-copulation for any pedagogic purpose; he has simply come upon hundreds of his drunken sailors ('[a] strange human cyclone', p. 27) engaged in an act of insubordination, the details of which are lingered on, albeit not for straightforward reasons of titillation:

It was on the bodies of the young negresses stretched out on their backs beside the filthy pool that hundreds of naked warriors were excitedly giving vent to their passions while others were forming a queue and awaiting their turn. . . . one could see the smooth and glistening bodies of the young women and their small breasts, the colour of roasted coffee, twisting with pain beneath the heavy hands of the males whose bronze loins rose and fell tirelessly among the cracking sounds of the green mass of putrefaction. (pp. 28–30)

The orgy continues with the sailors riding on the gigantic human raft formed by the bodies of the dying women, with necrophilia added to the diversions of a primitive nautical version of rape on a water-bed. At first, the incident is presented neutrally, with neither marked eroticism nor explicit moral condemnation, until Mafarka intervenes to berate his men in an outburst of 'anger, nausea and disgust' at what he has witnessed.

Mangy curs! festering wretches! lily-livered faint-hearted morons. . . . Beneath your thick skulls where your brains should be, you have nothing but a foul-smelling, cancerous growth, to disgorge in this way from your mouths and through the rotting slits of your eyes so much poisonous pus. . . . The sexual organs of women in chains—so that's the enemy you prefer to take on. (p. 35)

Mafarka's outraged rebuke is only comprehensible within a warped Futurist sense of values. In actual fact, the sailors are being censured for fighting mere women, for showing more interest in easy rape than in manly combat, and for responding to the bait (the negresses) which the enemy had deliberately set as a trap for them. As a consequence, what Marinetti's readers are likely to find disconcerting in this episode, no doubt provocatively positioned at the beginning of his first Futurist novel as a gesture against passéist values, is hardly going to be what Mafarka himself finds so contemptible in his men's excesses.

Partly because of the notoriety from the court-case brought against the work, 'The Rape of the Negresses' has tended to figure over-prominently in the work's reception. Yet such a sensational picture in many ways does an injustice to *Mafarka*. For gradually even Marinetti's ruthless hero wearies of battle and passionless slaughter and gradually turns his attention to the more positive challenge of the creation of the new Futurist man. Without necessarily elevating *Mafarka* to the status of a *Bildungsroman*, one can at least allow that the work's protagonist is far from static. He develops from being archaically primitive, in a series of regressive early scenes, towards at least some intimation of the new sensibility that so preoccupied Boccioni in his theoretical writings. Attention moves from the initial orgy of destruction and wallowing in cruelty to a final self-transcending act of creation—even if one may still have misgivings about just what activities the artificial son Gazourmah is being brought to life to perpetrate.

It would present a deceptively positive impression of Marinetti's primitivism to leave things with the end of *Mafarka*. In any case, one might, on the basis of this work, deduce that Marinetti's more reactionary primitivism largely pre-dates Futurism proper[74] and therefore comes some years before the genuinely modern Futurist proclamation of a new sensibility. Or, on the strength of a comparison between the battle-scenes in his two most famous works, *Mafarka* and *Zang tumb tumb*, one might be tempted to decide that Marinetti's fictive accounts of war were more bloodthirsty and barbaric than those rooted in actual battle-experience. An alternative possibility is even that his impressions of real war had in some respects tempered his enthusiasm for the subject. After all, there is no description in *Zang tumb tumb* or 'Battaglia peso + odore' to hold a candle to the barbarism of the set pieces in *Mafarka*. But Marinetti's war-writing unfortunately does not come to a standstill in 1918; there is no ceasefire for him. His two major works on the subject during the immediate post-war period, *L'alcova d'acciaio* (*The Alcove of Steel* (1921))[75] and *Gli indomabili* (*The Untamables* (1922)) both revert to something more akin to the unrestrained barbarities of *Mafarka*, rather than continuing the experimentation and the tone of *Zang tumb tumb*.

Whatever Boccioni, and sometimes even Marinetti, may have had to say about the primitivism of a new sensibility, things were more embroiled in the Italian Futurist camp than such innocuous theorizing might suggest. It would no doubt serve to contain the problem to make a distinction between the reactionary, even barbaric, primitivism of a Marinetti and Boccioni's modernist primitivism. It has, after all, been a standard ploy for dealing with some of the more distasteful aspects of Italian Futurism to hive off 'Marinettism' from Futurism proper and lay the blame for most of the movement's excesses squarely at the one man's door[76]—thus rather conveniently passing over the fact that he also happened to be the acknowledged spiritual and organizational leader of the entire movement.

Yet even such an approach will not do justice to the realities of the situation. Marinetti's barbaric primitivism and Boccioni's primitivism of a new sensibility cannot be kept so tidily distinct from one another. If we now turn to one of the earliest literary evocations of the new Futurist primitivism and the modified sensibility of 'multiplied man', Marinetti's 'La Guerre électrique' ('The Electric War'),[77] what we in fact find is one form of (Boc-

[74] It is clear from Giovanni Lista's account of the genesis of *Mafarka* (*Marinetti* (Paris, 1976), 43) that the seeds of the work reach back to the early years of this century, i.e. that they pre-date the birth of Futurism.

[75] For a detailed response to this novel, see Christopher Wagstaff, 'Dead Man Erect', in Holger Klein (ed.), *The First World War in Fiction* (2nd rev. edn.; London, 1978), 149–59.

[76] It was Palazzeschi, Papini, and Soffici, 'Futurismo e marinettismo' (*Lacerba* (14 Feb. 1915), 49–51) that launched the idea that Futurism should be distinguished from 'marinettism', with an iconically dismissive small 'm' (which has 'a different nature and different characteristics'). Since then, isolating Marinetti's Futurism from other manifestations (sometimes presented as the authentic Futurism) has been a common strategy among a large faction of the movement's apologists.

[77] The idea of man 'multiplied' by his technology, which Marinetti derives primarily from Mario

cionian) primitivism gradually veering off in the direction of Mafarka's world of reactionary values.

'La Guerre électrique' was originally published in Marinetti's introductory collection of Futurist manifestos and essays (Paris, 1911). For promotional reasons this volume was called simply *Le Futurisme*, although a more revealing title was the one under which most of the material[78] was republished in Italy four years later: *Guerra sola igiene del mondo* (*War, the World's Only Hygiene*)—more revealing, because the book contains a number of pieces extolling the virtues of combat and praising works of art able to convey the invigorating value of war, as well as including 'La Guerre électrique' itself, Marinetti's first major literary work since *Mafarka*. Because of its position as final piece among the essays of *Le Futurisme* and sole literary embodiment of the future they imply, 'La Guerre électrique' acquires a special contextual importance within the collection. The volume's dedication—'To the students of Paris, that they may love The Great Futurist Italy that we are preparing' (p. 75)—is a clear pointer in this respect, for nowhere is Marinetti's vision of a future 'Futurist' Italy more explicitly unfolded than in 'La Guerre électrique'. With an appropriate symmetry, the first piece of *Le Futurisme*, 'The First Battles' ('Les Premières Batailles'), arrives at what is described as 'our first Futurist conclusion'—namely, 'Long live war'—while 'La Guerre électrique' concludes Part One with a picture of Futurist combat *par excellence*.

A striking feature of the version in *Le Futurisme* is its length. 'La Guerre électrique' is more than twice as long as its Italian counterpart, the reason being that the Paris version is preceded by an extensive rationalizing and, given the embryonic nature of the movement, understandably defensive preamble.[79] Here, Marinetti sets out to disarm his critics by acknowledging that 'numerous objections have already accumulated in your[80] minds concerning our destructive and anti-traditional principle' (p. 133). Where, he assumes people will be by now asking, do the movement's real achievements lie? What are the goals that will justify its much-vaunted iconoclasm? Where are the literary and artistic

Morasso, has been popularized in recent years by Marshall McLuhan, *Understanding Media: The Extensions of Man* (London, 1964), and in the theories of cybernetics and robotics.

[78] With the exception of 'La Guerre électrique', the remaining contents of the volume consist mainly of reprinted manifestos and occasional pieces. Although the collection was published solely in Marinetti's name, the 'Manifestes et proclamations futuristes' that comprise the second part include a number of pieces where Marinetti was only the co-author.

[79] Most of this material does not appear in Italian as an integral part of *Guerra sola igiene del mondo*, but is hived off under the title 'Nascita di un'estetica futurista' ('Birth of a Futurist Aesthetic') (*TIF* 269–73).

[80] The way in which the concluding paragraphs of 'La Guerre électrique' directly address the reader is much more consistent with the French original's defensive buttonholing strategy, little of which remains in the Italian version.

embodiments of the Futurists' ideas? 'La Guerre électrique' is thus expressly presented as a counter to the idea that Futurism is simply a destructive aesthetic and largely a matter of theory, of manifestos inadequately underpinned by creative activity.

Having enjoined his readers to place their trust in progress ('Hope for everything from the Future, have confidence in Progress, which is always in the right, even when it is wrong, because it is movement, life, struggle, hope' (p. 136)), Marinetti tries to show how that trust will eventually be rewarded. The context is specifically a patriotic one. 'Les Premières Batailles' prefaces the volume not only with the slogan 'Long live war!', but with the more local rallying cry of 'Down with Austria!' (p. 80).[81] There is a clear hint that by acquitting herself heroically in war Italy will be both making amends for previous sins of passéism and taking the most momentous step possible towards becoming 'The Great Futurist Italy that we are preparing'.[82]

Later to be subtitled 'Futurist Vision–Hypothesis' when published in the Italian,[83] the part of 'La Guerre électrique' common to both Paris and Milan versions ostensibly belongs to a widespread genre of optimistically predictive visions of a national or global future which proliferated in the latter decades of the nineteenth century and continued to abound until the First World War shattered their essentially ameliorist assumptions once and for all.[84] However, on account of the peculiarly aggressive goals of Italian Futurism and the idiosyncrasies of Marinetti's view of the relationship between present and future, this genre-affinity soon turns out to be a looser one than the early pages might suggest; the work hardly represents the 'optimistic complement to Wells' that the Venice catalogue sees it as being (*FF* 568). To be sure, Marinetti begins on familiar enough (Boccionian) ground, proudly unfolding a picture of what the twenty-first century will look like in his 'beautiful peninsula, wholly vivified, shaken, and bridled by the new electric forces' (p. 138). Indeed, it is

[81] Various historical factors, including Austria's annexation of Bosnia and Herzegovina in 1908 and the fiftieth anniversary of the Risorgimento, had led, despite the Triple Alliance, to a marked upsurge in anti-Austrian sentiment in pre-war Italy.

[82] According to the words of the preamble: 'In the next, inevitable battles between peoples, the winner will be the one who is most fully aware of the difference [i.e. between the old and the new world]. It is the most forgetful, the most Futurist, the most knowledgeable, the most machine-orientated, and above all the richest who will win' (p. 137). Needless to say, this means the Italians.

[83] The work is presented as a 'vision–*hypothesis*', rather than just a vision of things to come, because its achievement depended, in Marinetti's eyes, on the realization of Futurist principles and in particular on Italy's victory in any forthcoming war. The Futurists' future is, therefore, not the guaranteed one of prophecy, but an opportunity. At the time of the work's republication in Milan under this new subtitle—with Italy in 1915 only about to enter the war—such success was by no means yet guaranteed.

[84] The genre is surveyed and examples analysed in Part Two of I. F. Clarke, *The Pattern of Expectation 1644–2000* (London, 1979). Of the work's many forerunners, Paul Adam, *Le Trust* (Paris, 1910), depicting a war between two gigantic trusts, Electric Standard and Ports du Méxique, and H. G. Wells, *War of the Worlds*, would appear to have had some influence on the details of Marinetti's piece.

because of the relative innocuousness of the first part of this vision that those anxious to present Futurism in a positive light tend to focus exclusively on it.[85] For what we are initially offered is little more than a conventional utopian vision of Nature harnessed.

> Here, along the entire shore, the immense grey-green sea, no longer idle, no longer lazy like a loved, prodigal, and unfaithful courtesan, finally shows itself mastered, finally made functional and productive. The immense grey-green sea, stupidly adored by poets, is now everywhere at work, with all its diligent, violent storms which set in motion countless iron pontoons that operate two million dynamos placed along the beaches and in a thousand working gulfs.
>
> Through a network of metal cables the double force of the Mediterranean and Adriatic Seas climbs to the crest of the Apennines to concentrate in great cages of iron and crystal, mighty accumulators, enormous nerve-centres set up here and there along Italy's mountainous backbone. The energy of distant winds, the rebellions of the sea, transformed by man's genius into many millions of kilowatts, will penetrate every muscle, every artery and nerve of the peninsula, needing no wires, controlled from key-boards with a fertilizing abundance that pulsates beneath the fingers of the engineers. (p. 138)

Although Marinetti's picture contains a handful of relatively forward-thinking prophecies, including the use of pontoons to harvest wave energy, the future he conjures up is by and large little more than an unimaginative elaboration of various ingredients of his own contemporary world. In this, 'La Guerre électrique' would appear to be in its first part an example of the same modernist primitivism one finds in Boccioni. (Tellingly, 'needing no wires' is one of its more ultra-modern descriptions.) For Marinetti and his Italian contemporaries, the 'new electric force' whose praise he sings here still radiates a magic aura of hyper-modernity, which is why at one stage he even contemplated the designation 'Elettricismo' for the embryonic movement.[86] On the whole, Marinetti's vision remains a somewhat lame and time-bound document, when judged by

[85] No doubt in order to create an ethically more acceptable picture of Marinetti, commentators have preferred to focus on the first part of the vision. Thus, Reyner Banham says of it: 'The tone of voice is that of science-fiction, and fairly clear debts to Jules Verne can be seen, but this is science-fiction of the rare type that comes true in detail. . . . It fixes a vision of a smart, glittering, businesslike technological life, and a corresponding architecture, which was to haunt the imagination of the next generation' (*Theory and Design in the First Machine Age* (London, 1960), 125). In fact, as anyone who perseveres with it until its grim and bitter end will appreciate, there are worse things in this vision to haunt the imagination than the glittering architecture. This is presumably why, in what looks like a similar attempt to pass over the more disturbing aspects of the piece, Christa Baumgarth's *Geschichte des Futurismus* (Reinbek, 1966), 215, which quotes extensively from the first half of 'La Guerre électrique', neither refers to the unattractively belligerent second part nor even mentions the work's telltale title. In Marinetti's scheme of things, however, the vision entails a logical progression: from the conventional, and thus more acceptable, depiction of electricity as a source of agricultural energy to the ecstatic portrayal of the electric war proper, which is presented as the only conceivable goal of such a transformed society.

[86] See E. Cavacchioli, 'I futuristi', *Attualità* (25 June 1911), 1.

the yardstick of his own contemporary technology or on the criterion of literary depictions of the future then available. After all, in the period between J. D. Falk's early vision of an *Elektropolis* in 1803 and André Bleunard's highly popular *La Babylone électrique* of 1888, the all-electric culture of future millennia had become something of a commonplace in predictive literature. Now, more inventive—and scientifically more plausible—visions based on newly discovered or rationally projected sources of power were gradually coming to the fore. Yet the Italy described in the first part of Marinetti's piece remains unrepentantly agrarian.[87] Although there are passing references to cities, the main emphasis is on farming techniques, seed-sowing, and harvesting, rather than on a world of efficient industrial production. No wonder Pontus Hulten recently felt justified in claiming that, while 'Marinetti understood that the new culture was to be of a technological nature ... to see a robotized automobile factory in modern-day Turin would nevertheless have been a shock to him' ('Future Prophecies', *FF* 15).

Marinetti's conception of the future also betrays a distinctly old-fashioned *quantitative* attitude to technology, with references to 'countless iron pontoons', 'two million dynamos', 'many millions of kilowatts' (even the linguistic reference here is inefficient), 'a thousand working gulfs', and such like. When set alongside the work of Marinetti's mentors Jules Verne, H. G. Wells, and Mario Morasso, this new 'vision–hypothesis' seems remarkably jejune, whether conceived of as science fiction or viewed as an exercise in futurology proper. There is little rationalized or miniaturized technology. Any sense of the future being more economic in its production of energy than a multiplied present is scarcely in evidence. The nearest one comes to such an idea is with the suggestion that the engineers of the future will avail themselves of 'books of nickel no thicker than three centimetres, costing no more than eight francs, and still containing one hundred thousand pages' (p. 139). When it comes to information-storage, Marinetti's new millennium remains quaintly book-orientated,[88] but

[87] This agrarian emphasis is untypical of Marinetti's work and of that of his fellow Futurists. But it does produce a picture of the peninsula as a self-servicing technological artefact and facilitates an image of Futurism as a 'natural' development which helps stage-manage the later parts of Marinetti's vision.

[88] The assumption that even in the twenty-first century books will still be in use may seem surprising, coming from a movement whose spokesmen at times talked so disparagingly about the book as a symbol of passéism. In 'The Futurist Cinema' ('La cinematografia futurista'), for instance, the book is dismissed as 'a wholly passéist means of preserving and communicating thought ... fated to disappear, like cathedrals, towers, battlements, museums and the pacifist ideal. The book, static companion of the sedentary ... cannot entertain or exalt the new Futurist generations intoxicated with revolutionary and bellicose dynamism' (*Futurist Manifestos*, p. 207). Either 'La Guerre électrique' represents early days in the evolution of a Futurist mythology, or else a transformed notion of the book will rescue it from oblivion as a means of information-storage. As someone who is by profession a writer, Marinetti may simply be clinging to a more sentimental relationship to the printed word than some of his non-literary fellow Futurists. Such allegiance can also be found in Khlebnikov's 'The Radio of the Future'

at least the book will have a certain handy compactness—unlike Depero's famous Futurist 'bolted book'.[89]

Admittedly, it is frequently not difficult to make fun of the past's images of the future, especially if that future is virtually our own present. One only has to compare the technical realities of current space exploration with the naïve ways in which it was pictured in the past. Yet this would be an unproductive approach to 'La Guerre électrique', since the work eventually reveals itself to be much more about the present than might have at first seemed to be the case. Reyner Banham proved as misleading in approaching this vision as a blend of science-fiction and futurology as a recent book was to assemble the Futurist architectural designs of Sant'Elia and Chiattone under the rubric 'Impossible Cities of Science Fiction and Fantasy'.[90] Ideally, Futurist architecture would have been the building style of an intensified present, not part of some world of mere fantasy or science-fiction.[91] (This is something nicely brought out in the juxtaposed images of Futurist plans and modern Italy in Lutz Becker's recent film *Vita futurista*.) Neither, by the same token, is a Futurist work like 'La Guerre électrique' to be primarily understood as a mere futuristic fiction, for its relationship is as much to the present as to the distant future which it is ostensibly foretelling. The Russian Ego-Futurist manifesto which declared that Futurism's prime concern was with 'the possibilities of the Future in the Present'[92] comes closer to the truth. Even those among Marinetti's detractors who were out to demonstrate that his world was fundamentally one of 'Presentism' rather than a Futurism of the distant future were not so wide of the mark.[93]

Much of the early part of 'La Guerre électrique' reads like a hymn to the transformed sensibility that Boccioni outlines in *Pittura scultura futuriste*. And because this new sensibility depends so heavily on the rejection of any vestiges of passéism, Marinetti's vision appears to be constantly looking over its shoulder. A great deal of rhetorical energy is expended on establishing that an ideal twenty-first century will have successfully enslaved and made productive

('Radio budushchego'), where the transmitting of non-verbal images remains subordinate to large-screen reproduction of the printed page.

[89] Depero's large 'bolted book', published by Dinamo-Azari (Milan, 1927), is contained within aggressively large steel bolts to emphasize the machine-quality of the work.

[90] Robert Sheckley, *Futuropolis: Impossible Cities of Science Fiction and Fantasy* (London, 1978), plates 128, 140, 141.

[91] Prampolini writes: 'The Futurist city is not a dream for us Futurists, but a stylistic command' to react to the 'dynamic aspects of contemporary life which imperiously await their architectural expression' ('Architettura futurista', *Futurismo*, 1 14 (11 Dec. 1932), cited in Crispolti, *Il mito della macchina*, p. 635).

[92] Cited in Markov, *Russian Futurism*, p. 75. The same point is made by Evgeny Zamyatin in 'Prezentisty', *Delo Naroda*, 9 (31 Mar. 1918), 4.

[93] See Wyndham Lewis, *Blasting and Bombardiering: An Autobiography (1914–1926)* (London, 1967), 34.

the 'lazy' objects which had previously been no more than the jaded symbols of a decorative nineteenth-century lyric poetry (attacked earlier in *Le Futurisme* in 'Nous renions nos maîtres les symbolistes, derniers amants de la lune' ('We abjure our Symbolist Masters, the Last Lovers of the Moon')). 'The immense grey-green sea, stupidly adored by the poets' of the past, 'the rebellions of the sea, transformed by man's genius', Nature's new 'fertilizing abundance', controlled by men at keyboards, now that it is electricity that determines crop-production, all make up the new miracle: 'the great miracle dreamed by the passéist poets is coming to life around us.... Great forests, immense woodlands spring up, matting the mountainside, higher and higher, obeying our Futurist wills and scourging the cadaverous old face, pitted with tears, of the ancient Queen of loves' (pp. 138–40).

What invalidates any slick identification of this vision with a foreward-looking 'primitive' sensibility à la Boccioni is the direction 'La Guerre électrique' takes in its second half. For here we are not just offered the reassuringly conventional millennial vision of electrically engendered peace and harmony such as even an unregenerate passéist poet might dream of, but a new variant form of 'utopia' in which war has a central role to play. For, in the words of the 'Foundation and Manifesto of Futurism', 'poetry must be conceived of as a violent attack on unknown forces, to reduce and prostrate them before men' (*TIF* 10). Christopher Wagstaff crisply sums up the implications of this during the course of his discussion of *L'alcova d'acciaio*: 'Many in Europe thought the war was a necessary regressive step to prepare for a better world. For Marinetti regression *was* that better world.'[94]

In its second part, 'La Guerre électrique' is consequently neither dystopia nor cautionary tale, but conjures up a positively longed-for new aggressive world dedicated to three aims: 'hygiene, pleasure, and struggle' (p. 140). In this sense, its reappearance in *Guerra sola igiene del mondo* is entirely logical. What, for all that, remains so untypical of the time, if one compares Italian Futurism with, say, German Expressionism, is the fact that, even after the horrors of early mechanized warfare have become apparent, Marinetti still feels able to stand by his naïve 1911 vision, re-offering it without any substantive modification. War is still seen in 1915 as the best form of hygiene (no doubt hygiene in the sense of obliterating both the stain of passéism and the enemies of Italy). And the three goals of the future society remain inextricably bound up with one another: *struggle* (domination of nature and one's passéist fellow men) is equated with *hygiene*, and this is in itself a source of intense *pleasure*.

Some parts of Marinetti's vision have a more reassuringly humanitarian ring to them, and as a result have occasionally enjoyed the dubious privilege of being quoted out of context, in order to make their author's thinking seem morally more palatable. For example: 'Hunger and poverty will disappear. The bitter

[94] 'Dead Man Erect', p. 157.

social question will be disposed of. The financial question will be reduced to a simple matter of production accounting. Freedom for all to make money. . . . Now the need for degrading manual labour will cease' (p. 140). But the other side of this coin must surely remain the question of what Marinetti assumes men will devote their energies to, indeed *should* confine themselves to, once more favourable material conditions have been brought about. 'No longer needing to struggle for his food, man finally conceives the pure idea of the ascensional record. His will and his ambition grow immense' (p. 140). Unfortunately, it is clear from the rest of the piece that the ascensional record is not expected to remain a 'pure' idea or an innocent goal for long. In practice, the growth of a Nietzschean will to power and of aggressive nationalist ambition becomes something inherently destructive, a situation made even muddier by the fact that such destruction is welcomed by Futurist Man in the name of hygiene.

War is not eradicated by man's new power over his agricultural environment; on the contrary, it emerges from the new situation in a heightened form: 'Because they can easily find enough to eat, everybody can perfect their lives in countless antagonistic exertions. An anarchy of perfections' (p. 141). When applied to production, Marinetti sees the principle of the ascensional record leading inevitably to wars of capitalism: 'Twenty-five syndicates govern the world, fighting over the markets of a superabundant industrial production. And this is why we finally arrive at the first electric war' (p. 141).

In the seldom-quoted[95] picture of battle in which 'La Guerre électrique' culminates, the stress is once again on man's mastery—through machines—over his environment. The war is electric, just as the harnessing of nature evoked at the beginning of the piece had been; and man, pictured in the cockpit of his Wellsian war-machines, is presented as if this were just another logical extension of the earlier image of the engineers controlling nature from their high-tension rooms, with their keyboards and the splendid flash of polished levers:

On the border between two nations steel elephants, bristling with shiny trunks pointed in the direction of the enemy, advance from two directions, enormous pneumatic machines, rolling down their tracks.

These monstrous drinkers of air are easily driven by small mechanics perched, like mahouts, in their glassed-in cabins. Their tiny figures are encased in a kind of diving-suit that allows them to make all the oxygen that they need to breathe.

The refined and subtle electric potential of these men is able to exploit the friendship and power of thunderstorms to conquer weariness and sleep.

[95] To his credit, R. W. Flint includes the entire 'La Guerre électrique' in *Marinetti: Selected Writings*, ed. Flint and Arthur A. Coppotelli (London, 1972), 104–8, although attributing it solely to *Guerra sola igiene del mondo*. The major requirement in English still remains a full translation of *Mafarka*, although I understand that one is at present being prepared for Calders by Dr Carole Diethe.

Suddenly, the cleverer of the two armies has thinned out its rival's atmosphere through the violent suction of its thousand pneumatic machines.

These slip away soon afterwards, to the right and left along their tracks, making way for locomotives armed with electric batteries. There they are, pointed at the border like cannon. Certain men, or rather masters of primordial forces, regulate the fire of these batteries that shoot great clusters of angry bolts through the gaps of a new sky, unbreathable and empty of all matter.

Can you see them rioting in the sky, those convulsive knots of thundering serpents? They strangle the countless chimney-pots that the worker cities display; they shatter the seaports' own jaws; they strike the white mountain peaks and sweep clean the bile-coloured sea that falls back and then rises madly to lay waste the maritime cities. Twenty electric explosions in the sky, a measureless glass chamber pneumatically emptied, have restarted the brave torments of two rival peoples with the fullness and splendour of their frightful interplanetary electric volleys. (pp. 141 f.)

For all the underlying fashionable cult of *agon*, Marinetti's future electric war entails a strangely depersonalized mode of combat, more a matter of antiseptic necessity than passion. Buildings and seaports are said to be destroyed, but not people. In this respect, the new clinical barbarities of 'La Guerre électrique' do differ markedly from those of *Mafarka*.

War shows man at his pinnacle, the warlord in his machine—'master of primordial forces'—godlike in his supremacy over nature. Yet in setting out his vision Marinetti is careful not to pursue the ethical implications of its picture too far. After all, Futurism is still needing to win converts. Hence, the sea is *swept clean*, the sky *pneumatically emptied*, the requisite acts of *hygiene* are scrupulously carried out on all fronts:

Between two battles, diseases have been attacked from every side and confined to two or three final hospitals that have become useless. The sick and weak, crushed, crumbled, pulverized by the vehement wheels of intense civilization. The green beards of provincial lanes will be shaved clean by the cruel razors of speed. Radiotherapists, their faces protected by rubber masks, their bodies encased in overalls made of lead, india rubber, and bismuth, will train their spectacles down through leaded windows on the piercing, healing danger of radium. (pp. 142 f.)

Such imagery not infrequently makes it difficult to tell whether one is encountering a healing or a killing machine. In fact, the latter is usually masquerading as the former. And if this is the primitivism of the new sensibility, then it comes painfully close to the eugenics and euthanasia policies practised by Italy's main ally during the Third Reich.

At one stage of his account of the future war, Marinetti, who had begun the piece with words of envy for 'the men who will be born a hundred years from now' (p. 138) and has talked repeatedly of the future Italians in the third person ('these men', 'the engineers', 'they', etc.), suddenly switches to the first person plural: 'the miracle . . . is coming to life around us' (p. 139). From here on, he continues to speak with the voice of the masters of the twenty-first century, no longer merely as their world's humble prophet. 'In our monoplanes we follow the fantastic growth of the forests towards the moon'; 'Great forests . . . spring up . . . obeying our Futurist wills' (p. 140), and so on, as if the act of envisaging in itself created such a brave new world. During the course of 'La Guerre électrique' we observe the author gradually immersing himself in the domain of his vision–hypothesis until he speaks to us from it. It is scarcely fortuitous that he has already begun to function as a representative of the twenty-first century before he begins to wax lyrical about the new electric conflict. It is as if it were *his* personally engineered war, and not that of his great-great-grandchildren, that he wants to introduce us to—hence, the way in which he takes his readers inside the cockpits of the war-machines and insulates them from the actual destruction on the ground. The perspective governing our experiences is no longer that of outside observer or victim, but of the victorious combatant. The narrative invites us to fight the war in an intoxicated state of identification with the victor, and, as was the case with the artillerist's bombardment in *Le Soir*, the result is an act of aesthetic creation rather than a conventional eyewitness account.

It could be argued that there are really two wars taking place in Marinetti's vision–hypothesis: the electric war between two of the great power syndicates of the next century, and an ideological conflict between what that Futurist world stands for and the supposedly passéist values of the majority of Marinetti's contemporary readership. From the outset, the presentation of the future is marked by a vitalist sense of struggle as a virtue: mastering nature, defeating one's foes, and vanquishing passéism are put on a par. It is therefore quite logical that, after the vision of the war has predictably concluded with the defeat of the enemy, Marinetti's attention should henceforth turn to the more crucial attack on the anachronism of his present-day world. The image of the hospital of the future where radio-therapists go about their work, their faces protected like those of the military combatants they so closely resemble, now becomes a metaphor for dealing 'hygienically' with twentieth-century passéism. For, according to Marinetti, our world is also in need of radio-therapy—the kind where, in the event of success, the patient is conveniently eradicated along with the disease. 'Treatment' here becomes a far more sinister variation on the idea of 'hygiene' than any Futurist notion of burning books or destroying museums ever was (it is closer to the Nazi euphemism of *Sonderbehandlung*). We hear Marinetti ask, now from his present-day viewpoint:

> when will we finally invent masks and overalls that will protect us from the deadly infection of imbecility, your imbecility, you passéists, who naturally dislike the cruel sincerity of my attacks on Old Italy? You say that everyone should wash his dirty linen in private. . . . Rubbish! We are no subtle and delicate-handed washerwomen. Today, we make a joyful bonfire of our infected, pestilential laundry on the loftiest peak of

human thought. We spare no one. After having insulted every stranger who adores our Italian past and despises us as singers of serenades, as ciceroni or beggars, we have asked them to admire us as the most gifted race on earth.

Thanks to us, Italy will cease to be the 'love-room' of the cosmopolitan world. To this end, we have undertaken to take up the battle on behalf of courage against the epidemic of cowardice, and to create an artificial optimism in the face of chronic pessimism. Our hatred of Austria; our feverish anticipation of war; our desire to strangle Pan-Germanism. This is the corollary of our Futurist theorem! (p. 143)

Marinetti's Futurist dream now finally reveals itself for what it really is: less a forecast of things to come than a beleaguered act of deliberately 'artificial optimism', a fanatically contrived Interventionist hypothesis *avant la lettre* to counter the threats the writer sees on all sides: from passéist Italy's culture and its admirers, from the political situation, and, most tellingly, represented in the persons of the majority of readers of 'La Guerre électrique' itself. In fact, it is against them that Marinetti ultimately adopts his most belligerent stance. The arsenal of hostile images that concludes the piece—'imbeciles! We hurl ourselves against you, holding our hearts in our fingers like revolvers, our hearts burdened with hate and courage' (p. 143)—is targeted on the archetypal passéist—'you', me, the reader. In contrast to this final vitriolic outburst, the combat of the twenty-first century possesses a disengaged, almost 'sporting' quality. For if the 'artificial optimism' of the vision is unmasked as merely a strategy for attacking the ills of the present, then Marinetti's contemporaries are the real enemies of the Futurists who want to create such a millennial Italy.

With us begins the violent strike of the young gravediggers. Away with graves! We let the corpses bury themselves and enter into the great Futurist City that points its mighty battery of smoking factory chimneys against the enveloping army of the Dead, marching along the MilkyWay. (p. 143)

In the name of 'hygiene', the path to the Great Futurist Metropolis will be strewn with unburied corpses. And while on the surface this might be no more than an ill-chosen metaphor[96] for the fact that Futurist Man will not waste his time with hypocritical tokens of respect for the degraded (and degrading) past, the figurative language of struggle that pervades even the early parts of 'La

[96] When asked in an interview about his exhortation to destroy the museums and libraries, Marinetti replied: 'It's simply a violent metaphor for the desire we all have at least to escape from the sanctification of the past' (*Poesia*, 5 (1905) 3 f.; a French version of the interview was published in *Comoedia* (2 March 1909)). Bengt Jangfeldt argues that 'it was never a question of contempt for the old culture as such, but a struggle against the *influence* of the old art on the creation of contemporary art' (*Majakovskij and Futurism, 1917–1921* (Stockholm, 1977), 11). Nevertheless, violent metaphors create violent climates; and, no matter how anxious Marinetti's defenders have been to stress the figurative nature of some of his incitements, this cannot detract from the fact that they contributed to the general atmosphere of irrationality that prepared the way for both Fascism and National Socialism, and in the former case the fact that the Futurists were very much involved with Mussolini's movement.

Guerre électrique' suggests there is a more insidious barbarism at work. 'The sick and weak, crushed, crumbled, pulverized by the vehement wheels of intense civilization. The green beards . . . shaved clean by . . . cruel . . . speed' is only 'optimism' by the standards of the most grotesquely distorted interpretation of the Nietzschean Superman concept. We are not so far from the primitivism of Mafarka's world of ritual sacrifices, of chieftains elevated in status by the number of people they have dispatched, and the weak and untouchables, the 'unhygienic', falling foul of the dominant warrior caste, as some of the futuristic trappings of 'La Guerre électrique' might lead us to hope. 'Unhygienic' here, it should be remembered, means little more than 'passéist', and in all likelihood it refers to us as readers. So Boccioni's 'higher barbarism' and the baser kind of primitivism that has always been all too present over the centuries, and especially in the first half of our own, are not as distinguishable as some more charitably inclined Futurist commentators might want them to be.

Primitivism in Futurist art and primitivism in literature are substantially different, though obviously related phenomena, as a juxtaposition of any representative Futurist paintings with these examples from Marinetti's work would confirm. Futurist art can offer no equivalent to the archaic barbarism of *Mafarka* or the final part of 'La Guerre électrique', though one can, admittedly, detect elements of Boccioni's 'primitivism of a new sensibility' in the subject-matter of many Futurist paintings. Russolo's *Dinamismo di una automobile* (*Dynamism of an Automobile* (1913)) combines striking use of bold colour with aggressive dynamic rhythms and a sense of thematic modernity that would scarcely discredit a 1980s advertisement for a high-performance car. Boccioni's own *Le forze di una strada* (*The Forces of a Street* (1911)) also conveys the exhilarating sense of dynamism, coupled with an exploitation of loud colour and references to the tokens of modern city life, features which characterize the Futurist interpretation of the modern street. And the force-lines of Depero's splendid *Grattacieli a New York* (*New York Skyscrapers* (1927)) (fig. 5.1) give a sense of the brutal vitality and building-machine analogy that many earlier Futurists, including Mayakovsky and Sant'Elia, had attributed to the modern city. Or to lessen the thematic gulf between the works under comparison by turning to paintings on the subject of war: Severini's *Il treno blindato* (*Armoured Train* (1915)), depicting a military train with mounted guns and riflemen shooting out from it, as if in some mobile version of trench warfare, and Nevinson's *The Twentieth Century* both contain that mixture of technology, combat, and dynamism that for Boccioni marks out the new primitivism from the neo-primitive and which was for many a Futurist the essence of modern warfare. Yet Marianne Martin can talk of 'the icy colours and clear-cut forms' of *Il treno blindato* that 'convey the cutting,

Fig. 5.1. Fortunato Depero, *Grattacieli a New York*, 1927 (Collection Prof. Mary Vieira Belloli, Milan-Basilea)

unemotional precision of gunfire',[97] with none of that hint of compassion to be found in the companion piece *Il treno dei feriti* (*Red Cross Train*). And 'unemotional' is the right word here, for one finds little of the mindless aggression, the parading of carnage and excess, or the flaunting of power that permeate *Mafarka*.

Such differences cannot be considered without reference to Marinetti's personality, of course. For wherever one looks in his works, from early titles containing such words as *Déstructions*, *La Conquête* . . ., 'Uccidiamo . . .', 'Abbasso . . .', and 'Bombardamento' through to the post-war novels, one is confronted by a wallowing in violence and in the more bestial aspects of vitalist primitivism. Whether his setting is the past, the present, or the future, Marinetti manages to create a world that invariably radiates the same archaic quality. Yet even if this seems to single Marinetti out as different in his excesses from most of the Futurists around him, one should not over-emphasize this difference. As we have seen, they too sang their hymns to cruelty and violence. And, more importantly, generic differences are also at work, not just matters of personality. It would be difficult to imagine just what painting's equivalent to *Mafarka* could possibly be. It is certainly not Boccioni's war collage *Carica dei lancieri* (*Charge of the Lancers*). And in sculpture, similarly, it would not be the hieraticism of Boccioni's *Forme uniche della continuità nello spazio* (*Unique Forms of Continuity in Space*) or a work like Epstein's *Rock Drill*, even if this is more archaically primitivist in spirit than anything Boccioni ever produced. If the Futurist film had been developed further, one suspects that only there could a real visual counterpart to the regressive primitivism of *Mafarka* and 'La Guerre électrique' have been found.[98] And the original *Vita futurista*, the main Futurist film of the time, does contain scenes of violence.

Fortunately, as some of the remarks about aesthetic renewal quoted earlier suggest, primitivism is not simply a matter of reactionary content; it embraces a whole spectrum of technical possibilities.

Robert Goldwater takes the line that 'primitivism is not the name for a particular period or school in the history of painting, and consequently no description of a limited set of objective characteristics which will define it can be given'.[99] It is at best a tendency, a cluster of formal experiments, rather than a single style or '-ism'. Once, however, one restricts one's attention to the primitivism of a particular period, then the picture becomes a little more sharply focused than Goldwater's scepticism would suggest. Confining himself to primitivist aspects of Expressionism and Dada, and then only to 'the stages by

[97] *Futurist Art and Theory*, p. 198.

[98] The closest suggestion to this comes in Point 4 of the manifesto 'The Variety Theatre' ('Il teatro di Varietà'), where it is argued that the new theatre will be able to present 'otherwise unrealizable spectacles (battles, riots . . .)' (*TIF* 71).

[99] *Primitivism in Modern Art*, p. xxiv.

which the value "primitive" became positive and desirable', Christopher Middleton still warns that 'primitivism has many facets', something borne out by the fact that his typology of the most significant for his subject runs to as many as ten items. In fact, the characteristics itemized go some way towards offering a clear picture of the nature of the tendency in this century. The forms of 'primitivism' considered are:

1. 'prelogical' or 'savage' art
2. 'naïve' European art
3. alogism in poetry
4. infantilism
5. resuscitation of folk forms
6. anti-aestheticism
7. feigned illiteracy
8. exoticism
9. 'das Elementare'
10. atavism

Middleton's survey of 'the word's semantic range'[1] covers (with the possible exception of the second category) forms of primitivism which also occur in Futurist literature, even if they are not all being considered in the present context. (Some of them, including 'alogism' and 'anti-aestheticism', have been discussed in earlier chapters.) And yet this list does not take account of other features which have been associated with the Futurist primitive. To start with an eccentric pair of examples: in 'Les Exposants au public' the painters even saw force-lines as the application of a primitive technique. The way in which 'the general violence of the picture', the 'sheaves of lines corresponding with all the conflicting forces' of the work, draws the observer into the chaos of the canvas was for them 'the foundation of *Futurist primitivism*' (*Archivi*, i. 105 f.). In addition, as if to give the term an even wider framework of reference, they stress how fundamentally this differs from what they term 'Impressionist primitivism', with its concentration on 'reflections of light'. Here, the primitive is seen solely as a matter of technique, with no thematic implications at all.

Christa Baumgarth is even prepared to interpret the Futurists' general enthusiasm for the experimental as a form of primitivism.[2] And no doubt if one sees the compulsive aesthetic experimentation, the sense of the in-built ephemeral in much Futurist theorizing and in the quest for new non-anachronistic forms as symptoms of the new anti-traditional sensibility, then this is not beyond the bounds of acceptability, at least in Boccioni's sense of the word. Nevertheless, it is doubtful whether, *outside the Futurist context*, anyone would think of positing a link between extensive literary experimentation *per se* and the primitive.

[1] *Bolshevism in Art*, pp. 24 f. [2] *Geschichte des Futurismus*, p. 133.

Not all technical primitivism is a vehicle of aesthetic renewal, however. One has to allow for the fact that primitive elements can be there purely for their shock potential, as a 'slap in the face of public taste', as the Russian manifesto puts it. Moreover, there have been instances, for example the case of the wallpaper used in the first *Sadok sudey*, where critics have been unable to agree about whether they were dealing primarily with a mere provocation or a radical act of artistic revitalization.[3] Certainly there have been occasions when *épater le passéiste* has been as attractive a pastime for the budding Futurist as the regeneration of art through systematic recourse to features of the primitive. One need not go as far as Peter Bürger's *Theory of the Avant-garde*, with its view of Futurism as a rejection of the institution of art itself,[4] to allow for the provocative element in the deployment of the primitive.

Michael Bell sees the primitivist urge in much modern writing as 'born out of the interplay between the civilized self and the desire to reject or transform it'. As a result it becomes 'the projection by the civilized sensibility of an inverted image of the self'.[5] On the level of provocation, that interplay is dramatized: it is a clash between the expectations of the (civilized, or passéist) public and the adversary Futurist artist. More specifically, it is a contrast between the aspects of intellectualized civilization within the psyche of the Futurist himself and his new creations. This interplay between the civilized persona and its inverted image, between old and new sensibilities, partly accounts for the fact that the Futurist cult of the primitive is accompanied by so much frenetic manifesto-production, self-conscious theorizing, and attempts at justification. There is, at least in this instance, no simple contrast between modern cerebrality and a quest for its negation in some regressively simple form of primitivism.

In a letter to Vico Baer of 21 June 1914, Boccioni presents Futurist primitivism as 'the extreme point of a form of *complexity*', whereas he views the primitivism of earlier times as a matter of *simplicity* (*Archivi*, ii. 49). He elaborates the point in *Pittura scultura futuriste*:

it is a basic mistake to declare that man is distancing himself from nature! If this were so, one could naïvely believe that an animal was closer to nature than a chemist. . . . We all possess a new instinct: the instinct for the complex. . . . In our opinion, science had

[3] For Vasily Kamensky (*Put' entuziasta* (Moscow, 1931), 113), the publishing of poetry on the reverse side of cheap wallpaper was no more than a 'protest against luxurious bourgeois publications'. But for Krystyna Pomorska, it represents a more serious focus on the 'formal, technical side' of book-production (*Russian Formalist Theory and its Poetic Ambiance* (The Hague, 1968), 80).

[4] Cf. *Theory of the Avant-garde*, trans. Michael Shaw (Theory and History of Literature, 4; Manchester, 1984), especially the section 'The Avant-garde as Self-criticism of Art in Bourgeois Society' (pp. 20–7).

[5] *Primitivism*, p. 80. Bell warns of the dangers of what he calls '"armchair" primitivism' (p. 77), but, in the case of the Futurists, one worries more about what they were likely to get up to whenever they decided to leave their armchairs.

led man back to a sort of barbarism, to the most marvellous higher form of barbarism. (*Gli scritti*, p. 87)

Two aspects which are crucial, if one is trying to understand the connection between Futurist primitivism and the movement's literature, are its *anti-intellectualism* (which is not at odds with what Boccioni says about the 'complexity' of the new primitivism) and its quest for *the essential*—the desire to strip away from reality its misleading surfaces and penetrate to true reality, which in Italian Futurism tends to mean the material dimension.

In her exploration of what she calls the movement's 'poetics of matter', Carmine Chiellino analyses the role played by 'matter' and 'intuition' in early Italian Futurism.[6] While Marinetti's insistent championing of 'intuition' (as opposed to 'intelligence') demonstrably owes much to the Bergson of the *Introduction à la métaphysique* (1903) and *L'Intuition philosophique* (1912), Chiellino argues that there is a more radical antinomy between intuition and intelligence in Marinetti's thought. Marinetti's espousal of intuition is deliberately anti-intellectual, in a way that it is not in Bergson. (Whether this is a 'tendentious interpretation' of Bergson, as Chiellino suggests,[7] or merely a parting of the ways, is open to debate.) But the fact remains that whatever Marinetti derives from Bergson is taken in a new, and often more insidious direction. 'We believe along with Bergson', he writes in 1915, 'that *la vie déborde l'intelligence*' (*TIF* 285). But Bergson's critique of intelligence, as it is outmanœuvred by life, readily widens in Marinetti's case to a full-scale primitivist anti-intellectualism. In a tautological flaunting of his lack of respect for intellectual argument, Marinetti declares: 'By *intuition* I mean a stage of thought which is almost entirely intuitive and unconscious' (the source of the analogies of the 'wireless imagination', in fact), and 'by *intelligence* I mean a state of thought which is almost entirely intellectual and willed'. While he concedes that the two cannot always be tidily distinguished, he is anxious to show where his interest lies and why. This is the context of the much-quoted remarks: 'I taught you to hate museums and libraries in order to prepare you *to hate intelligence*, reawakening in you divine intuition. . . . Through intuition we will conquer the seemingly unconquerable hostility that separates out human flesh from the metal of motors' (*TIF* 47).

Apart from the *intuitive* doubling of nouns to create bold analogies, one of the chief results of this attitude is the attention paid to the senses, a deliberate

[6] *Die Futurismusdebatte: Zur Bestimmung des futuristischen Einflusses in Deutschland* (Europäische Hochschulschriften, Ser. 1, vol. 252; Berne, 1978), 94–105. On the centrality of 'matter' in Futurist theory, see also Blumenkranz-Onimus, *La Poésie futuriste italienne*, pp. 130f. Peter Drews surveys Bergson's importance for literary Futurism in *Die slawische Avantgarde und der Westen. Die Programme der russischen, polnischen und tschechischen literarischen Avantgarde und ihr europäischer Kontext* (Forum Slavicum, 55; Munich, 1983), 119–37.

[7] *Die Futurismusdebatte*, p. 98.

compensation for modern man's debilitating intelligence. According to Marinetti's 'Technical Manifesto of Futurist Literature' ('Manifesto tecnico della letteratura futurista', 1912), 'three elements hitherto overlooked in literature must be introduced:

1. Sound (manifestation of the dynamism of objects)
2. Weight (objects' faculty of flight)
3. Smell (objects' faculty of dispersing themselves)

To force oneself, for example, to render the landscape of smells that a dog perceives. To listen to motors and reproduce their conversations. (*TIF* 45)

Sound, weight, and smell are here ingeniously defined in terms intended to relate them to matter: they are all viewed as properties of objects, rather than directly related to human concerns. 'Man tends to foul matter with his youthful joy or elderly sorrows,' the argument continues, whereas 'matter has an admirable continuity of impulse towards greater warmth, a greater sub-division of itself. Matter is neither sad nor gay. Its essence is courage, will power, and absolute force.' The problem is, in Marinetti's view, that hitherto *materia* ('matter' or 'material') has always been contemplated by an 'I', a self 'too preoccupied with itself, full of preconceived wisdom and human obsessions' (*TIF* 45). Far more important than the overstressed recommendation that 'the *I*' in literature should be 'destroyed' is this desire to bring about a new matter-orientated aesthetics, for it accounts for so many features of Futurism, whereas the Futurists were seldom able to eradicate the self from their writings.

The need outlined here is to break with an ego-/anthropocentric way of presenting the world and give matter its autonomous place, i.e. the goal that some of the French *nouveaux romanciers* were to set themselves some forty years later. The poem 'Battaglia peso + odore' appended to this manifesto represents one such attempt. Even a consideration of the title reveals differences between Marinetti's view and a traditional approach, but also some of the problems inherent in the new aim.

A more conventional account of battle in terms of smells and weight would probably stress the load of equipment being carried, the smell of ammunition, possibly even the stench of putrifaction, and, what is more important, it would register these details through the medium of a participating soldier's sense-impressions. Indeed, recalling the genesis of this poem, Marinetti speaks of the occasion when, 'coming away from the maddening mixture of smells and perfumes of the Souk, I wrote in my mind words-in-freedom telling of the murderous weight the colonial infantry carried under the burning sun that made even heavier their rifles knapsacks blankets ammunition machine guns cannons mules and the sand in their boots unfit for desert dunes' (*GM* 254). Marinetti, on the other hand, does try to depersonalize such data. The work's *leitmotif*, *Tintinnio zaini fucili zoccoli chiodi cannoni* ('Tinkling knapsacks rifles clogs nails cannon') is deliberately put in the plural so as not to tie the impressions down to one experiencing person on the battle-field. The references to 'ammonia', 'excrement', or even 'orange-blossom nausea' (*fiore-d'arancio nausea* (*TIF* 52)) are also made autonomous, in the sense that there is no indication of a subject experiencing them. But smells, weight, sounds, and many of the other 'material' characteristics that make up the poem really only have a significance in terms of a notional perceiver. They may be properties of matter itself, but it is difficult to conceive of a smell or a sound without at the same time automatically supplying the context of someone receiving such sense-impressions. In other words, the 'I' cannot be destroyed; at most one can depersonalize the 'focalizer' (Genette) of such impressions.

An indication that Marinetti was well aware of the problem comes in his poem 'paesaggio d'odore del mio cane-lupo (lirica olfattiva—parole in libertà)' ('my alsatian's landscape of smells. (olfactory lyric. words-in-freedom)'), published in *I nuovi poeti futuristi* of 1925. The poem manifestly embroiders on the idea that literature could eliminate the human subject by 'rendering the landscape of smells that a dog perceives'. As the canine man-surrogate savours the 'delicious variety from the immense ocean of odours', ranging from the 'green smell of tepid soggy grass' and 'smells of damp cold earth' through to the 'plump musky vanilla-flavoured thick rich intimate MAN smell', it becomes clear that the mere substitution of a canine subject for a human one represents an unsatisfactory solution to the problem. Admittedly, the dog's world is not ours, therefore the poem's eccentric perspective functions as a defamiliarization of the passéist's concern with landscape (a target in Boccioni's *Pittura scultura futuriste*). But in the final analysis, one form of subjectivity has simply replaced another. Instead of getting closer to autonomous matter, we are still experiencing matter through the medium of language—and via a human's projection of what a dog's world must be like, at that. Russolo's noise-machines, his *intonarumori*, may be able to approximate the acoustic properties of the world of matter, but for literature, contained, as it is, within the medium of language, such a goal must remain a more remote eventuality. However, there were other respects in which Futurism was pushing its experiments closer to this goal.

Touch is not a sense given much prominence in early Futurist writings, but it did come into its own later in the manifesto 'Tactilism' ('Tattilismo' (1924)). Just as Russolo had furnished a typology of noises, so here Marinetti envisages the creation of tactilist works intended to bypass intelligence (language, even) and evince a more intuitive response. The resultant assemblages might be construed as a form of sculpture—after all, touchable sculptures have enjoyed something of a vogue in recent years—but the way in which Marinetti presents them suggests that such works can be just as meaningfully thought of as a new form of theatrical *sintesi* or even concrete poetry.

The description of *Sudan–Paris* (illustrated *FF* 338) uses the generic label *tavola* which began life as a term for some of the free-word layouts of *Zang tumb tumb*:

I created the first[8] abstract suggestive tactile table, which was entitled *Sudan–Paris*. In its *Sudan* part, this table has: spongy material, sandpaper, wool, pig's bristle, and wire bristle. (*Crude, greasy, rough, sharp, burning tactile values that evoke African visions in the mind of the toucher*.)

In the *sea* part, the table has different types of emery paper. (*Slippery, metallic, cool, elastic, marine tactile values*.)

In the *Paris* part, the table has silk, watered silk, velvet and large and small feathers. (*Soft, very delicate, warm and cool at once, artificial, civilized*.)

After careful differentiation of tactilism from other plastic arts and what he calls 'morbid erotomania' (evidently a passéist preoccupation, though no culprits are named), Marinetti seeks to legitimize the experiment in terms once more relating to the aesthetics of matter:

We are aware of the hypothesis of material essence. This provisional hypothesis considers matter to be a harmony of electronic systems, and through it we have come to deny the distinction between matter and spirit.

When we feel a piece of iron, we say: this is iron; we satisfy ourselves with a word and nothing more. Between iron and hand a conflict of preconscious force–thought–sentiment takes place. Perhaps there is more thought in the fingertips and the hand than in the brain that prides itself on observing phenomena.

With Tactilism, we propose to penetrate deeper and outside normal scientific method into the true essence of matter. (*TIF* 151 f.)

In the manifesto 'Tactilism' ('Il tattilismo'), not to be confused with the one entitled simply 'Tattilismo' from which the above extract is taken, Marinetti lists various possibilities for the creation of tactilist works: tactile clothing, tactile furniture, tactile rooms, and tactile poems (*TIF* 140), although it is not clear how many of these experiments were actually carried out.

However, it is not in this literal form of tactilist creation, but rather in the

[8] In *Una sensibilità italiana nata in Egitto*, Marinetti was later to state that *Paris–Sudan* had been made by his wife Benedetta, even though he claims to have created it in his tactilist manifestos; see 'Le tavole tattili costruite da Benedetta sulla spiaggia di Antignano' (*GM* 263–7). Possibly the idea was Marinetti's and the construction itself was left to Benedetta. In any case, an article in *Comoedia* (Feb. 1921) by Picabia disputes the paternity of tactilism, pointing out that similar experiments had already been conducted some years before in New York. No doubt he was thinking of Clifford Williams, *A Plaster to Touch* (1918), an account of which was given in Apollinaire, 'La Vie anecdotique' (*Mercure de France* (Feb. 1918), also in *Anecdotiques*, pp. 251 f.). For details of the work, see Buckley, *Guillaume Apollinaire as an Art Critic*, p. 310. The author of the entry on 'Tactilism' in the Venice exhibition catalogue (*FF* 584), confronted by a rare instance of Marinetti having been pipped at the post, concludes: 'Marinetti certainly deserves credit as the only artist to have produced interesting results in the field'! It is quite conceivable that Marinetti may have been set thinking in the direction of tactilist tables by Apollinaire's report of Williams's work, but there is already a complex theoretical underpinning for tactilism in Futurist aesthetics, whereas Williams's experiment was a far less systematic affair.

works in *I nuovi poeti futuristi* that one finds the most detailed series of attempts at applying such principles aesthetically, i.e. in the creation of 'tactile poetry'.[9]

Apart from the Alsatian's landscape, various other poems, all subtitled in a manner emphasizing the sense-impressions entailed, are given in both Italian and French, an indication of Marinetti's desire to publish the genre as widely as possible. The poems are *navigazione tattile (lirica tattile—parole in libertà)* (pp. 279f.), *ritratto olfattiva di una donna (lirica olfattiva—parole in libertà)* (pp. 281 f.), and *temperature del corpe del nuotatore (lirica termometrica—parole in libertà)* (pp. 284 f.). Hardly any of the other poets in the collection follow Marinetti in this direction, although the reader is offered an interesting 'SINFONIA TATTILE' by Loris Catrizzi, striking in its by then old-fashioned use of telegraphic noun-pairings but also in its attempts at rendering the tactile through neologistic formations:

muro-cartavetrata
vetri-cartavelina-cartargentata
marciapiedi-porcellana
paletots-garza
pelliccie-cotoneidrofilo
marmi-celluloide
fazzoletti-cartascuigante
manisenzaguanti-legnobagnato
temperatura-ghiaccata al limone
l'atmosfera e spugna di champagne[10]

wall–sandpaper
panes of glass–flimsy paper–silver-paper
pavement–china
coats–gauze
fur-coats–cotton wool
pieces of marble–celluloid
handkerchiefs–blotting paper
gloveless hands–wetwood
temperature–iced lemon sweet
the atmosphere is impregnated with champagne

Catrizzi was to publish further matter-orientated works, including his molecular poem 'Vibrazioni di fosforo', and the exercises by Marinetti were republished in 1932 in his collection *Parole in libertà futuriste—Tattile—Termiche—Olfattiva*. Moreover, his *Il fascino dell'Egitto* (1933) contains a section 'Tattilismi rissanti del grassume fecondo edella vetrosità sterile' ('Brawling Tactilisms of Fertile Earth and Barren Vitrescence' (*TIF* 982)). Another Futur-

[9] When the adjective 'tactile' is attached to Futurist poetry, it tends to mean relating *thematically* to the sensation of touch, rather than referring to the material quality of any poem itself.

[10] *F. T. Marinetti presenta I nuovi poeti futuristi* (Rome, 1925), 27 f.

ist, Angelo Rognani, also went on to create a number of tactile poems.[11] But the effect of most of these works is merely to translate the tactile or some other sensation back into the verbal medium and hence bring about a 're-intellectualization' of it to some degree. That is to say, as long as language is being used to take literature back in the direction of matter, the result will always be some degree of compromise.

On a more literally sense-dominated plane, Futurist theatre experimented in a related direction. Fillia's *Sensualità meccanica* (1927), for example, makes use of various temperatures in the theatre to control the audience's aesthetic response. But rather than an exploration of the essence of matter, what such experiments are moving towards is a series of new sense-orientated strategies for audience-manipulation.

Marinetti's manifesto 'Dynamic and Synoptic Declamation' ('La declamazione dinamica e sinottica'), first published in Cangiullo's *Piedigrotta* (1916), points to another way in which the man–machine symbiosis could be brought about. The art of Futurist declaiming could introduce mechanical and non-human characteristics into performance:

> the Futurist performer must completely dehumanize his voice, systematically doing away with every modulation or nuance . . . Completely dehumanize his face, avoiding every facial expression or trick of the eyes. . . . Metallize, liquefy, vegetablize, petrify, and electrify his voice. . . . Gesticulate geometrically, thereby giving his arms the sharp rigidity of semaphore signals and lighthouse beams, to indicate the direction of forces, or of pistons and wheels, to express the dynamism of words-in-freedom. . . . Gesticulate in a draughtsmanlike, topographical manner, synthetically creating in mid-air cubes, cones, spirals, ellipses etc. . . . In all ways imitate motors and their rhythms . . . (*TIF* 106–8)

In making his style of delivery more mechanical, the Futurist performer described here will not only become more mechanical himself, but in doing so will bring out the machine-like and materialist properties of the new literature he is mediating. Yet one can imagine from Marinetti's description just how difficult some of these values would have been to communicate effectively.

While the development of the *intonarumori* was of great significance within the development of modern music, the fetichizing of sounds, tactile sensations, and the depiction of weights and smells (in both thematic and technical ways) frequently remains of little more than peripheral importance within the sphere of literary Futurism. Only onomatopeia, the least revolutionary of all these various sense-experiments, strengthened the repertoire of poetry to any substantial degree. But there were other ways in which the aesthetics of matter, which can be seen as the most productive of all aspects of Futurist primitivism, entered into the mainstream of Futurist writing to produce significant results.

[11] For details, see Blumenkranz-Onimus, *La Poésie futuriste italienne*, p. 130.

For ultimately, underlying the stress on matter and the need to make more of the senses and less of intelligence, is something more revolutionary than the idiosyncratic desire to reproduce the sense-impressions of a dog. (As suggested by the scatological note on which Marinetti's poem on that subject ends, this is little more than a provocation.) A far more significant goal was the narrowing of the gap between man and matter: or, more specifically, between man and machine. 'Through intuition we will conquer the seemingly unconquerable hostility that separates our human flesh from the metal of motors,' as Marinetti puts it (*TIF* 47). In some lines from Malevich, quoted in the notes to the translated *Essays on Art*, we hear

> Meat and the machine are the muscles of life.
> Both are the bodies in which man lives.
> Here two worlds have collided.
> The world of meat and the world of iron.[12]

But in Italian Futurism the result is less a collision than an attempt at a new synthesis, one which, if ever achieved, would engender not only a new conception of art but even a new kind of man.

'You Wops insist too much on the machine', Wyndham Lewis is supposed to have grumbled to Marinetti. 'You're always on about driving-belts, you're always exploding about internal combustion. We've had machines here in England for donkey's years. They're no novelty to us.' To which Marinetti is alleged to have retorted: 'You have never understood your machines. You have never known the *ivresse* of travelling at a kilometre a minute.'[13] He might have added: you have not been adequately changed by your machines. Or even: you have not succeeded in *becoming* your machines.

Renato Poggioli would have sided with Lewis, for in his *Theory of the Avant-garde* we encounter the complaint that all too often

> modernism . . . cheapens and vulgarizes modernity into what Marinetti called, encomiastically, modernolatry [in fact, the term is Boccioni's, coined at the end of *Pittura scultura futuriste*]: nothing but a blind adoration of the idols and fetiches of our time. . . . To exemplify the degeneration of modernity into modernism, all we have to do is refer to the failure of the attempt to express what the surrealists . . . called the *modern marvellous* [the 'Variety Theatre' manifesto also talks of a 'Futurist marvellous' (*TIF* 71)]. Why not look ingenuously and freshly, with sympathy and enthusiasm, at certain aspects of modern life and draw from them a new poetic magic, new fables and fantasies?

Although this is not the spirit of Poggioli's piece, one could scarcely ask for a better description of the direction in which many of the Futurists were hoping

[12] *Essays on Art*, i. 28. [13] *Blasting and Bombardiering*, p. 34.

to take their modernist primitivism. But Poggioli is sceptical of how far such a development ever progressed: 'The attempt, anticipated by Walt Whitman and Verhaeren, and already begun by the futurists . . . ended up in almost complete failure. This failure suggests the hypothesis that the modern imagination is congenitally impotent when it comes to mythic and legendary creation.'[14]

It is certainly not difficult, in the Futurist context, to think of examples of a blind adoration of the fetiches and idols of the time. The movement's manifestos are peppered with illustrations of such debased *modernolatria*. Futurism's most hyperbolic and least profound response to modern technology came in the series of lapidary manifesto comparisons that furnish us with some of the most memorable of all avant-garde pronouncements.

The series was initiated by Marinetti's declaration that 'a roaring car that seems to ride on grapeshot is more beautiful than the *Victory of Samothrace*'[15]—an auspiciously provocative juxtaposition which was soon to spawn a whole range of ingenious variations, both within Italian Futurism and beyond its confines. Nearest to home was Boccioni's remark, in *Pittura scultura futuriste*, that 'it is this passionate love of Reality that makes us prefer the American *cake-walk* to hearing the *Valkyrie* and makes us prefer the newsreel in the cinema to a classical tragedy' (*Gli scritti*, p. 87). Many other formulations soon followed in the wake of Marinetti's first manifesto:

The Morse Code transmitter is a 1000-times greater and more essential a work of art than Byron's *Don Juan*
an American shoe is superior to the Venus de Milo
Today the tatters of the signboards are more beautiful than an evening dress
You like the lightning in the sky, and I in an electric iron.
any laundry bill is superior to Pushkin
a racing car is superior to the paintings of Thorwaldsen

In 'The Futurist Universe 1918' ('L'universo futurista 1918') Balla elaborates a similarly outrageous set of contrasts:

Any store in a modern town, with its elegant windows all displaying useful and pleasing objects, is much more aesthetically enjoyable than all those passéist exhibitions which have been so much praised everywhere. An electric iron, its white steel gleaming clean

[14] *The Theory of the Avant-garde*, p. 218. That the term *modernolatria* was generally acknowledged as coming from Boccioni is clear from Marinetti's introduction to *I nuovi poeti futuristi*, p. 14, where there is a reference to that 'typical passion for life today which Boccioni called *modernolatria*'.

[15] The phrase 'che sembra correre sulla mitraglia' has occasioned a diverse array of interpretations. Is it 'grapeshot' or 'machine-gun fire', is the car running physically *across* grapeshot, or is it under fire? Clearly, the phrase is intended to unite associations of speed with those of danger, but these various readings (most of them tenable) make the image seem either more or less anachronistic and vary the quality of the danger suggested. Eventually, Marinetti dropped this part of the contrast and, by the time of 'Nascita di un'estetica futurista' (1913), was content simply to play 'a racing car' off against 'the *Victory of Samothrace*'.

as a whistle, is more attractive to the eye than a nude statuette, stuck on a pedestal, hideously tinted for the occasion. A typewriter is *more architectural* than all those building projects which win prizes at academies and in competitions. The windows of a perfumery, with little boxes and packets, bottles and . . . phials. . . . The clever and gay modelling of ladies' dancing-shoes, the bizarre ingenuity of multi-coloured parasols. Furs, travelling bags, china—these things are all a much more rewarding sight than the grimy little pictures nailed on the grey wall of the passéist painter's studio.

Or, in the words of Mayakovsky: 'heavy steel monsters are more necessary than *Eugene Onegin* . . . *crème de violettes* more profound than Dostoevsky' (*PSS* i. 368). Eventually, not to be outdone, Marinetti came back with:

In my first manifesto you read the following declaration which called forth a storm of protest: 'A racing car is more beautiful than the Victory of Samothrace.' I now leave you, as an explosive parting gift, with another image that expresses our Futurist thinking better: 'Nothing is more beautiful than the framework of a house that is still in the process of being built.' We prefer the scaffolding of a house under construction to a well-built house.[16]

Scaffolding is doubtless preferred, both on account of its machine-like primitivism, as opposed to the aesthetically rounded, and also because it points in the direction of the future, rather than of some edifice completed in the past.

This naming of Futurist cult-objects (polemically juxtaposed with some of the sacred cows of passéism) soon degenerated into a form of ritualistic fetichism.[17] A variation on the technique is the kind of enumerative paean to modernity that concerned us at the beginning of this chapter: with Boccioni's list of tokens of modernist primitivism, Marinetti's 'modifiers' of the new sensibility, and Mayakovsky's rapturous account of life in New York. Enumeration takes the cult of fetich-objects in the direction of a reverential form of chanting. But, even in the initial stages of the movement's history, the cult of the machine went further than simply the act of naming, even though

[16] These juxtapositions are taken, respectively, from: Bruno Jasieński, 'Manifest w sprawie natychmiastowej futuryzacji' (*Polska awangarda poetycka. Programy lat 1917–1923*, ii, ed. Andrzej Lam (Cracow, 1969), 205); an undocumented remark attributed to Zdanevich (reported by Markov, *Russian Futurism*, p. 185); Paolo Buzzi, *Versi liberi* (Milan, 1913), 220; Shershenevich, *Zelenaya ulitsa* (Moscow, 1916), 134; Mayakovsky, 'Avto-biograficheskiy ocherk', *Novoe russkoe slovo*, 49 (23 Jan. 1959), 4; Marinetti, *Le Futurisme*, pp. 115 f.; Kruchenykh, *Zaumnaya gniga* (Moscow, 1915, cited in Markov, *Russian Futurism*, p. 335); Buzzi, *Versi liberi*, p. 220; and the catalogue *Balla*, 1918 (cited in Crispolti, *Il mito della macchina*, p. 278). In similar vein, see also Buzzi's 'I don't love music any more, Naxar! The only thing that exhilarates me is the sound of gunfire' (*L'ellisse e la spirale* (Milan, 1915), 25), and Marinetti's 'taxis are just as beautiful and just as proud as the stars' (*Le Monoplane du Pape* (Paris, 1912), 223). This latter is the only example of which I am aware that predates the founding manifesto's juxtaposition.

[17] The charge of fetichism was a familiar one to the Futurists. Marinetti is, for example, at pains to point out that he published in French, 'not because Paris is a fetich [non per feticismo di Parigi], but in order to spread [his] revolutionary Italian ideas throughout the world' ('Piazze e Angoli di produttività interventismo cazzotti revolverate simultaneità politiche artistiche evasioni nell'Assoluto', *GM* 107).

there are many embarrassing instances of collections of Futurist poetry paying greater tribute to technology in their titles than in the body of the ensuing works.

One of the most prominent objects of such Futurist *modernolatria* was the motorcar; indeed, 'Automobilism' was a term of abuse occasionally applied to the entire movement. An early hymn to the car comes in the form of Marinetti's 'A mon Pégase', which begins with the characteristically rhapsodic *Dieu véhément d'une race d'acier | Automobile ivre d'espace* ('Vehement god of a race of steel | Automobile drunk with space'). When originally published in 1905 in *Poesia*, the work was entitled 'A l'Automobile'; then in *I poeti futuristi* this was changed to 'A l'Automobile de course', but this too was rejected in favour of 'A mon Pégase' in the 1908 version in *La Ville charnelle*.[18] This final one is a revealing change, for during the poem the racing-car does gradually metamorphose into a winged beast, as we progress from evocations of *la danse | que tu mènes sur les blanches routes du monde* ('the dance | you lead on the white roads of the world') to the final

> Hurrah! plus de contact avec la terre immonde . . .!
> Enfin, je me détache et je vole en souplesse
> sur la grisante plénitude
> des Astres ruisselants dans le grand lit du ciel!
>
> Hurrah! no more contact with the unclean earth . . .!
> At last, I disengage myself and fly supply
> over the inebriating fullness
> of the Stars streaming into the vast bed of the heavens!

Driver now transforms into rider (even, symbolically, pilot), and modern car turns into mythological beast. This poem, written well before the founding manifesto, is nevertheless not so much at odds with the spirit of the programmatic racing car/Victory of Samothrace contrast as its recourse to Greek mythology might imply. There is, to be sure, as so often in early Futurism, a heavy weight of anachronism to both imagery and diction. And the poet's dithyrambic lucubrations also reveal a central uncertainty about whether the machine is a superior god to be worshipped ('vehement god . . . formidable monster') or something to be controlled ('*My* Pegasus'): *Je lâche enfin tes brides métalliques* ('I finally loosen your metallic reins'). Yet, compared with a statue of Victory, this is a dramatic image, with the stress being very much on the dynamic interaction between man and beast. Indeed, there are places, including the poem's first line, where one could go further and assume that the poet is seeking to identify fully with his means of transportation (to become a new mechanical centaur). The description of the beast as belonging to a steel race

[18] In the Italian translation in *Lussuria velocità* (Milan, 1921), 'elastic spelling' has been added to intensify the acoustic aspect of certain words in the poem.

can be interpreted at the same time as symbolic of what Futurist man himself needs to become. (It is this drawing closer together and eventual symbiosis of man and machine that Marinetti in all probability had in mind when he told Wyndham Lewis that the English had never really 'understood' their machines.) A metaphor such as *je vais éperonnant ma fièvre et mon désir* ('I go spurring on my fever and my desire') goes some way towards internalizing the machine-beast image. In this respect, for all its fetichism, Marinetti's poem at least points in the direction of the machine-man image which was to emerge in later Futurism; and it is closer to the cult of a new primitivism than many of the movement's more gauche early hymns to violence, iconoclasm, and the irrational.

In Paolo Buzzi's *Il premio di Futurismo* (*The Futurist Prize* (1916)), a jury is sitting around with its individual members declaring who should in their opinion receive the grand Futurist Prize (an aeroplane) for most fully embodying the new sensibility. There are various candidates. One proposes a painter, another a sculptor, a third a musician, a fourth an architect. But at this point the Artificial Man arrives and announces that he wishes to enter for the competition. His credentials: 'a cork leg, a plaited-rope arm, a rubber ear, a glass eye, and, if that is not enough, a wig'.[19] He wins, the point being that the artificial man, a prefiguration of machine-man, is a greater Futurist work of art than any sculptor, painter or musician could produce.

The 'Manifesto of Futurist Playwrights' ('Manifesto dei drammaturghi futuristi' (January 1911)) spoke of the necessity 'to introduce into the theatre the feeling of the domination of the machine'.[20] Unfortunately, the fruits of this injunction often tip over into banality and the ludicrous, as they range from such whimsies as the ballet *The Love of Two Locomotives for the Station Master* (1924, music by Casavola, costumes by Depero) and Folgore's *I tre momenti* (1927), with no actors, only performing machines, to Marinetti's *Dance of the Shrapnel* (*Danza dello shrapnel*), *Dance of the Machine Gun* (*Danza della mitragliatrice*), and *Dance of the Aviator* (*Danza dell'aviatrice*) (first published under this trio of titles in *L'Italia futurista* and subsequently as part of the 'Manifesto of Futurist Dance' ('Manifesto della danza futurista')[21]). Balla's *Macchina tipografica* (1914) works more in the mainstream mode, with actors imitating machines, though so does one of the sillier plays in this idiom, Mario Scaparro's *Un parto* (*A Birth* (1920)). Dramatis personae include a female seaplane and a female biplane ('stripped of any modesty, she dashes resolutely on the blue-green surface of the water. She performs a voluptuous dance based on spirals').

[19] Quoted from Kirby, *Futurist Performance*, pp. 242 f.

[20] Quoted from ibid. 27.

[21] *L'Italia futuriste* (8 July 1917), 12. The 'Manifesto of Futurist Dance' appeared in *I manifesti del Futurismo* (Milan, 1919), 199 ff. See *TIF* 127–30 for a more accessible reprint of Marinetti's description of the three dances.

Eventually, the seaplane becomes pregnant, 'her belly opens and four AVIATORS jump out completely equipped'.[22]

Italian Futurism's puerile obsession with machines and gadgetry readily opened itself up to ridicule, and that response came very quickly. In an article in *La voce* of, appropriately enough, 1 April 1909, Ardengo Soffici offered his recipe for such pseudo-modernism:

> Take a kilo of Verhaeren, two hundred grammes of Alfred Jarry, a hundred grammes of Laforgue, twenty of Laurent Tailhade, five of Vielé-Griffin, a handful of Morasso ... a pinch of Pascoli, a mouthful of Acqua Nunzia. Also take: fifteen cars, seven aeroplanes, four trains, two steamers, two bicycles, various electric accumulators, some ingredients of your own: ... beat in all the ingredients into a lake of grey matter and aphrodisiac liquid. Boil the mixture in the vacuum of your mind, on the fire of American charlatanism, and then serve up to the Italian public.[23]

Turning to the machine in the wrong spirit, as Soffici's lampoon implies, could result in even greater imbecilities than passéism ever perpetrated. Fortunately, the Futurist response to the machine amounted to more than such thematic high jinks.

In *La nuova arma: la macchina* Marinetti's mentor Mario Morasso had prophesied a time when man 'will become one with his machine and form a single entity, together with it. He will become a strange being, with which legend is not yet familiar ... a being half man and half iron construction, a monster made of components, a centaur, a siren, with whom no myth is conversant, a phantom that belongs, not to the past, but to the future.'[24] We have already encountered an intimation of this new hybrid creature in 'A mon Pégase', but the work in which it is first most clearly articulated is *Mafarka le futuriste*, specifically in the section near the end where the hero constructs his artificial son, Gazourmah.

'I want to surpass myself in creating, with the effort of my own heart, a youth more radiant than my own was, an immortal youth' the warrior Mafarka tells his followers (p. 208). Gazourmah, described by his progenitor–constructor as 'the child worthy of my spirit' (p. 211), is no mere resurrection of Frankenstein's monster. He is the apotheosis of the Futurist vision of a new man in at least three respects: he has superhuman physical abilities, he is immortal, and he is an artificial construct demonstrating what the Will is able to accomplish without the help of despicable woman. The novel's preface had already announced that 'the hour is near when men with large foreheads and brows of steel will prodigiously give birth to children, with a single effort of

[22] The piece appeared in *Roma futurista*, 3 (1 Feb. 1920), here quoted from *Teatro italiano d'avanguardia: Drammi e sintesi futuriste*, ed. Mario Verdone (Rome, 1970), 106 f.

[23] 'La ricetta di Ribi buffone', *La voce*, 1 (1 Apr. 1909), 63.

[24] Turin, 1905, 286.

their rich will-power they will produce giants. . . . the spirit of man is an unused ovary. . . . It is we who will fertilize it for the first time' (p. xi).[25] Now Mafarka makes this prophecy come true by singlehandedly effecting the birth of the new artificial giant:

I have concluded that it is possible, without the help of the disgusting complicity of the female womb, to express from one's flesh an immortal giant with infallible wings!

You must believe in the absolute and definitive power of the Will, it is necessary to cultivate it, to intensify it, in following a cruel discipline, until the moment when it leaps out of our nervous centres and launches itself beyond the limits of our muscles, with inconceivable force and speed. (pp. 214 f.)

Although such an artificial man has been prefigured in the various anthropomorphic references to war-machines earlier in Marinetti's novel,[26] Gazourmah is the first machine-man to be created 'without recourse to the female vulva' (p. 215). As such, he represents 'the Will externalized . . . the superior manifestation of organized and vital matter' (pp. 216 f.).

The main importance of this new superman, as much a cousin of Jarry's 'surmâle' as of Nietzsche's 'Übermensch', lies in his ability to surpass his creator. Mafarka is still earthbound; Gazourmah can fly. Mafarka is afraid of impending death; Gazourmah, as a machine, is immortal. Mafarka fights mere earthlings; Gazourmah attacks and defeats the sun (as symbol of reason), thereafter to live under the aegis of Mars. And yet, despite all this, it is Mafarka who is designated 'the Futurist' and upon whom the novel's attention is lavished, for he is the text's real equivalent of the Italian Futurists, whereas Gazourmah (like the Electric War) remains a distant dream, a 'vision–hypothesis'.

Although a milestone, this incident near the end of *Mafarka* is but an early stage in the evolution of Futurism's machine-man myth. For a start, Mafarka does not transform *himself* into Gazourmah, he simply creates an artificial son in his own counter-image.

Later, Marinetti was to claim that 'wings are asleep in the flesh of man' (*TIF* 255), and no doubt by then he meant aeroplane wings not Pegasus's or some mythological vestiges of a Daedalian heritage. The image suggests, much more than the creation of Gazourmah did, that man will eventually transform himself into a new future being, not simply create more complex machines or artefact-children to compensate for his own mortal limitations. Perhaps the only way in which to see in Gazourmah a more genuinely Futurist embodiment of what Morasso and later Marinetti predicted is to interpret him as an

[25] 'The Will' was, according to Carrà, regarded in Futurist circles as 'something analogous to electricity, which we use to our advantage without knowing exactly what it is. The Will is determined and given expression in the facts which are the body, the action and instinct of Reality. Reality and Will have for us the same law of reciprocity in actions and facts' (*Guerrapittura*, p. 27).

[26] e.g. the references to 'huge machines with the shape of animals' (p. 92) and the 'giant machine' (p. 98) which point forward to Gazourmah's size and automaton status.

extension of his creator. But this would mean that the ideal figure would have to be an amalgam of man (Mafarka) and his machine-extension (Gazourmah), rather than just the one figure.

The notion of the machine as an 'extension' or 'multiplication' of man's previous powers is central to the new role assigned to the machine. It is explored most fully in a 1910 manifesto entitled 'Multiplied Man and the Reign of the Machine' ('L'homme multiplié et le règne de la Machine').

Marinetti begins by enunciating the Futurist aim of the 'abolition in literature of the seemingly unchallengeable fusion of the two ideas *Woman* and *Beauty*' (*TIF* 255). Beauty is now to be acknowledged as the true attribute of the machine, consequently woman (like syntax) will be 'abolished' from literature, along with sentimentality for her. The new love-story will be that of the mechanic and his machine:

> we are developing and proclaiming a great new idea that runs through contemporary life: the idea of mechanical beauty; we therefore exalt love for the machine, the love that we notice burning on the cheeks of mechanics scorched and smeared with coal. Have you never observed a mechanic lovingly at work on the powerful huge body of his engine? His is the minute, knowing tenderness of a lover caressing the woman he adores.

There are often moments, and this is for me one of them, where it is difficult to distinguish between calculated provocation and downright silliness, on Marinetti's part. To replace one sort of sentimentality with another (and there was relatively little in *Mafarka*) may seem perverse enough. To claim for the modern mechanic the nimbus of a Barbara Cartland hero and set up this *macchinolatria* as the hallmark of a world of exciting technology seems juvenile. Later in the manifesto, however, Marinetti successfully redeems himself by making points that are of importance to the mythology of 'the Futurist marvellous'.

Revealingly, the machine gradually changes roles during the course of Marinetti's argument. It begins as a surrogate woman, with Marinetti displaying a form of patronizing machine-sexism in passages such as: 'his beautiful machine that glowed with pleasure beneath his ardent caresses', and 'you must fondle them, treat them with respect, never mishandle or overtire them'. Then, as in *Mafarka*, it becomes the artificial construct of the new Futurist man (a 'motor built to exact specifications'), until it eventually becomes man's double and finally an extension of his self. It is at this point that the myth of the machine begins to emerge as a major component of the Futurist view of the world. Marinetti's manifesto encapsulates—in partly realistic, partly fantastic terms—the role of the machine in the twentieth century. And in promoting a man–machine analogy, it prepares the ground for many of the techniques of Futurist literary experimentation, as well as its more dehumanizing features.

Expatiating on the 'coming discovery of the laws of a true sensibility of machines', Marinetti explains that

> it is . . . necessary to prepare for the imminent, inevitable identification of man with motor, facilitating and perfecting a constant interchange of intuition, rhythm, instinct, and metallic discipline. . . . we are looking for the reaction of a non-human type in whom moral suffering, goodness of heart, affection and love, those sole corrosive poisons of inexhaustible vital energy, sole interrupters of our powerful bodily electricity, will be abolished.

In using a word like 'abolish', Marinetti favours the kind of discourse that implies the repeal of antequated laws, as if some inexorable process of historical evolution were involved. In addition, he avails himself of a dubious image of scientific progress to underwrite the myth.

> It is certain that, if we grant the truth of Lamarck's transformational hypothesis, we must admit that we are looking for the creation of a non-human type. . . . On the day on which it is possible for man to extend his will and make it into a huge invisible arm, Dream and Desire, which are empty words now, will become master and reign over time and space. (*TIF* 256)

In Boccioni's words, 'Man is evolving towards the machine, and the machine towards Man' (*Gli scritti*, p. 84). If the machine is given meaning by man, man is multiplied by the machine, which extends his physical *and mental* capacities.

Marinetti's notion of *multiplied* man signifies two things. First, and more familiarly, it suggests a picture of man multiplied by the new dimensions and capabilities that his machines give him (a connotation that owes much to Marinetti's reading of Morasso).[27] But it also means man multiplied by a process of self-transformation, to the point where he is no longer man, but has become man–machine. Marinetti appears to be in no doubt that this new 'mechanical being . . . will be naturally cruel, omniscient and combative'; to be 'non-human', as he envisages it, will entail becoming *in*human. As 'La Guerre électrique' shows, when the time comes when 'future man will reduce his heart to its true distributive function' (*TIF* 257), the result will be the callousness of perpetual global warfare.

Marinetti nevertheless deserves credit for recognizing at such an early stage that we are 'already being educated by the machine and . . . coming into some

[27] On Marinetti's indebtedness to Morasso, see Pär Bergman, *'Modernolatria' et 'simultaneità': Recherches sur deux tendences dans l'avant-garde littéraire en Italie et en France à la veille de la première guerre mondiale* (Uppsala, 1962), *passim*, and his 'L'Esthétique de la vitesse' in Jean-Claude Marcadé (ed.), *Présence de F. T. Marinetti: Actes du Colloque International tenu à l'UNESCO* (Lausanne, 1982), 13–25, Carlo Gentili, 'A proposito di F. T. Marinetti e M. Morasso', *Atti e Memorie dell'Accademia Clementina di Bologna*, 11 (1974), 1–16, and V. Vercelloni, *Macchinolatria e modernolatria di Mario Morasso* (San Lazzaro di Savena, 1985). Apart from Morasso, a further source for this image of the machine as man's 'multiplier' is E. Montfort, 'La Beauté moderne', *La Plume*, 300 (15 Oct. 1901), 822.

relationship with motors'. But the second, potentially far more primitivist association of the idea of 'multiplied man' has man becoming *physically*, rather than just in sensibility, a machine. Into this historical situation, the manifesto then injects the element of literature: 'Art and literature exercise a determining influence' in this process of Lamarckian transformation, it is posited. 'We can . . . speed up or retard the movement of humanity towards this form of life liberated from sentimentality.' Literature is thus to have its part to play in preparing for the time when people will have 'methodically learnt to destroy . . . all the sorrows of the heart' (*TIF* 258)[28] and will have become the genetic successors to their own machines.

If one re-examines 'La Guerre électrique' in the light of the notion of future 'multiplied man' elaborated in Marinetti's manifesto (which was originally published alongside it in *Le Futurisme*), certain features of the 'vision–hypothesis' assume a fresh significance. There is, for example, symptomatically little in Marinetti's picture of the Italy of the new millennium to suggest any great degree of automation. We do hear of 'polyhedral garages from which motorized ploughs incessantly leap out to the fields, to till, plough, and irrigate the lands and crops electrically' (p. 139), but these the narrator is careful to relegate to the status of 'centres of secondary activity'. The main emphasis is on the technologist Zarathustras, in their surveillance planes or esconced in their Apennine lairs controlling the power supplies. These men live 'in high tension rooms where a hundred thousand volts palpitate between the plate-glass windows. They sit before switchboards, with dials to right and left of them, keyboards, regulators, and commutators, and everywhere the splendid flash of polished levers' (p. 138). These privileged men of the future 'have finally gained the pleasure of living between iron walls. They have steel furniture, twenty times lighter and less expensive then ours. They are finally emancipated from wood, with its lesson of weakness and debilitating softness, and from fabrics with their rustic ornaments' (pp. 138 f.). They are encased, as if in—or part of—a machine. (Le Corbusier was later to give wider circulation to the Futurist idea of the building as a living-machine, but it is already suggested here in Marinetti's vision and it also plays an important role in Sant'Elia's writings.) A juvenile gadget-fixation is again not lacking here, but the focus, as during the eventual electric war, is less on the machines as such or on any accompanying technological expertise than on the image of the men at the controls, men who 'by means of wireless telephones regulate the lightning speed of the . . . trains' (p. 139). Freight trains, we are informed, are the only kind left on the earth's surface, because man, 'having become airborne, only sets foot there once in a while' (p. 140).

[28] In Marinetti's eyes, such a world 'neatly challenges the pessimism of Schopenhauer' (*TIF* 258). However, there appears to be a misreading of Schopenhauer's concept of the Will here, and possibly an attempt at introducing a more Nietzschean interpretation of it as well.

On the subject of Marinetti and machines, Giovanni Lista once observed:

Already in his pre-Futurist period, the racing-car was preferred by Marinetti to the train, which is evidently still too collectivist and too utilitarian a type of machine. It is here, moreover, that one should look for the most salient differences between Marinetti's brand of Futurism and Russian Futurism. For Marinetti, it is not so much a matter of the working machine, the factory or, to paraphrase a famous title, the machine as such, but it is a matter of a concrete metaphor for vitalist energy and the will to power. What Marinetti retains above all else from the machine is its dynamic efficiency, its capacity for acceleration: its speed and its Dionysian surpassing of its limits, things which contain all the spirit of Marinetti's decadence.[29]

Yet, in the light of 'La Guerre électrique', one could modify these observations in one or two respects. For here the essential distinction is not between the useful and the non-utilitarian; after all, the emphasis throughout Marinetti's vision, both in the peacetime section and during the electric war, is on *functional technology*. The grain-sowing trains and the combat machines are both 'useful', in their different ways. The stress is more specifically on the machine dominated by a single operator.[30] This is also true of other Futurist cult-machines: the racing-car, the single-seater aeroplane, the 'bicyclettes aggressives' (as they are called in the manifesto 'The gods depart, D'Annunzio remains' ('Les dieux s'en vont, D'Annunzio reste')) or the war-machines of the twenty-first century with their echoes of *The War of the Worlds*. The train, by contrast, even granting Lista's point about its collectivist nature, has the further disadvantage of rendering man passive. Even the driver (and it was Buzzi who once announced that, if he were to travel by train, then the driver's cab was the only place for him[31]) merely follows the lines and obeys signals. Marinetti's primary vision, on the other hand, is of man as an active controller of machines, his extensions, to the point where eventually the two would even transcend symbiotic co-existence.[32]

29 *Marinetti* (Paris, 1976), 68.

30 In *L'alcova d'acciaio: Romanzo vissuto* (Milan, 1921), the operator is described in similar terms as the 'tamer of primordial forces' (p. 176) and as being the 'ruler of the heart-motor' which is 'totally obedient' (p. 79), to the point where the two coalesce in a new unity.

31 *Omaggio a Paolo Buzzi* (Milan, 1958), 94. In a recent article on Futurist *macchinolatria*, Helmut Meter singles out the technical triad of train, car, and aeroplane as the focus of attention ('Maschinenkult und Eschatologie: Zum historischen Ort des italienischen Futurismus', *Sprachkunst*, 17 (1986), 276). But there are sound reasons why the humble bicycle looms as large as the locomotive within Futurist mythology. The Futurists' vision of a symbiosis of man and machine, or at least of the machine as the 'extension' or 'multiplier' of man, accounts for the emphasis on certain kinds of machine. Despite its relative slowness, the bicycle becomes important within Futurist symbolism in this respect. Cf. a remark on the bicycle, made in connection with Fillia's *Bicicletta*, that 'the machine and the rider represent a completely indivisible fusion of mechanical forces' (Fillia, Curtoni, Caligaris, 'L'idolo meccanico', *L'Antenna* (16 May 1926), 2).

32 In *L'alcova d'acciaio*, where one still finds Marinetti predicting that 'one day we shall have electric war', there is, by contrast, more emphasis on automation: 'ghost planes laden with bombs and pilotless, guided from a distance by command planes . . . bombs, guided even from the ground with an electric

Each waggon of the seed-sowing train mentioned near the beginning of 'La Guerre électrique' carries 'a gigantic steel arm on its roof that sweeps horizontally throwing fecundating seed in all directions' (p. 139). This image gives at least a foretaste of the new 'fusion of instinct with the efficiency of motors and conquered forces' promised in 'Destruction of Syntax' (*TIF* 59). The train, with its gigantic overhead arm, is a clumsily explicit genuflexion in the direction of Morasso's *La nuova arma*. In this agrarian version, Italy itself is presented anthropomorphically—with repeated references to its muscles, arteries, nerves, and circulation—while the taming of nature is translated into the allegorical picture of 'the earth . . . squeezed in man's vast electrical hand'. And, while Marinetti tries to give some measure of technical plausibility to the concept of the new 'electric' farming,[33] the topic's main role is to suggest a naturalness that can prepare the way for the electric war which is in turn meant to be seen as a 'natural' human activity. The prophetic image of the tank used in the warfare of the twenty-first century is stylized in terms which suggest both the same man–machine synthesis as in the agrarian section and a vision of a general technologizing of nature: 'On the border between two nations, steel elephants, bristling with shiny trunks pointed in the direction of the enemy, advance from two directions enormous pneumatic machines, rolling down their tracks' (p. 141). Such an evocation of elephants-of-war may make us feel that we are looking back over our shoulders at the world of Mafarka, if not that of the Punic Wars. But Marinetti's aim is to suggest the new mechanical *animal*, a natural image to imply a just and logical companion to the man–machines of the new century.

Balla's and Depero's 'Futurist Reconstruction of the Universe' ('Ricostruzione futurista dell'universo') develops further the idea of the 'Metallic Animal' (the title of one of its sections):

> Fusion of art and science, chemistry, physics, continuous and unexpected pyrotechnics all incorporated into a new creature, a creature that will speak, shout and dance automatically. We Futurists . . . will construct millions of metallic animals for the vaster war (conflagration of all the creative forces of Europe, Africa and America, which will undoubtedly follow the current marvellous conflagration . . .). (*Archivi*, i. 51)

Eventually, as Verdone and Kirby have shown, the idea was to come to its most fruitful, and perhaps least barbaric, realization in the theatre: in Vasari's and

panel of buttons . . . aerial missiles' (p. 175). Again, though, the primary stress is on the technician behind these remotely controlled weapons.

[33] From their planes, men 'survey the entire irradiating circulation of electricity across the limitless chessboard of the plains. Everywhere plants grow in an abnormal fashion, from the effects of artificial electricity at high tension. Electric irrigation and drainage. By means of electrolysis and the multiple reactions it sets off, electricity stimulates assimilation elsewhere, in vegetal cells, in the nutritional properties of the soil, and directly excites vegetative activity' (pp. 139f.). There is in fact no scientific basis to what Marinetti describes, nor was there assumed to be any at his time.

Depero's various mechanical figures for the ballet and the stage. 'La Guerre électrique' though, despite the early date at which it was written, contains many more of the seeds of Futurism's subsequent development than *Mafarka* does, no doubt largely because its setting in the twenty-first century predisposes it towards a more radical confrontation with the movement's ultimate goals.

To reinforce the 'naturalness' of his vision of the future, Marinetti not only juxtaposes the agrarian use of machines with their role in the electric war and employs animal imagery to describe war-machines; he also devotes much energy to presenting electricity as a *natural* source of energy, culminating in his picture of the Futurist warriors 'making use of the friendship and force of thunderstorms' in order to achieve their aggressive ends. The resultant 'electric explosions' and 'electric volleys' are at times described in terms more appropriate to nuclear than electrical energy. At all events, whether derived from hydro-electric sources or borrowed from a passing thunderstorm, the electrical power behind the future war is firmly associated with nature.

In *Mafarka* a new man is actually created. His wings are constructed of canvas, the same material that aeroplane wings would have been made of at this time. And even if he does not rejoice in a body of steel, as Marinetti's 'Pegasus' does, he nevertheless displays those steely virtues that were to underscore later Futurism's cult of the metallic. In 'La Guerre électrique', on the other hand, the idea is more hinted at than actually embodied in the vision, even though we have come much further into the future and presumably closer to the Futurists' ideal world. Man is at one point described as having liberated himself from his passéist environment of soft wooden furniture and decadent fabrics. Yet, although these images might seem to prophesy the imminent coming of Metal Man, the rest of the vision remains in a world of 'examples' only, without the transformation taking place. Perhaps Marinetti envisaged such 'Lamarckian' transformations as requiring a longer time-scale.

Yet, even if the vision of the man–machine is postponed until an indefinite future, the cult of the machine and the machine aesthetic make themselves felt throughout Italian Futurism, but specifically after the Interventionist question, which took up so much of the writers' energies, had been solved.

In 1923 Vinicio Paladini issued a manifesto called 'The Mechanical Aesthetic' ('Estetica meccanica') which was revealingly prefaced by Boccioni's pronouncement: 'We are the primitives of a new sensibility.'[34] This was one of a whole series of Second Futurism manifestos linking the machine with new aesthetic preoccupations, including Prampolini's, Pannaggi's, and Paladini's 'Manifesto of Mechanical Art' ('L'arte meccanica' (1922)), Fillia's, Curtoni's, and Caligaris's 'Sacred Mechanical Art' ('L'arte meccanica sacra' (1926)), Marinetti's 'The Landscape and Futurist Aesthetic of the Machine' ('Paesaggio e

[34] *Noi*, 2nd ser., 1 (May 1923), 21.

l'estetica futurista della macchina' (1931)), and Cascadalluminio's 'Machinolatry' ('Macchinolatria' (1935)).[35] As Enrico Crispolti has shown, the aim was to derive an aesthetic of purity, of 'machine' qualities—including a form of quasi-religious machine-spirituality—from technology. The machine had ceased to be a source of mere thematic interest, and was beginning to have a key aesthetic role to play in the Futurists painters' awareness of the way they should create.

It is 'not a question of new subject-matter', Paladini emphasized, in 'The Mechanical Aesthetic', but what he calls 'the myth of the machine': the machine as the repository of certain values of precision, abstraction, clarity, and beauty which were to make it the guiding Muse of Second Futurism.[36] In the 'Manifesto of Mechanical Art', Prampolini, Pannaggi and Paladini stress associations of decisiveness, mechanicalness, and precision and point out that one should distinguish between the exterior of a machine and its spirit; for it is above all the latter that interests them.[37] 'A Futurist work ... should be constructed like a machine', the preface to the 1927 Galleria Pesaro exhibition declares.[38] And in a similar vein, Severini writes:

> The precision, the rhythm, the brutality of machines and their movements have without doubt led us towards a new form of realism which we can express without painting locomotives. There is in fact an analogy between the machine and the work of art. All the material components which make up a machine, for instance, are ordered according to a unique act of will, the will of the inventor who adds to the integral vitality of these different materials another life or action or movement. The process of constructing a machine is analogous to the process of constructing a work of art.[39]

Severini's remarks bridge the gap between some of the thematic forms of primitivism looked at during the first half of this chapter and the eventual consciousness of how important the machine was to Futurist aesthetics. The phrase 'the brutality of machines' well sums up the direction in which 'La

[35] The originals of some of these manifestos are extremely difficult to locate. Crispolti offers extracts from them in *Il mito della macchina*, but the best source of reprinted material at present available is the facsimile *Manifesti, Proclami, Interventi e documenti teorici del Futurismo*, ed. Luciano Caruso (4 boxes; Florence, 1979).

[36] Of the two studies on the subject, both entitled *Il mito della macchina*—Enrico Crispolti's (Trapani, 1972) and Roberto Tessari's (Milan, 1973)—the former is exclusively concerned with Futurism and contains a wealth of useful perspectives from the theoreticians of Second Futurism. Tessari's volume offers a more general survey.

[37] The manifesto, issued in Rome in Oct. 1922, was also published in *Noi*, 2nd ser., 1 (May 1923), 1 f., and in the *Bulletin de l'Effort Moderne*, 15 (May 1925), 13–15.

[38] Catalogue to the Mostra di 34 Pittori Futuristi, held at the Galleria Pesaro, Milan (Nov.–Dec. 1927). The preface is signed by Marinetti, Benedetta, Azari, and Fillia. Cf. also in the 'Manifesto of Futurist Art' ('Manifesto d'arte futurista'): 'we feel mechanically and we feel ourelves to be constructed of steel' (Crispolti, *Il mito della macchina*, p. 392).

[39] From the section 'Le Machinisme et l'art' of 'Avant-garde Painting' ('La Peinture d'avant-garde'), originally published in the *Mercure de France* (1 Jan. 1917), here cited from the reprint in *Archivi*, i. 211 f.

Guerre électrique', parts of *Mafarka*, and the cult of the machine-man were moving. But the comparison between the construction of the machine and the Futurist interpretation of aesthetic creation is relevant to many of the other features of Futurist technique which we have been exploring.

Some of the things Severini says about painting echo Shershenevich's remark of the year before, when, in the context of Russian Futurism, he rejected the idea that 'it would suffice to write about the city, the automobile, the skyscraper to become urbanist . . . the crucial thing is not to write about the city, but to do so in a city-like fashion'.[40] The contrast between writing 'about the city' (*o gorode*) and in a 'city-like fashion' (*po gorodskomu*) is very much akin to that between a thematic and technical use of the machine myth, in general. Machines are now, in the context of Second Futurist visual art, being put forward as constructional models. In 'The Mechanical Aesthetic' it is 'the Diesel engine, the flywheel, the wireless transmitter' which, we are assured, 'have an immense value in the structuring of the self', and by extension in the formation of the new aesthetic. And in 'The Mechanical Idol' ('L'idolo meccanico' (1926)), the *idols* listed, 'AEROPLANE—AUTOMOBILE — BICYCLE — TURBINE — SKYSCRAPER — LINOTYPE — LOUDSPEAKER — ELECTRIC LIGHT—HYDRAULIC PRESS—BOILER—REFRIGERATOR',[41] are cited as aesthetic paradigms.

In *Il mito della macchina* Enrico Crispolti has chronicled just what this machine aesthetic meant in theoretical terms to the artists and sculptors of Second Futurism, and he rightly points out that we are in need of some kind of *Archivi del Secondo Futurismo* comparable to Gambillo's and Fiori's compilation on the earlier period.[42] What is happening in the area of painting and sculpture in the 1920s is in some respects similar to the developments Christopher Middleton traces in his survey of primitivism, in particular the process by which primitivism gradually becomes allied to an abstractionist tendency. Quoting Kandinsky's 'the freer the abstraction of form is, the purer and more abstract a ring a work has',[43] Middleton sees here the first attempt at linking primitivism with abstraction in art. (In his manifesto 'Abstract Art' ('Arte astratta'), Julius Evola talks in similar terms of a 'plunge into brutality for the sake of purification' (see *FF* 474).) Middleton's subsequent argument that 'primitivism has to

[40] *Zelenaya ulitsa* (Moscow, 1916), 45. Cf. the later remark: 'When there is rhetoric—there is no urbanism, because the logic of reasoning is alien to the city' (p. 46), which hints at a connection between urbanism and the telegraphic dimension of Futurist language.

[41] 'Estetica meccanica', *Noi*, 2nd ser., 1 (May 1923), 2; 'L'idolo meccanico', *L'antenna* (16 May 1926), 2.

[42] See, in particular, the sections 'Il costruttivismo "meccanico" di Pannaggi' and 'La poetica dell'"arte meccanica"' in Crispolti, *Il mito della macchina*, pp. 382–408, 450–63. Crispolti's account of the links between Second Futurism and Le Corbusier's L'Esprit Nouveau also has a strong bearing on this subject.

[43] *Bolshevism in Art*, p. 27. Kandinsky's remark is taken from *Über das Geistige in der Kunst*[2] (Munich, 1912), 60.

do with plane surfaces in painting and in poetry'[44] is convincingly illustrated in respect of Expressionist and Dada poetry, but it only accounts for one form of stylistic primitivism. The alliance between primitivism and abstraction that occurs in Second Futurist visual art has no counterpart in the literature of the period. While the machine-analogy pushes the visual arts more and more in the direction of purity and abstraction, in literature the most significant developments in the area of the mechanically primitive were there, it could be argued, practically from the beginning.

As if consciously taking up the imagery of 'La Guerre électrique', Marinetti's manifesto 'Geometric and Mechanical Splendour and the Numerical Sensibility' ('Lo splendore geometrico e meccanico e la sensibilità numerica' (1914)) declares that 'nothing is more beautiful than a huge humming electric station', which explains why technology is then presented as a model for Futurist poets:

> the electric station . . . holds the hydraulic pressure of a mountain chain and the electric power of a vast horizon, synthesized in marble distribution panels bristling with dials, keyboards, and shining commutators. *These panels are our only models for the writing of poetry*. For precursors we have gymnasts and high-wire artists who, in their evolutions, their pauses, and the cadences of their musculature, achieve the sparkling perfection of precise gears, and the geometric splendour that we want to attain in poetry with words-in-freedom. (*TIF* 86)

Many of this manifesto's images are mere rhetoric, of course. References to 'the enthusiastic imitation of electricity' or the poet's need to acquire 'the happy precision of gears and well-oiled thoughts' give little away in pragmatic terms. But the underlying literature–machine analogy (in a manifesto which, incidentally, was first published in *Lacerba* of 15 March 1914, with the more revealing title of 'Geometric and Mechanical Splendour in Words-in-freedom' ('Lo splendore geometrico e meccanico nelle parole in libertà')) showed a very early recognition of the technical implications of the machine model for a new form of literature. (When reprinted in *Spagna veloce e toro futurista* (Milan, 1931), the manifesto received the even more transparent title 'Geometric Splendour and the Aesthetic of the Machine' ('Lo splendore geometrico e l'estetica della macchina')).

The forms of telegraphic writing examined above marked an attempt to introduce a new machinist, functional economy into the language of Futurist poetry. Giuseppe Ungaretti even sees in words-in-freedom an expression of Marinetti's 'blind faith in raw matter . . . in chaotic matter'.[45] The poet himself used various machine analogies: from steam passing through pipes ('Destruction of Syntax') to 'the lyric initiative of electricity flowing through the sheaths

[44] *Bolshevism in Art*, p. 32.

[45] 'Commemorazione del futurismo', in his *Saggi e interventi*, ed. M. Diacono and L. Rebay (Milan, 1964), 172.

of the quadruple turrets' of a Dreadnought; hence a liberated sentence would offer an uninterrupted flow of intuitive ideas 'without arresting the current of the words-in-freedom' ('Geometric and Mechanical Splendour' (*TIF* 88)), the main point being that these are only figurative expressions of a new view of artistic constructs, not literal applications of technological ideas. Just how loose the metaphor is can be seen by the fact that the *Lacerba* version refers to 'dynamism' instead of 'electricity'.

The collage principle that was explored in Chapter 2 is very much a matter of the assembling of parts into a new component whole—indeed, the alternative term for the process, 'montage', is actually derived from the machine workshop. Even the quest for the iconic, with which this study began, is related to a machine aesthetic, inasmuch as the machine is understood as an archetypal embodiment of essential matter. Hence the attempt to return (via iconicity) to material essences represents part of the same primitivist urge to bypass the intellect and return to 'the brutality' of objects unhampered by a surfeit of cerebrality. If the Futurist painters and non-verbal collagists were groping their way forward to a machine aesthetic during the inter-war years, then arguably the writers had been closer to such an aesthetic from the very outset of the movement's history. Indeed, some of the most notable achievements of literary Futurism were governed by such a principle.

Glossary

THE following glossary is intended to offer an outline explanation of some of the main technical vocabulary used by, and on the subject of, the Futurists and their allies. Particular attention is devoted to concepts relating to national literary groupings and to key terms which do not form part of general critical discourse. Concepts which have already been explored in some detail in the individual chapters of the present study have been omitted. For supplementary information of a similar nature, the reader is referred to the 'Dictionary of Futurism' in *Futurismo & Futurismi* (*FF* 409–613), although the entries there are less concerned with the *terminology* of either literary Futurism or Futurism in the visual arts than with brief biographical sketches of leading Futurists, the main events in the history of the movement, and a miscellany of rather generalized Futurist 'topics'.

AERO-POETRY *Aeropoesia*, a relatively late Futurist phenomenon, is the child of aero-painting (*aeropittura*), a new genre conceived by the Italian Futurists in the late 1920s. The guiding principle informing such works was the new Olympian, dynamic perspective that flight brought with it. Not only did Marinetti's and Mino Somenzi's 1929 manifesto 'Futurist Aero-painting' ('Aeropittura futurista') proclaim that a fresh visual synthesis would result in painting, if such a pilot's-eye perspective were applied to a work's organization, they also associated the loftiness of aero-painting with a new spirituality. Examples of Futurist aero-painting include Azari's *Flight Panorama* (1926), Dottori's *Landscape seen from a High Flight*, and Balla's *Balbo and the Transatlantic Fliers*. Various second-generation Futurist poets sought to transfer such perspectival innovations from aero-painting to descriptive sequences of poetry, according to principles outlined in Marinetti's 'Manifesto of Aero-Poetry' ('Manifesto della aeropoesia' (30 October 1931)). The background to these experiments is given in Chapter Fifteen of Marinetti's *La Grande Milano* ('Il primato della aeropoesia della aeropittura e della aeromusica', *GM* 247–50). Marinetti's own *Aeropoema del Golfo della Spezia* is one of the best-known examples of aero-poetry, a genre which, *pace* Marinetti, is by no means exclusively Italian and which also embraces such works as Kamensky's 'Polet Vasi Kamenskogo na aeroplane v Varshave' ('Vasya Kamensky's Aeroplane Flight in Warsaw') and many of Konstantin Olimpov's aeroplane poems. Other Italian Futurist examples include Krimer's *Aviatori* (1935), Ignazio Scurto's *Aeropoesia del paracadute* (1936), and Bruno Sanzin's *Aerobizzarria* (1935) and *Fiori d'Italia: Aeroprofumi futuristi* (1942). This eccentric chapter in the eccentric history of literary Futurism is impressively documented in Felix Philipp Ingold, *Literatur und Aviatik: Europäische Flugdichtung 1909–1927* (Basle, 1978).

CUBO-FUTURISM A term generally assumed to have been coined by David Burliuk *c.*1913 to distinguish the main group of Russian Futurist writers and painters from their closest rivals among the Ego-Futurists and various other contemporary avant-gardes abroad. As the term implies, there is more of a sense of continuity between the Russian Futurists of this grouping and the painters of French Cubism. Matyushin had by this time translated Gleizes's and Metzinger's *Du cubisme* into Russian and lectured on the subject. Malevich, Goncharova, Larionov, Rozanova, and the Burliuks were the leading representatives of Russian Cubo-Futurism in painting, although the group remained relatively fluid and was continually splitting and realigning, often because of internal feuds. Of the writers associated with Cubo-Futurism, the most prominent were Khlebnikov, Kruchenykh, Kamensky, and Mayakovksy. In their concern with the applicability of quasi-'Cubist' principles of word-fragmentation and deformation to verbal innovation and with autotelic, non-referential language (as outlined in the manifesto 'The Word as such' ('Slovo kak takovoye')), they were closer in spirit to the poetic experimenters of the German Expressionist *Sturmkreis* than to the Italian Futurists, whose influence they in any case always hotly denied. A general introduction to the group is given in Vahan D. Barooshian's *Russian Cubo-Futurism* (The Hague, 1974), a survey which covers the period 1910–30, although the heyday of Cubo-Futurism as a group, before the writers went their various ways, was confined to the years 1913–15.

EGO-FUTURISM A far less significant grouping, in terms of literary achievement, than that of the Russian Cubo-Futurists, and one consisting solely of writers, whereas one of the main achievements of Cubo-Futurism lay in the sheer variety of ways in which painting and literature were being encouraged to interact. Ego-Futurism was the name for a post-Symbolist coterie centred on the figure of Igor Severyanin, the poet of urbanism and, as the term 'Ego-Futurism' itself suggests, the cult of the self. The concern with the city and with aspects of the modern world (witness: Severyanin's own *Elektricheskie stikhi* (*Electric Poems*, 1911), and Olimpov's *Aeroplannye poezy* (*Aeroplane Poetry*, 1912)) inclined the Ego-Futurists to be more receptive to the influence of the Italians. Vadim Shershenevich translated a number of Marinetti's main manifestos (*Manifesty ital'yanskogo futurizma* (Moscow, 1914)), as well as *Bitva u Tripoli* (*The Battle of Tripoli*, (Moscow, 1916)) and *Mafarka le futuriste* (*Futurist Mafarka* (Moscow, 1916)); and in *Futurizm bez maski* (*Futurism without a Mask* (1913)) he sought to bring out (and even over-played) Ego-Futurism's general indebtedness to the inspiration of Marinetti and his Italian fellow Futurists, when in fact none of the other Ego-Futurists was as taken with Marinetti's ideas as Shershenevich was. Anna Lawton, *Vadim Shershenevich: From Futurism to Imaginism* (Ann Arbor, 1981), contains much material on the links between Italian Futurism and Russian Ego-Futurism.

HYLAEA An essentially primitivist forerunner of Cubo-Futurism, which Markov dates from the publication of Kamensky, *Zemlyanka* (*The Mud Hut*), in 1910. The name 'Hylaea' (Russian *Gilea*), derived from the word used by the ancient Greeks for the area around Kherson near the Black Sea coast where the group was based at the estate of Chernyanka, already gives some indication of the deliberately backward-looking aspect of this early form of Futurism, especially when compared with the urbanism and technological slant of Russian Ego-Futurism and Elena Guro's and Mayakovsky's later work, not to mention the *modernolatria* of Italian Futurism. The organizing forces behind the group were the brothers David and Nikolai Burliuk, who together with Benedikt Livshits (its chronicler in *Polutoraglazyy strelets* (*The One and a Half-Eyed*

Archer)) founded the group. The divide between Hylaea and Cubo-Futurism is difficult to be precise about, either chronologically or in terms of membership, except to note that the majority of Hylaea writers appear later as self-proclaimed 'Cubo-Futurists'. In Markov's words, 'in 1913 members of the Hylaea Group became known as "cubo-futurists"; however, the designation "Hylaea" was not abandoned. It continued to appear on the covers of futurist publications, sometimes together with the new name, "futurists"' (*Russian Futurism: A History* (Berkeley, 1968), 117). In recent years, there has been a tendency to treat the major writers involved, notably Khlebnikov and Kruchenykh, under the generic label 'Cubo-Futurists', but, as Markov has shown, the Hylaea grouping does tend to be more generally concerned with the primitive than the later Cubo-Futurist one, which assumes many more of the attributes of contemporary avant-gardism.

INTONARUMORI (NOISE-INTONERS) The Italian Futurists' neologistic term for a series of noise-making machines designed largely by Luigi Russolo and Ugo Piatti. Their primary function was the introduction of 'the art of noises' into the repertoire of Futurist music, noise being specifically associated with the modern world of the machine and in this instance actually being produced by machines rather than (passéist) instruments. Michael Kirby writes: 'Externally, all the *intonarumori* were quite similar. . . . rectangular wooden boxes with funnel-shaped acoustical amplifiers, or megaphones, projecting from the front. The boxes, averaging about two-to-three feet in height, and the megaphones varied in size, but the general appearance was the same. They were "played" by means of a protruding handle that moved in a slot on the top or side of the instrument. . . . Although the quality of the sound remained constant, its pitch was variable. Just as the pitch of an electric motor changes as its rotors spin faster or slower, each "noise-intoner" had to create a variable sound' (*Futurist Performance* (New York, 1971), 37). Further technical details are given in Russolo, 'The Futurist *Intonarumori*' ('Gl'intonarumori futuristi') (*Lacerba* (1 July 1913), 140–1). Pieces such as *The Awakening of a Great City* and *A Meeting of Motorcars and Aeroplanes* were played by whole banks of differentiated *intonarumori*. Although Russolo's manifesto 'The Art of Noises' ('L'arte dei rumori') makes it clear that his researches with sounds are considerably indebted to the acoustic effects in works like Marinetti's *Zang tumb tumb*, he is also anxious to stress that 'THE ART OF NOISE MUST NOT LIMIT ITSELF TO IMITATIVE REPRODUCTION' and simply content itself with 'a succession of noises that imitate life'. On the other hand, Russolo and his fellow cacophonists did not restrict themselves to the production of non-iconic sounds or *musique concrète*. However, Russolo's axiomatic point that noise, in contrast to music, 'loses its character of *effect* by transforming itself into *element* and *primary material*' ('The Futurist *Intonarumori*', p. 141) was to have important repercussions for both Futurist theatre and poetry. The interaction between the musicians' 'art of noises', on the one hand, and poetic onomatopoeia, *onomalingua*, and the large-scale exploration of the word's materiality by the Futurist poets, on the other, is certainly not a matter of a one-way influence but a complex of reciprocal cross-fertilizations. Marinetti's importance for Russolo is matched by Russolo's subsequent influence on Depero and the *sintesi* playwrights and creators of Futurist ballet. Stage-works like Marinetti's *Il tamburo di fuoco* and Balla's *Macchina tipografica* would be inconceivable without the earlier experiments of Pratella, Russolo, and Piatti. The subject is well covered in Michael Kirby's *Futurist Performance*.

IRREDENTISM The longstanding political campaign to bring about the assimilation of the *terra irredenta* (i.e. the unredeemed territory) into Italy. The Italians considered a number of regions beyond their then borders, including Trento, Trieste, Bolzano, and Merano, to be ethnically theirs and consequently in need of incorporation into the new Italian state. In the nineteenth century a number of societies were founded to further this cause, the most famous being the Pro Italia Irredenta (Naples, 1877); in essence their aim was the completion of the Risorgimento. During the pre-war period the Italian Futurists allied themselves particularly vociferously with the intense anti-Austrianism that Irredentism implied. One of the main items on the 'Futurist Political Programme' ('Programma politico futurista') signed by Marinetti, Boccioni, Carrà, and Russolo (11 Oct. 1913) reads: 'IRREDENTISM.—PAN-ITALIANISM.—ITALY'S SUPREMACY'. However, the coming of the more pressing Interventionist campaign in the latter part of 1914 gave their political efforts a new direction. Documentation of Futurist Irredentist sympathies during the early period can be found in Marinetti's preface to Palazzeschi's *L'incendario* (Milan, 1910), entitled 'Report on the Futurist Victory of Trieste'; the 1913 manifesto 'Irredentism and Pan-Italianism', signed by Marinetti, Boccioni, Carrà, and Russolo; and Buzzi's volume of poetry *Aeroplani* (Milan, 1909), dedicated to 'the flag of Trieste that we will win back'. One of the side-effects of Irredentism was the way in which the Futurists tended to single out the Austrians as their natural enemies from among those with whom they were at war during the 1915–18 period.

ORPHISM The name given by Apollinaire to a group of Paris-based painters, including Delaunay, Kupka, Léger, and Picabia, who were active in the years just before the First World War. Orphism came into existence largely as a reaction to the colour-limitations and rigid formal preoccupations of Cubist painting. Inasmuch as it was in dispute with Cubism, it shared a measure of common ground with Italian Futurism. But in certain areas—the concern with simultaneity, the development of collage, the evolution of dynamic forms of painting, the boldness of its colours—there was, nevertheless, intense rivalry between the Orphists and Futurists, who repeatedly charged each other with plagiarism. Boccioni's article 'The Futurists Plagiarized in France' (*Lacerba* (1 Apr. 1913), 66–8) is the leading document in this respect, identifying the main bones of contention. Virginia Spate suggests that, while the Orphists 'were unimpressed by the Futurists' paintings, they were interested in their ideas' (*Orphism: The Evolution of Non-Figurative Painting in Paris, 1910–1914* (Oxford, 1979), 30), ideas which were to some considerable degree mediated by the Orphists' spokesman Apollinaire, in *Les Soirées de Paris*, and by Severini's presence in Paris. Although the poet Blaise Cendrars is sometimes thought of as a literary Orphist, Orphism is first and foremost a chapter in the history of painting. The points of contact relevant to literary Futurism are above all, as can be seen from Apollinaire's commentaries, at the level of highly generalized aesthetic goals. In realization, Orphist and Futurist painting differ substantially, Orphism and literary Futurism even more so, with the result that the charges of plagiarism made at the time frequently seem contrived in retrospect.

PASSÉISM The Italian Futurists' principal term of abuse in their campaign to denigrate and 'mock everything consecrated by time' (*TIF* 14). Passéism (*passatismo*) was a pejorative label for the debilitating cult of the past, particularly in Italy, and in that sense assumed the function of a central contrast-term to Futurism. Targets included all forms of cultural historicism (history seen as that 'miserable collector of stamps, medals, and counterfeit coins', *TIF* 260), antiquarian locations (Venice: 'the bordello of history', Florence, Rome, museums, libraries), Symbolism, Wagner, D'Annunzio, sentimentalism, even Cubism—in fact almost everything in contemporary and past culture that is not Futurist or proto-Futurist in spirit. Based on the assumption that 'the past is necessarily inferior to the Future' (*TIF* 260), the onslaught against all forms of passéism inevitably doubles as ground-preparing for Futurism, although the attacks on anachronistic phenomena are carefully kept at such a level of generality as not to deflect attention from the Futurist antidote. In a letter to Gian Pietro Lucini of Feb. 1913, Marinetti describes *passatismo* as 'a word invented by E. A. Butti, one which is already widely in circulation in Milanese circles where the debate about Futurism is raging' (*La Voce*, 5 1905, 1050). Marinetti's own manifestos 'Against Past-loving Venice' ('Contro Venezia passatista' (*TIF* 30–3)) and earlier 'Against Past-loving Spain' ('Contro la Spagna passatista' (*TIF* 34–40)), and Papini's 'Against Past-loving Florence' ('Contro Firenze passatista' (*Archivi*, i. 180–3)) gave much impetus to the idea. Summing up the campaign in retrospect, Marinetti refers succinctly to Futurism's 'struggle against the old passéist (Romantic, Symbolist, and Decadent) aesthetic' and its 'immense bric-à-brac of a sick sensibility' (Dinamo-Azari, *Depero futurista* (Milan–Paris, 1927), 151).

The Italian noun *passatismo* and related words (*passatista*, both noun and adjective, and *antipassatista*) have generated two parallel sets of English words: *passéism* and *passéist*, on the one hand, and *passatism* and *passatist*, on the other—the former being preferred by the Futurists themselves when launching their ideas in English manifestos. Much of the idiosyncratic loan-vocabulary associated with this complex of ideas consists of a strange semi-anglicizing of foreign words (hence the lack of French endings on *passéism* and *passéist*, from the French, and the hybrid formations *passatism* and *passatist*, from the original Italian terminology).

SECOND FUTURISM Not a very precise period-concept, but a necessary one, if the Futurist achievements of the 1920s and 1930s are in any way to be meaningfully distinguished from the Futurism of the early or, as it is often called, 'heroic' phase of the years 1909–15. In the visual arts, the paintings and assemblages of the Second Futurists, including Depero, Farfa, Prampolini, Fillia, Diulgheroff, Paladini, Pannaggi, Rosso, *et al.*, appear markedly more abstract and more in tune with developments in Dada and Surrealism than with early (First) Futurist styles. Enrico Crispolti, who has single-handedly done more for the rediscovery of Second Futurism than any scholar, has put forward the view that one does not find the same historical watershed in literary Futurism that characterizes the visual arts. Nevertheless, a comparison of the poetry in the first anthology *I poeti futuristi* (1912) with that in *I nuovi poeti futuristi* (1925) readily reveals that literary Futurism is no more to be equated with the well-known classic works of the 'heroic' period than the later paintings and collages can be. At present, the periodization of Italian literary Futurism is still in its infancy. In spite of works like Glauco Viazzi's informative commentary to *I poeti del futurismo, 1909–1944* (Milan, 1983) and Noëmi Blumenkranz-Onimus's *La Poésie futuriste italienne: Essai d'analyse esthétique* (Paris, 1984) which extend their coverage into the inter-war years and beyond, literary Futurism is still as likely to be identified with the early phase as Futurist painting was until the last two decades.

SIMULTANEITY One of the more recalcitrant of Futurist terms, since it can be applied to a number of phenomena: from the juxtapositional or perspectival structure of various kinds of Futurist writing (the 'wireless imagination', aero-poetry, collage, *sintesi*, and forms of synoptic reception and synchronized declamation) to the mixture of present and past images in *stato d'animo* paintings and poems. An additional complication, as Pär Bergman's detailed study of the subject (*'Simultaneità' et 'modernolatria': Recherches sur deux tendences dans l'avant-garde littéraire en Italie et en France à la veille de la première guerre mondiale* (Uppsala, 1962)) has shown, is that the Italian Futurists did not usually mean by *simultaneità* what the Orphists or Blaise Cendrars or Henri-Martin Barzun meant by *simultanéité*, *simultanéisme*, or *simultanisme*. Boccioni devotes the whole of Chapter Sixteen of *Pittura scultura futuriste* (Milan, 1914) to an attempted defence of the Futurists' conception of simultaneity against Cubist plagiarism, but without adequately defining the main differences. Earlier in the volume, he more revealingly describes the 'dynamism' of an object as its '*simultaneous action* of absolute and relative movement' (*Gli scritti*, p. 144), a 'simultaneity of centrifugal + centripetal forces' which characterizes the dynamism of the object and its ambience. If such a concept of simultaneity is, as Bergman suggests (p. 156), virtually synonymous with *velocità*, then it is in terms of a dynamism of a distinctly Bergsonian kind. Yet even within the Italian Futurists' own interpretation of the idea, 'simultaneity' was not simply derived from Bergson; it could also mean the kind of globalism associated with Whitman's and Delaunay's 'univers simultanique' or the Cubist manner of viewing objects 'from various sides at once' (Apollinaire, 'Simultanisme—Librettisme', *Les Soirées de Paris* (15 June 1914), 324).

SINTESI Literally: 'synthetic plays'. The manifesto 'Futurist Synthetic Theatre' (signed by Marinetti, Settimelli, and Corra, 11 Jan. 1915) states: 'Our Futurist theatre will be: SYNTHETIC. That is, very brief.' The resultant plays are called *sintesi* because, it was claimed, they 'synthesize fact and idea in the smallest number of words and gestures'. 'Synthetic', along with 'laconic' and 'telegraphic', thus became one of the principal Futurist 'in'-group terms for the new elliptical literature and was invariably present in the form of a generic sub-title to the works involved. But, whereas the micro-structural compression of 'telegraphic lyricism' was primarily intended as a highly organized aesthetic reform with the single goal of liberating writing from the constraints of conventional syntax, dramatic *sintesi* displayed a greater variety of aesthetic functions. As the structural equivalent to the new pared-down syntax, they could be geared to the streamlining of art (previous theatre being rejected in the 'Futurist Synthetic Theatre' ('Il teatro futurista sintetico') as 'too prolix, analytical, pedantically psychological, explanatory, diffuse, finicky, and static'); but usually the counter-measure of compression comes only at the macro-structural level, whereas the *sintesi*'s actual syntax was usually surprisingly conventional. As an alternative to such a minimal-art function, they could be designed to provoke traditional audiences by offering ridiculously truncated versions of the revered (passéist) classics or an anti-climactic theatrical experience. And often they could develop inconsequentiality in a Dada or Surrealist direction. With the *sintesi* the Futurists claimed to have invented a new genre involving no sentiments, no psychological development, and no illusionism. The classic study of the subject remains Michael Kirby's *Futurist Performance* (New York, 1971).

STATO D'ANIMO A term derived from the Symbolist phrase *état d'âme* to describe a certain type of Futurist painting or poem depicting a state of mind or mood. The single most important impetus for such works comes in theoretical form in Boccioni's *Pittura scultura futuriste*, and in practice from his triptych *Stati d'animo: Gli addii* (*The Farewells*), *Quelli che vanno* (*Those who go*), and *Quelli che restano* (*Those who stay*) of 1911. Since *Pittura scultura futuriste* did not appear until 1914 (and even the pre-published material in *Lacerba* only dates from approximately a year before), it is obvious that Boccioni's theory of *stato d'animo* painting is a retrospective account of his earlier practice, not part of some manifesto programme (it only becomes this for the Futurist poets who followed in his wake). In *Pittura scultura futuriste* Boccioni defines the thematic focus of the *stato d'animo* style as 'an object, observed in the mirror of memory', a 'synthesis' of inner and outer worlds, of past and present (*Gli scritti*, pp. 191, 194). In general, Futurist painting aims to break down the barrier between the subject and the object world ('the interior and the exterior world appear in a simultaneous compenetration' (*Gli scritti*, p. 110)). Thus, the *stato d'animo* genre might be thought of as one particular instance of this general pattern of 'compenetration', receiving, because of the subject involved, more of a slant in the direction of memory-organization. In both painting and literature there are marked similarities between what the Italian Futurists think of as a *stato d'animo* work and the techniques of collage and other simultaneous structuring principles ('compenetration', *carta sincrona*). But, whereas in collage it is the autonomous act of creation which creates new juxtapositions, in this instance it is a particular state of mind (e.g. Viviani's character on the verge of suicide in 'Stati d'animo di un suicida') which brings objects from the past and the present together.

STURMKREIS (STURM CIRCLE) The group of writers associated with the German Expressionist periodical *Der Sturm*, edited by Herwarth Walden from 1910 until 1932. Within the Futurist context, the pertinent phase of the journal's history is less that of the founding period (dominated by the work of Else Lasker-Schüler, Paul Scheerbart, Albert Ehrenstein, and Alfred Döblin) than the time when Walden began publishing the Italian Futurists and the work of August Stramm in 1913. Stramm and his emulators, including Lothar Schreyer and the early Kurt Schwitters, took *Der Sturm* much more in the direction of Futurist language innovations, even if few of the Expressionist poets shared the obsession with war and cultural iconoclasm that so blighted much Italian Futurism. Walden was, in any case, never content to let *Der Sturm* become simply the German organ of Futurism (even though Stramm once referred to Walden's brand of 'Futurism'); instead he preferred to publish a cosmopolitan spectrum of avant-garde works, theories, and illustrations, with Kandinsky's influence certainly as strongly in evidence as Marinetti's during the pre-war period. The Futurist vein within the *Sturmkreis* writings was influenced by Italian Futurism, but not to any marked degree by Russian Futurism, although in many respects the language experiments of the Sturm writers were closer to the work of the Cubo-Futurists than to the Italians' *parole in libertà*. Although considerable attention has been paid to the Futurist influence on the

Sturmkreis during the period 1912–18, the phase of further influence, particularly associated with the Futurist dramatist Ruggero Vasari, remains to be explored. *Der Sturm*, as Vasari's presence suggests, was influenced by Second Futurism, as well as the Futurism of the first, 'heroic' period.

VORTICISM An English literary and artistic movement, contemporaneous with European Futurism and the subject of various Italian attempts at highjacking it. Among their number, the Vorticists included Wyndham Lewis, David Bomberg, Henri Gaudier-Brzeska, C. R. W. Nevinson, William Roberts, and Edward Wadsworth, a heterogeneous group of whom Lewis was less the Marinettian leader-figure than he in retrospect liked to pretend. Vorticism generally derives much of its vocabulary from both Cubism and Futurism; indeed the dynamic vortex image itself owes much to the Futurist concern with *velocità*. The term 'Vorticism' was coined by Ezra Pound in 1914, at a stage where he was seeking a new focus for his literary endeavours following growing disillusionment with Imagism. The two issues of *BLAST* mark the theoretical high point of literary Vorticism, which was arguably never to attain the heights that Vorticism did in the visual arts. Charles Nevinson was the main link between Italian Futurism and the Vorticists. Not only were paintings by him like *The Departure of the Train de Luxe* and *A Bursting Shell* inordinately indebted to Russolo's and Severini's work; he was also the co-signatory with Marinetti of the manifesto 'Vital English Art' and the Italian's main informant about developments in the country. (By the time he came to write his autobiography *Paint and Prejudice* in 1937, however, Nevinson found these skeletons in his cupboard somewhat compromising and he tends to minimize the role he had played as Italian Futurism's main representative in England.) Even at the time, Wyndham Lewis, in particular, was unwilling to be annexed by the Italian avant-garde and objected strongly to aspects of Futurist ideology: Futurism, he declared, 'means the Present, with the Past rigidly excluded, and flavoured strongly with H. G. Wells' dreams of the dance of monstrous and arrogant machinery'. If this sounds like a welcome, pragmatic English response to Latin excesses, it must be borne in mind that the animosities between the Vorticists and the Futurists (documented in Lewis's *Blasting and Bombardiering: An Autobiography (1914–1926)* (London, 1967)) were also matched by just as much infighting among the Vorticists themselves, especially as Lewis struggled to assert his hegemony over the disparate group. Richard Cork describes the main aims of the Vorticists as 'generalized to the point of vagueness', but he does identify certain salient factors: 'a demand for the rigidity and stability which Futurism ignored, a need for a degree of abstraction that would liberate the rebel artists from any distracting reliance on representational subject-matter, and a willingness to arm themselves with the attributes of a mechanical weapon which could effectively destroy all opposition' (*Vorticism and its Allies* (London, 1974), p. 12). Vorticism's points of contact with Futurism are either over-derivative—as in the case of some of Nevinson's paintings (e.g. *The Arrival* and *Tum-Tiddly-Um-Tum-Pom-Pom*)—or have often been overrated—as is the case with the supposed indebtedness of *BLAST* to Futurist typography. The 'AGAINST' and 'WE WANT' sections of 'Vital English Art' certainly recall the structure of *BLAST*, whose name derives from Nevinson, but shared features with Futurist manifestos are little more than part of the newly evolving avant-garde propaganda techniques. In fact, Futurism and Vorticism are no closer than Futurism and Cubism, and very little actual

Futurist literature was known in England at the time, apart from a limited repertoire of pieces performed publically by Marinetti. The Futurism–Vorticism relationship in the visual arts is surveyed with appropriate scepticism in Richard Cork, *Vorticism and Abstract Art in the First Machine Age* (London, 1975), and in literature in William C. Wees, *Vorticism and the English Avant-garde* (Manchester, 1972). The strong representation of English Vorticism in *Futurismo & Futurismi*, however, suggests that the original Futurists' claim that Vorticism was but Futurism by another name is still not completely out of favour.

WORDS-IN-FREEDOM (*PAROLE IN LIBERTÀ*) Although other English translations do exist (e.g. 'words-at-freedom', 'liberated words'), the ungainly hyphenated phrase 'words-in-freedom' was established early on by the Italian Futurists as the authorized English term for their new, allegedly a-syntactical poetry: *parole in libertà*—i.e. the noun *libertà* ('freedom') refers primarily to the emancipation from the hold of traditional syntax. The phrase was no doubt also chosen to suggest an analogy with *vers libre*, since Marinetti liked to imply that the development from free verse (which he had already been campaigning for while editor of *Poesia*) to a poetry emancipated from syntax was a logical one. ('After Free Verse Words-in-freedom' ('Dopo il verso libero le parole in libertà') is the title of an article by Marinetti in *Lacerba* of 15 Nov. 1913.) But, whereas the crusade for free verse was largely a matter of publicizing Gustave Kahn's ideas in Italy (and, in any case, Lucini was as much at the forefront of this campaign as Marinetti himself), the introduction of words-in-freedom could be presented as an exclusively Futurist phenomenon. The unfamiliar phrase *parole in libertà* (French: *mots en liberté*) in turn soon generated a series of cognate neologisms in Italian, English, and French Futurist writings, including the nouns *mot-libriste* and *free-wordist* and such adjectives as *parolibero*, *free-word*, and *parolibriste*.

Index

Acquaviva, Giovanni, 249 n. 59
Adam, Paul, *Le Trust*, 322 n. 84
Adler, Jeremy D., 142, 187 n. 96, 198 n. 20, 240 n. 36, 248 n. 57
Aergeter, Emmanuel, *Le Voilier aux diamants*, 154 n. 30
Aksenov, Ivan Alexandrovich, *Eiffelaya*, 155 n. 36
Neuvazhitelnye osnavaniya, 155 n. 36
Albert-Birot, Pierre, 'Derrière la fenêtre', 236 n. 29
Albertini, Luigi, 90
Almanacco purgativo, 92 n. 58
Altenberg, Peter, 135 f.;
Bilderbogen des kleinen Lebens, 193
Was der Tag mir zuträgt, 193
'Alyagrov' (i.e. Roman Jakobson), 287;
(with Alexei Kruchenykh), *Zaumnaya gniga*, 87 n. 33
Andersen, Troels, 301 n. 38
Annenkov, Yuri, *Rolls Royce*, *Exlibris*, 139
Ansani, Antonella, 45 n. 73
Apollinaire, Guillaume, 9, 14, 22, 24 f., 64, 74, 99 f., 109 n. 92, 116, 145 f., 152, 155, 157, 168 f., 180, 203, 283, 291, 294 f., 362;
L'antitradizione futurista, 58 n. 93
Calligrammes, 13, 22 n. 21, 41, 44, 84, 142, 155 n. 36, 156 fig. 3.1
'L'Esprit nouveau et les poètes', 29 n. 40, 60
'Quelconqueries II', 262 ff.
Lettre-Océan, 15, 72 n. 18, 80, 84, 88, 96, 153, 158
'La Peinture nouvelle', 306 n. 49
Le Poète assassiné, 306 n. 49
'Simultanisme-librettisme', 61, 364
'La Vie anecdotique', 339 n. 8
'Zône', 155 n. 36, 158, 176
(under the pseudonym 'Gabriel Arbouin'), 'Devant l'idéogramme d'Apollinaire', 15 n. 13
Apollon of Rhodes, 47
Apollonio, Umbro, 21 n. 19, 171 n. 67, 225 n. 13, 295 n. 24
Aragon, Louis, 85
Arcari, P., 298 n. 30
Archipenko, Alexander, 288, 305
Arishima, Ikuma, 94 n. 63
Arnheim, Rudolf, 53, 56
Arnold, Armin, 177, 204 n. 32
Auden, W. H., 145, 168, 176
Azari, Fedele, *Flight Panorama*, 359;
(with Marinetti, Benedetta, and Fillia), 'Manifesto d'arte futurista', 355 n. 38

Baer, Vico, 305, 335
Balan, Jaroslaw, 199 n. 22, 200 n. 26
Baldissone, Giusi, 56 n. 90
Ball, Hugo, 28, 316;
'gadji beri bimba', 254
Balla, Giacomo, 21 n. 20, 109, 217 f., 227, 344 n. 16;
'canzone di maggio', 160
'Dialogo di un passatista con un futurista', 152
Dimostrazioni interventiste, 88 n. 39
Guinzaglio in moto, 51
Linea di velocità + paesaggio, 49
Linea di velocità + spazio, 49
Linee di velocità + paesaggio, 50
Macchina tipografica, 260 f., 346, 361
Ritmi dell'archetto, 51
'Technical Manifesto of Men's Clothing', 225 f.
'L'universo futurista 1918', 343
Velocità astratta, 60
(with Depero) 'Ricostruzione futurista dell'universo', 353
Balland, André, 145 n. 6
Ballerini, Luigi, 71, 181
Bally, Charles, 25 n. 26
Baltz-Balzberg, Regina, 291 n. 11
Banham, Reyner, 323 n. 85, 325
Barooshian, Vahan D., 138 n. 37, 141 n. 47, 287, 360
Barzun, Henri-Martin, 364
Bassani, Ezio, 307 n. 51
Baumgarth, Christa, 178, 230 f., 323 n. 85, 334
Bazhan, M., 154
Beauduin, Nicolas, 147, 153;
'La Poésie de l'époque', 148, 220
'Vision de la Cité Future', 158
Becher, Johannes R., 166 n. 58, 242;
'Panzerwagenballade', 211

Becker, Jim, 194
Becker, Lutz, *Vita futurista*, 325
Begot, Jean-Pierre, 228 n. 16
Behrens, Franz Richard, 147, 213, 230;
 B=C. Roman der Lyrik, 211 f.
 'Tannentod', 271
Bell, Michael, 298, 335
Bellanova, Piero, 70 n. 15
Belloli, Carlo, 68
Bellonzi, Fortunato, 293 n. 20
Benjamin, Walter, 83 n. 23
Benn, Gottfried, 'Rede auf Marinetti', 7
Benveniste, Emile, 26
Bergman, Pär, 22 n. 21, 94, 108 n. 91, 109, 153 n. 25, 155, 171, 173 n. 72, 350 n. 27, 364
Bergson, Henri, 49 n. 81, 109, 290, 364;
 Introduction à la métaphysique, 336
 L'Intuition philosophique, 336
Bètuda, Mario, 11;
 'Looping the Loop', 19, 20 fig. 1.8, 45
Bihalji-Merin, Oto, 289, 316
Blast, 11 n. 7, 203, 303, 365
Der blaue Reiter, Almanac, 288 ff.
Blesh, Rudi, 75, 87 n. 37, 98 n. 71, 104
Bleunard, André, *La Babylone électrique*, 324
Blumenkranz-Onimus, Noëmi, 38 f., 68 n. 11, 162 n. 52, 302 n. 41, 336 n. 6, 341 n. 11, 363
Blümner, Rudolf, *Ango laïna*, 254;
 Der Geist des Kubismus und die Künste, 245 f.
Boccioni, Umberto, 37 n. 52, 67 n. 9, 73, 75, 111–14, 116, 140, 168, 196, 295, 297, 300, 304–7, 310, 316, 319, 323, 334, 336, 344, 350;
 Antigrazioso (la madre), 107
 Carica dei lancieri, 86 n. 32, 121 ff., 333
 Dinamismo di un ciclista, 50, 54
 'Fondamente plastico della pittura e scultura futuriste', 291
 Forme uniche della continuità nello spazio, 333
 Le forze di una strada, 331
 Fusione di una testa e di una finestra, 76 n. 8, 107
 'Il cerchio non si chiude', 112 f.
 'I futuristi plagiati in Francia', 362
 'Manifesto tecnico della scultura futurista', 107, 113, 115 n. 6, 317
 'Moto assoluto + moto relativo = dinamismo', 110
 Pittura scultura futuriste, 6, 49, 92 n. 58, 113, 153, 295, 303, 310, 325, 335, 338, 343, 364 f.
 Rissa in galleria, 49
 'La scultura futurista', 21
 Stati d'animo (Gli addii, Quelli che vanno, Quelli che restano), 365
 Still Life with Glass and Siphon, 49 n. 81
 La strada entra nella casa, 107 f.
 Sviluppo di una bottiglia nello spazio, 107
 Testa + casa + luce, 107
 'UOMO + VALLATA + MONTAGNA', 93, 107, 108 fig. 2.6
 Voglio fissare le forme umane in movimento, 49 n. 81
 Voglio sintetizzare le forme uniche della continuità, 49 n. 81
 (with Marinetti, Carrà, and Russolo), 'Irredentismo e Pan-Italianismo', 362
 (with Marinetti, Carrà, and Russolo), 'Programma politico futurista', 362
Bohn, Willard, 10 n. 5, 23 n. 23, 44 n. 71, 88 f., 91 ff., 158 n. 39
Bomberg, David, 366
Booth, Wayne, 44
Borgeaud, Georges, 74 n. 5
Borges, G. A., 167 n. 59
Bosworth, Richard, 89 n. 44, 90 n. 48
Bot, Oswaldo (Osvaldo Barbieri), 217
Bouissac, Paul, 25 n. 25
Bowler, B. B., 9, 11, 67 n. 9
Bowlt, John E., 241 n. 38, 311 n. 60
Bozzetti, Elmar, 247
Bozzolla, Angelo, 64 n. 4, 72, 97 n. 70, 171 n. 67, 172 n. 70, 203
Bracci, T. A. (with Fillia), *alfabeto spirituale*, 292 n. 13
Bragaglia, Anton Giulio, 51
Braque, Georges, 74, 114;
 Bottle, Newspaper, Pipe and Glass, 81 n. 16
 Glass, Carafe and Newspaper, 101
 Le Portugais, 99
Breunig, L. C., 295 n. 26
Bridgwater, Patrick, 6, 207, 237, 242 f., 246, 266, 277
Briosi, Sandro, 33 n. 46
Brown, Edward J., 2, 5
Die Brücke, 288 f.
Bryusov, Valery, 46
Brzeska, Sophie, 306
Buccafusca, Emilio, *Realradioculla*, 153
Buckley, Harry E., 306, 339 n. 8
Bürger, Peter, 83 n. 23, 335
Burks, Arthur, 69
Burliuk, David, 311 f., 359, 360
 'Bez *R*', 213, 283
 'Bez *R* i *S*', 213, 283

'Kubizm', 245 n. 47
'Razdol'e Vremya', 209 f.
'Die "Wilden" Russlands', 289
Burliuk, Nikolai, 46 f., 311 f., 315, 360;
'Poeticheskie nachala', 35, 140, 276
Butti, E. A., 363
Buzzi, Paolo, 6, 140, 299, 352;
Aeroplani, 299, 310, 362
'L'artigliera', 51
'Bombardamento aereo', 123 n. 27, 124, 125 fig. 2.9
Conflagrazione, 73, 104, 124, 134 f., 135 fig. 2.15, 136 fig. 2.16
L'ellisse e la spirale, 15, 17 fig. 1.4, 41, 203, 344 n. 16
'Inno alla guerra', 297
'La notte di Francoforte sul Meno', 301 n. 36
Poema di radio-onde, 153
'Al porto d'Amburgo', 301 n. 36
Il premio di futurismo, 346
'I tempi di Poesia', 203
Byron, Lord, *Don Juan*, 343

Caligaris (with Curtoni and Fillia),
'L'arte meccanica sacra', 354
'L'idolo meccanico', 352 n. 31, 356 n. 41
Callimachus, 47
Calvesi, Maurizio, 49 n. 82, 97
Cangiullo, Francesco, 11, 26;
'Addiooo', 108 n. 91
'Caffè', 153 n. 26
Caffèconcerto, 73, 136
Canzone pirotecnica, 39
'Le coriste', 123 n. 27
'La declamazione dinamica e sinottica', 33
'Fumatori II.', 15, 17 fig. 1.5, 23, 24 fig. 1.10, 77
'Golfo estivo', 39, 45 n. 73
Milano-Dimostrazione, 14 fig. 1.2, 136 f., 142
Palpavoce, 73
Piedigrotta, 39, 218, 341
Poesia pentagrammata, 17, 31 f., 45 n. 73
'Serata in onore di YVONNE', 93 ff.
Le serate futuriste, 37, 39, 91 n. 53
Cantarelli, Gino, 'Direttissimo', 38
Cardile, Enrico, 9 n. 4;
'Oda alla violenza', 297
Carducci, Giosuè, 160;
Odi barbare, 310 n. 59
Carli, Mario, *Stati d'animo*, 256
Carmody, Francis J., 109 n. 92
Carpenter, Bogdana, 1, 188 n. 97, 315
Carrà, Carlo, 11, 21 n. 21, 36 n. 52, 61, 74–82, 83 fig. 2.2, 85–99, 115, 121, 136, 294 f., 348 n. 25;
'Cielo di guerra', 129, 130 fig. 2.13
Collage, 101
Festa patriottica, 10 n. 6, 64, 73, 76–99, 104, 134, 158 n. 39
'Guerra nell'Adriatico', 134
Guerrapittura, 38, 75, 104, 124, 133
Inseguimento, 104, 105 fig. 2.5
Manifestazione interventista (see *Festa patriottica*)
'La notte del 20 Gennaio 1915 sognai questo quadro (Angolo penetrante di Joffre sopra Marna contro 2 cubi germanici)', 74, 131, 132 fig. 2.4
'Piani plastici come espansione sferica nello spazio', 49
'La pittura dei suoni, rumori e odori', 37
'Rapporto di un NOTTAMBULO milanese', 10 n. 6
'Ritmi d'un cannone trainato al galoppo', 134
Ritratto di Boccioni, 306
Ritratto di Marinetti, 86
Ritratto di Rémy de Gourmont, 306
'Scoppio d'un obice: Espansioni atmosferiche', 133 f.
Sintesi circolare di oggetti, 101, 103 fig. 2.4, 115 n. 6
Sintesi futurista della guerra, 58, 59 fig. 1.23, 60
'Ufficiale francese che osserva le mosse del nemico', 134
'Vita moderna e arte popolare', 303
'Volontario garibaldino, ventre a terra (sparando)', 134
'1900–1913. Bilancio', 43, 93 f.
(with Marinetti, Boccioni, and Russolo), 'Irredentismo e Pan-Italianismo', 362
(with Marinetti, Boccioni, and Russolo), 'Programma politico futurista', 362
Carrà, Massimo, 89
Carrieri, Raffaele, 74 f., 99 f., 101 n. 79, 120 n. 20, 162 n. 50, 171 n. 67, 295 n. 26, 304
Carroll, Lewis, 256 f.
Caruso, Luciano, 10 n. 5, 104, 142, 181
Casavola, Franco, 346
Cascadalluminio, 'Macchinolatria', 355
Catrizzi, Loris, 'Sinfonia tattile', 340
'Stradaseragennaio', 218
'Vibrazioni di fosforo', 340
Cavacchioli, Enrico, *Cavalcando il sole*, 297 n. 27;
'I futuristi', 323 n. 86
'Inno alla crudeltà', 296

Celant, Germano, 226
Cendrars, Blaise, 5, 159, 173 n. 72, 362, 364;
'Crépitements', 155
'Dernière heure', 147 n. 16
Du Monde entier, 153, 154 n. 29
'La Tour Eiffel', 154, 156 f.
Cézanne, Paul, 312
Chagall, Marc, 312
Chiattone, Mario, 8, 325
Chiellino, Carmine, 109, 204 n. 32, 336
Chiti, Remo, 21 n. 20, 109
Chukovsky, Kornei, 143, 149 n. 18, 256 f., 289 f., 297 f., 310, 315
Cianci, Giovanni, 145 n. 7, 214 n. 54
Ciclone, Vasto, 'In auto', 38 f.
Cinti, Decio, 317 n. 73
Clark, I. F., 322 n. 84
Clavel, Gilbert (with Depero), *Plastic Dances*, 307 n. 51
Clough, Rosa Trillo, 22 n. 21, 97 n. 68, 115 n. 6, 155 n. 32, 229 f.
Coady, Robert, 308 f.
Cobbing, Bob, 11
Cohen, Arthur, 10 n. 5, 11 n. 8
Compton, Susan, 10, 47, 60 n. 94, 139, 198, 315
Cooke, Raymond, 35 f., 175 f. n. 76, 232, 235, 254, 275
Cooper, Douglas, 81 n. 16
Coppatelli, Arthur A., 327 n. 95
Cork, Richard, 306 n. 47, 366 f.
Corra, Bruno (also Bruno Corradini), 21 n. 20, 109, 217;
(with Marinetti and Settimelli), 'Il teatro futurista sintetico', 364
(with Settimelli), 'Pesi, misure e prezzi del genio artistico', 114 f., 309 f.
Correnti, Dinamo, 217
Coserik, E., 240
Crali, Tullio (with Marinetti), 'Manifesto dei parole musicali', 302
Cravan, Arthur, 308 f.;
'Sifflet', 309
Crispolti, Enrico, 60 n. 95, 88, 97 n. 68, 103 n. 85, 192 n. 7, 292 n. 13, 308 n. 55, 309 n. 57, 325 n. 91, 344 n. 16, 355 f.
Cruttwell, C. R. M. F., 104 n. 86
Curi, Fausto, 22 n. 21
Curtoni, Pino (with Caligaris and Fillia), 'L'arte meccanica sacra', 354
(with Caligaris and Fillia), 'L'idolo meccanico', 352 n. 31, 356 n. 41
Czyżewski, Tytus, 313
Daily Mail, 168
D'Alba, Auro, 11, 205;
Baionette, 51, 55 fig. 1.20, 58
'Il puro lirismo nella sensibilità futurista', 189
'Simultaneità', 189 n. 99
D'Annunzio, Gabriele, 9, 183 n. 92, 306, 347, 363
Dasenbrock, Reed Way, 306 n. 48
Däubler, Theodor, 156 n. 36
Davie, Donald, 186
Décaudin, Michel, 145 n. 6, 147 n. 16, 176
De Concini, Ennio, *Aeropoesie futuriste di bombardamente*, 227 n. 15
'Paracadute', 227
Delaunay, Robert, 74, 80, 152 f., 156 n. 36, 173 n. 72, 362, 364;
Du Cubisme à l'art abstrait, 156, 159
Delmarle, Felix, 299
Demetz, Peter, 204 n. 32
De Michelis, Cesare G., 2 n. 4
Depero, Fortunato, 225, 306, 325, 354, 363;
'antenne', 159 f.
Grattacieli, 13 fig. 1.1
Grattacieli a New York, 331, 332 fig. 5.1
Libro imbullonato, 325
Liriche radiofoniche, 51, 56 fig. 1.21, 153
New York—Film vissuto, 14 n. 11
Notes on the Theatre, 203
onomalingua, 257–60, 361
'Tramvai', 139, 258 f.
'The Great Savage', 307 n. 51
The Love of Two Locomotives for the Station Master, 346
'La voce dell'antenna', 153 f.
(with Balla), 'Ricostruzione futurista dell'universo', 353
De Saussure, Ferdinand, 25 f.
Diacono, Mario, 65, 71, 357 n. 45
Diethe, Carole, 327 n. 95
Diulgheroff, Nicolay, 363
Döblin, Alfred, 205, 299, 365;
Berge, Meere und Giganten, 204
Berlin Alexanderplatz, 204
Die drei Sprünge des Wang-lun, 204
'Offener Brief an F. T. Marinetti', 176 n. 77, 178, 204
Wallenstein, 204
Döhl, Reinhard, 'Apfel', 142
Dokhlaya luna, 148 n. 19
Dolfi, Emilio Mario, 'Wisky and Soda', 218
Dostoevsky, Fedor, 344
Dottori, Gerardo, *Landscape seen from a High Flight*, 359

Douglas, Charlotte, 270 n. 92
Drews, Peter, 1, 336 n. 6
Dujardin, Edouard, *Les Lauriers sont coupés*, 191 f.
Dunlop, Ian, 169
Dürsteler, Heinz Peter, 239, 241, 243

Eco, Umberto, 28, 40 f., 43, 178
Ehrenstein, Albert, 365
Eimert, Dorothea, 204 n. 32
Einstein, Albert, 149
Einstein, Carl, *Negerplastik*, 288
Eng, Jan van der, 138 n. 37
Enzensberger, Hans Magnus, 235 n. 29
Epstein, Jacob, 288;
  Rock Drill, 306, 333
Erbsloh, Gisela, 255 n. 71
Erlich, Victor, 234 n. 27, 287
Ernst, Max, 85 f., 93
Evola, Julius, *Arte astratta*, 38, 356

Fagioli dell'Arco, Maurizio, 260 n. 79
Falk, J. D., *Elektropolis*, 324
Farfa (Vittorio Tommasini), 186, 203, 217, 363;
  Marconia, 153
  'nervi', 198 f.
  'Newyorkcocktail', 218
Fenollosa, Ernest, 315 n. 69
Férat, Serge, *Still Life* (also *Still Life with 'Lacerba'*), 101
Fet, Afanasy, 'Shepot, robkoe dykhan'e', 146 n. 9
Fieguth, Rolf, 235 n. 29
Filla, Emil, 'O ctnosti novoprimitivismu', 289 n. 3
Fillia (Luigi Colombo), 103 n. 85, 217, 302, 363;
  Bicicletta, 354 n. 31
  Lussuria, 153
  Sensualità meccanica, 341
  'telegrafo', 194 f.
  (with T. A. Bracci), *alfabeto spirituale*, 292
  (with Caligaris and Curtoni), 'L'arte meccanica sacra', 354
  (with Caligaris and Curtoni), 'L'idolo meccanico', 252 n. 31, 356 n. 41
  (with Marinetti, Benedetta, and Azari), 'Manifesto dell'arte futurista', 355 n. 38
Fiori, Teresa, 356
Fivé, Gustave, 11;
  'Sports', 93, 95 f.
Flint, F. S., 214
Flint, R. W., 327 n. 95
Flora, Francesco, 115 n. 6
Folejewski, Zbigniew, 1, 22 n. 21, 144 n. 4, 210 n. 44
Folgore, Luciano, 11, 203;
  'Avanscoperta', 190
  'Le antenne', 159 n. 44
  'La barbarie', 297 n. 27
  'Caffè notturno', 190 n. 3
  Il canto dei motori, 297 n. 27
  'Cena', 190 n. 3
  'Città ferma', 190 n. 3
  Città veloce, 190
  'Lirismo sintetico e sensazione fisica', 189 f.
  Ponti sull'oceano, 190
  'Sobborgo', 190 f.
  I tre momenti, 346
Foster, Stephen, 10 n. 5
Foucault, Michel, 44 n. 71
François, F., 26 n. 30
Freud, Sigmund, 66, 192
Fuchs, Wolfgang J., 51
Fuller, Loie, 83 n. 20

Gambillo, Maria Drudi, 356
Gaudier-Brzeska, Henri, 288, 366;
  Hieratic Bust of Ezra Pound, 306
  Red Stone Dancer, 306
Gauguin, Paul, 303, 312
Genette, Gérard, 338
Gentile, Carlo, 350 n. 27
Gerbino, Giovanni, 'Romanza dell'attesa', 272
  Telegrafo e telefono dell'anima, 133, 272 n. 95
Gerhardus, Maly and Dietfried, 114 n. 4
Gga, 291
Gibian, George, 2 n. 2
Ginestier, Paul, 154 n. 30
Ginna, Arnaldo (Arnaldo Ginnani Corradini), 21 n. 20, 109, 217
Gisuti, Franco, 44, 168 n. 61, 170 n. 64, 221
Gleizes, Albert (with Metzinger), *Du Cubisme*, 360
Gnedov, Vasilisk, 35
Golding, John, 307 n. 51
Goldsmith, John, 288 n. 1
Goldwater, Robert, 291, 316, 333
Goll, Iwan, 143, 147, 153, 159, 167, 203;
  'An den Eiffelturm', 155 n. 36
  Der Eiffelturm, 155, 157 fig. 3.2
  'Paris brennt', 155 n. 36
Goncharova, Natalya, 289, 331 f., 360;
  The Cyclist, 138
Goretti, Maria, *La donna e il futurismo*, 279;
  'Volteggi acrobatici del mio apparecchio sulla lettera A', 278 f.
Gourmont, Rémy de, 120

Govoni, Corrado, 11, 17, 26;
Poesie elettriche, 153
'Il palombaro', 123 n. 27
Rarefazioni, 17, 41, 42 fig. 1.15, 47, 153
Grammont, M., 29 n. 39
Gramsci, Antonio, 123 n. 26
Graves, Robert, 'Wordsworth by Cable', 176
Gray, Camilla, 2, 312
Greimas, A. J., 27 n. 33
Grigor'ev, V. P., 227 n. 14
Gris, Juan, *The Table*, 101
Gropius, Walter, 8
Gumilov, Nikolai, 290
Grygar, Mojmir, 138 n. 37
Gryphius, Andreas, *Cardenio und Celinde*, 282 n. 12
Guiraud, Pierre, 51
Guro, Elena, 3 n. 7, 360;
(with Khlebnikov and Kruchenykh), *Troe*, 47

Haftmann, Werner, 101, 103 n. 84
Hage, Volker, 117 n. 16
Hamilton, Alastair, 63
Hamann, Richard, 193 n. 13
Hanson, Anne Coffin, 45 n. 73, 97 n. 67, 162 n. 53
Hartshorne, Charles, 26
Hatvani, Paul, 'Spracherotik', 217
Hausmann, Raoul, 74, 99
Heartfield, John, 93
Hemingway, Ernest, 168
Herf, Jeffrey, 297
Hermand, Jost, 193 n. 13
Hermans, Theo, 44, 147, 170, 203
Herzfeld, Michael, 25 n. 25
Heynicke, Kurt, 213
Hight, Arthur, 315 n. 68
Hillis Miller, J., 180
Hinton Thomas, Richard, 211
Hinz, Manfred, 92 n. 54, 165 n. 56
Hofmann, Werner, 116 n. 12
Hohoff, Curt, 193 n. 11, 205 n. 35
Höllerer, Walter, 166 n. 58
Holz, Arno, 161, 193, 227;
Phantasus, 225 n. 12, 226, 227 n. 14
Hopster, Norbert, 211 n. 47, 242
Huelsenbeck, Richard, 316 n. 70
Hulme, T. E., 315 n. 68
Hulten, Pontus, 324
Humesky, Assya, 241 n. 39
Hutton, Helen, 86 n. 33, 87

Ilnytzkyj, Oleh S., 199 n. 22
Ingold, Felix Philipp, 359
L'Intransigeant, 161 ff., 175, 185
Iser, Wolfgang, 114 n. 4, 180
L'Italia futurista, 104, 123

Jakobson, Roman, 163, 192 f., 271; *see also* 'Alyagrov'
Jameson, Frederic, 293
Janecek, Gerald, 10, 45, 47, 139 f., 142, 154 f. n. 30, 202 n. 29, 253, 265, 315
Jangfeldt, Bengt, 330 n. 96
Janis, Harriet, 75, 87 n. 37, 98 n. 71, 104
Jannelli, Guglielmo, 11;
'Messina', 43, 153
Jarry, Alfred, 347 f.
Jasieński, Bruno, 313;
But w butonierce, 159 n. 44
'Manifest w sprawie natychmiastowej futuryzacji', 344 n. 16
'Morse', 159 n. 44
Prologue to *Pieśń o głdzie*, 143 f., 148
Jensen, Kjeld Bjørnager, 330 n. 96
Joffre, Marshall, 65, 104, 123, 126 f.
Jones, M. S., 213 n. 51
Jones, Peter, 214 n. 53
Joyce, James, 110 f., 191 f.;
Finnegans Wake, 226
Juillian, René, 89
Jürgens-Kirchhoff, Annegret, 75, 83 n. 23, 104, 114 n. 4

Kahn, Gustave, 367
Kaiser, Georg, 161, 211;
Die Bürger von Calais, 206
Gas, 206
Gas 2, 206
Kamensky, Vasily, 10, 46, 196, 203, 251, 290, 360;
'Dvorets S. I. Shchukin', 202
'Konstantinopol'', 47 n. 78, 139 f., 141 fig. 2.17, 265 ff.
'Ot ieroglifov do A', 148
'Polet Vasi Kamenskogo na aeroplane v Varshave', 45, 359
Put' entuziasta, 335 n. 3
'Radio-Telegraf', 155 n. 30
Rykayushchiy Parnas, 265
Tango s korovami, 202, 265
Zemlyanka, 360
Kanbara, Tai, 93;
'Tai Kanbara's First Manifesto', 94 n. 63

Kandinsky, Wassily, 365;
Über das Geistige in der Kunst, 49 n. 82, 289 n. 3, 356
Kern, Gary, 149 n. 18, 151 n. 22, 313 n. 66
Khardziev, N., 138 n. 37
Khlebnikov, Velimir, 2, 22 n. 21, 36, 46 f., 138, 186, 196, 209, 227, 245, 250, 252 f., 257, 259, 262, 284, 311, 360 f.;
'Dva i tri', 276
'Eto parus rekacha ...', 235 n. 28
'Kakim obrazom v *so* est′ oblast′ sna', 276
'Khudozhniki mira', 175 n. 76
'Koe-chto pro Peterburg', 275
'Lelyanoyu nochi ...', 235 n. 28
Mirskontsa, 47
'Morskoy bereg', 235 n. 28
'Nasha osnova', 275
'O stikhakh', 253 n. 66
'Pereverten′', 281 f.
'Perun', 280, 312 f.
'Ploskost′', 235 n. 28
'Ra', 280
'Radio budushchego', 149–52, 324 f. n. 88
Razin, 281, 283
'Slovar bukvi El', 275
'Slovo o El', 276
'Truba marsian', 290
'Uchitel′ i uchenik', 238 f., 241, 273
Vnuchka Malushi, 291
'Ya videl', 274 f., 283
'Zaklyatiye smekhom', 232–6
Zangezi, 210 f., 225, 268 ff., 278
'Zhuravl′', 235 n. 28
(with Kruchenykh), 'Bukva kak takovaya', 225
(with Kruchenykh), 'Slovo kak takovoye', 45, 248, 252, 268, 360
(with Guro and Kruchenykh), *Troe*, 47
Kirby, Michael, 194 n. 14, 202 f., 307 n. 51, 346 n. 19, 353, 361 f., 365
Klemm, Wilhelm, *Die Satanspuppe*, 288
Knoblauch, Adolf, 213
Kobzar, 199 f.
Köhler, Wolfgang, 40
Kovtun, E. E., 28 n. 34
Kozloff, Max, 89, 92
Krampen, Martin, 43
Kriger, L., 267 n. 91
Krimer (Cristoforo Mercati), 217;
Aviatori, 359
Kruchenykh, Alexei, 10, 12 n. 10, 35, 87 n. 33, 140, 196, 239, 248, 249 n. 58, 257, 360 f.;
'Deklaratsiya zaumnogo yazyka', 253, 259, 262 f.
'Dyr bul shchyl', 251–5
'Novye puti slova', 250 f., 312
Pomada, 251
Starinnaya lyubov′, 35
'Vysoty (Vselenskiyy yazyk)', 252 f., 255
Vzorval′, 47 n. 79, 51, 54 fig. 1.19, 58
(with 'Alyagrov'), *Zaumnaya gniga*, 87 n. 33
(with Khlebnikov), *Bukh lesinyy*, 47 n. 79
(with Khlebnikov), 'Slovo kak takovoye', 45, 248, 252, 268, 360
(with Matyushin), *Pobeda nad solntsem*, 255 f.
Kuenzli, Rudolf, 10 n. 5
Kulbin, Nikolai, 58
Kunisch, Hermann, 193 n. 11
Kupka, Frank, 362
Küpper, H., 155 n. 36

Lacerba, 8, 11, 168, 186
Lach, Friedhelm, 7
Lam, Andrzej, 344 n. 16
Lamarck, Jean-Baptiste de, 350 f., 354
Landis, Linda, 97 n. 67, 117, 131, 162 n. 53
Lanne, Jean-Claude, 235
Larionov, Mikhail, 289, 311 f., 360
La Rochelle, Drieu, 'Fond de cantine', 154 n. 30
Lasker-Schüler, Else, 365
Laude, Jean, 316
Lawton, Anna, 231 f., 249, 360
Lecat, Jacques, 145 n. 6
Le Corbusier (Charles-Edouard Jeanneret), 8, 351, 356 n. 42
Léger, Ferdinand, 'Les Réalisations picturales actuelles', 164 n. 55
Lehnert, Herbert, 248
Leonhard, Rudolf, 'Futuristenabend in Berlin', 189, 204
Leramo, Sibilla, 121 n. 23
Lessing, G. E., 127
Levenson, Michael H., 315 n. 68
Lewis, Wyndham, 28 f., 293, 306, 325 n. 93, 342, 366;
Enemy of the Stars, 145
'One-Way Song', 291
Lichtenstein, Alfred, 147 n. 13
Liddell Hart, B. H., 104 n. 86
Liebmann, Kurt, 213
Liede, Alfred, 264 n. 85, 279 n. 8
Lista, Giovanni, 98 n. 71, 121 n. 23, 148 n. 19, 172 f., 191 n. 6, 236 n. 29, 352

Livshits, Benedikt, *Polutoraglazyy strelets*, 140, 241 n. 38, 311 f., 360
Longati, A., 169 n. 62
Longfellow, Henry Wadsworth, *The Song of Hiawatha*, 257
Lucini, Gian Pietro, 363, 367
Lucio, Venna, 'Caffè + Arno', 116 n. 10
Lyons, John, 29, 230

Maclean, H., 247 f.
McLuhan, Marshall, 11, 55, 302;
Counterblast, 11 n. 7, 55, 298 n. 30
The Mechanical Bride, 110 f.
Understanding Media, 11 n. 7, 321 n. 77
MacNeice, Louis, 145, 176, 231
Mac-Orlan, Pierre, 'Villes', 154 n. 30
Mainardi, Enzo, 'Le molecule del suono', 249
Makovsky, Sergei, 312
Malevich, Kasimir, 310, 342, 360;
An Englishman in Moscow, 138
'From Cubism and Futurism to Suprematism', 300 ff.
'On New Systems in Art', 303
Mallarmé, Stéphane, *Un coup de dés*, 9, 12, 24 f., 84, 142
Manca, 92 n. 58
Marc, Franz, 'Die "Wilden" Deutschlands', 289
Marcadé, Jean-Claude, 3 n. 6, 10 n. 6, 350 n. 27
Marchesi, Oreste, 'Un bel sogno', 40
Marchis, Giorgio de, 218 n. 3
Marconi, Guglielmo, 148 f., 152
Marcus, Susan, 99 n. 74
Maria, Luciano De, 22 n. 21, 190, 307 n. 52
Marinetti Barbi, Luce, 17 n. 5
Marinetti, Benedetta, 339 n. 8;
(with Marinetti, Azari, and Fillia), 'Manifesto d'arte futurista', 355 n. 38
Marinetti, Filippo Tomasso, 3–7, 11 ff., 16 f., 20 f., 24 f., 28–37, 41, 51, 54, 61–72, 74, 80 f., 83 f., 89, 91 n. 53, 96, 99 f., 108 f., 112 f., 115, 121, 136 f., 139 f., 142, 146 f., 152, 155, 160 f., 176, 188 f., 192 f., 194 f., 202–6, 208 f., 212, 216 f., 227–32, 236, 253, 259, 261 f., 278, 292, 294 n. 20, 299 f., 302, 310 f., 336, 342 ff., 367;
'Adrianopoli assiedo orchestra', *see* 'Bombardamento'
L'Aeropoema del Golfo della Spezia, 215, 221, 226, 359 *L'alcova d'acciaio*, 320, 326, 353 n. 32
8 anime in una bomba, 18 fig. 1.6, 221, 224, 277
'Après la Marne, Joffre visita le front en auto' (also called 'montagne + vallate + strade × joffre'), 2 n. 21, 61, 65, 71, 73, 123 n. 27, 124 f., 126 fig. 2.10, 127, 185
'Une Assemblée tumultueuse', 64, 71, 73, 128 fig. 2.11
'Battaglia di Tripoli', 175
'Battaglia nella nebbia', 66
'Battaglia peso + odore', 106, 161, 163, 172, 174 f., 178 ff., 180, 186, 204, 320, 337
'Bataille à 9 étages du Mont Altissimo', 61
'Bombardamento', 37, 106, 162, 172, 178, 180–7, 185 fig. 3.4, 188, 196, 208, 213, 333
'Bombardamento di Adrianopoli', *see* 'Bombardamento'
'Con Boccioni a Dossa Casina', 221 f.
La Conquête des étoiles, 333
'Contro i professori', 307
'Contro la Spagna passatista', 363
'Contro Venezia passatista', 296, 363
'Correzione di bozze + desideri in velocità', 93
'Corvée d'acqua sotto i forti Austriaci', 18, 19 fig. 1.7
'Il cracracracranio della notte', 222 ff.
'Danza dell'aviatrice', 346
'Danza della mitragliatrice', 346
'Danza dello shrapnel', 346
Déstructions, 333
'Les Dieux s'en vont, D'Annunzio reste', 352
'Distruzione della sintassi—Immaginazione senza fili—Parole in libertà', 9, 30, 36, 48, 104, 110, 146 n. 12, 162, 164, 167, 170–4, 178, 181, 209, 211, 213, 298, 357
'Dune', 93, 221
'L'elica', 97 n. 67
'Dopo il verso libero le parole in libertà', 367
'Il fascino dell'Egitto', 340
'Fondazione e Manifesto del Futurismo', 10, 166 n. 58, 231, 296, 309, 317, 326
Le Futurisme, 92, 98 n. 71, 321, 351
'I futuristi primi interventisti', 90 n. 45
Gli indomabili, 224, 309 n. 57, 320
Guerra sola igiene del mondo, 321, 326
'La Guerre électrique', 148, 307, 316 n. 71, 320–31, 324 n. 88, 331 ff., 348, 350–7
'L'Homme multiplié et le règne de la Machine', 349 f.
'Italianità dell'arte moderna', 192 n. 7
'Joyce e le parole in libertà', 191 f.
'Lettre d'une jolie femme à un monsieur passéiste', 67, 68 fig. 1.25
Lussuria velocità, 345 n. 18

'macchina lirica', 221 f.
Mafarka le futuriste, 158, 165, 185, 297, 307, 316–21, 328, 331, 347, 348 f., 353 f., 356
'Manifeste du Futurisme' (*see* 'Fondazione e Manifesto del Futurismo')
'Manifesto dell'aeropoesia', 97 n. 68, 155 n. 32, 226, 359
'Manifesto della danza futurista', 346
'Manifesto tecnico della letteratura futurista', 106, 162, 164, 170 n. 65, 175, 177, 183, 204, 337
'A mon Pégase', 345 f., 354
Le Monoplane du pape, 155 n. 31, 344 n. 16
'Morbidezze in agguato + bombarde italiane', 63, 66
Les Mots en liberté futuristes, 11, 61, 65, 67, 90 n. 46, 104, 127 f., 135, 137, 181, 183 n. 91 and n. 93
'Nascita di un'estetica futurista', 321 n. 79, 343 n. 15
'navigazione tattile', 340
Noi futuristi, 174 f., 177
I nuovi poeti futuristi, 11, 18, 162, 178, 182, 194, 197, 218, 221, 312, 340, 343 n. 14, 363
'Paesaggio e l'estetica futurista della macchina', 354 f.
'paesaggio d'odori del mio cane-lupo', 222
'Parole in libertà (Bombardamento sola igiene)', 108
Parole in libertà futuriste—Tattile—Termiche—Olfattive, 338
'Piazze e Angoli di produttività interventismo cazzotte revolverate simultaneità politiche artistiche evasioni nella Assoluto', 344 n. 17
'Poema non umano dei tecnicismi', 224, 299
I poeti futuristi, 297, 363
'ritratto olfattivo di una donna', 221 f., 340
Runio clacla, 296
Una sensibilità futurista nata in Egitto, 339 n. 8
Simultaneità di guerra volutta, 66 f.
'Le soir, couchée dans son lit, elle relisait la lettre de son artilleur au front', 61, 62 fig. 1.24, 63–73, 104, 122, 329
Spagna veloce e toro futurista, 357
'Lo splendore geometrico e meccanico e la sensibilità numerica', 15, 29 f., 77, 106, 163, 173 n. 74, 182, 195, 357 f.
Sudan-Paris, 338 f.
Il tamburo di fuoco, 361
'Tattilismo', 338
'Il tattilismo', 339
'Il teatro di Varietà', 92 n. 58, 167 f., 333 n. 98, 342
'temperature del corpe del nuotatore', 340
'Uccidiamo il Chiaro di Luna!', 98 n. 71, 296, 333
La Ville charnelle, 345
'Vive la France', 124 n. 29
War Diary, 124
Zang tumb tumb, 11, 15, 16 fig. 1.3, 22, 23 fig. 1.9, 30 fig. 1.12, 31 fig. 1.13, 34 f., 51, 53 fig. 1.18, 54–8, 60 f., 69, 92, 100, 107, 133, 153, 161, 163, 165, 172, 174, 177 n. 80, 178, 179 fig. 3.3, 180 f., 186 ff., 191, 200 f., 213, 221, 316 n. 72, 320, 339, 361
(with Benedetta, Azari, and Fillia), 'Manifesto d'arte futurista', 355 n. 38
(with Boccioni and Carrà), 'Irredentismo e Pan-Italianismo', 362
(with Boccioni and Carrà), 'Programma politico futurista', 362
(with Corra and Settimelli), 'Il teatro futurista sintetico', 70, 190 n. 3, 193, 364
(with Corra, Settimelli, Ginna, Balla, and Chiti), 'La cinematografia futurista', 21 n. 20, 109, 324 n. 88
(with Tullio Crali), 'Manifesto dei parole musicali', 302
(with C. R. W. Nevinson), 'Vital English Art', 305, 366
(with Mino Somenzi), 'Aeropittura futurista', 359
(with Alfredo Trimarco, Luigi Scrivo, and Piero Bellanova), 'L'arte tipografica di guerra e dopoguerra', 70 n. 15
(with 19 Futurist co-signatories), 'Manifesto dei drammaturghi futuristi', 346
Markov, Vladimir, 1 f., 140, 147 n. 15, 159 n. 45, 199 n. 22, 228 n. 17 and n. 18, 233 f., 236, 249 n. 58, 253, 289, 311, 325 n. 92, 344 n. 16, 361;
Isskustvo negrov, 288
Marsh, Edward, 28 f.
Martin, Marianne, 83 n. 20, 87 n. 35, 121 n. 23, 122, 331
Martini, Stelio M., 10 n. 5, 104, 142, 181
Martov, E., 46
Masnata, Pino, 11;
La radia, 153
Massin, 26
Materer, Timothy, 203
Matyushin, Mikhail, 28 n. 36, 29, 360;
(with Kruchenykh), *Pobeda nad solntsem*, 255 f.

Mayakovsky, Vladimir, 2, 10, 46, 147, 186, 196, 209, 238, 241 n. 39, 250, 331, 360;
'Avto-biograficheskiy ocherk', 344 n. 16
'Dostizheniya futurizma', 299
'Iz ulitsy v ulitsu', 198
'Kak sdelat' stikhi', 187
'Kaplya degtya', 291
'Khorosho', 144 f., 187
'Moe otkrytie Ameriki', 300 f., 310, 344
'Nash marsh', 275
'Puteshestvie s mamoi', 299 f.
'Skazka o Krasnoy Shapochke', 68
'Utro', 197, 271, 281
Mayer, Andy, 194
Mayer, Peter, 45
Mazza, Armando, 'Altare', 229 f.
'antenna marconica', 153 n. 26
'compenetrazioni', 108 n. 91
'Linea Palermo-Messina', 17
Merian, Mattaeus, *Todten-Tantz*, 282 n. 12
Meriano, Francesco, 11
Meter, Helmut, 213, 352 n. 31
Metzinger, Jean (with Albert Gleizes), *Du Cubisme*, 360
Meyer, Richard, 'Paris', 158 f.
Michelsen, Peter, 205 n. 35
Middleton, Christopher, 291 n. 10, 292, 316 n. 70, 334, 356
Minot, Claude, 121 n. 23
Młodożeniec, Stanisław, 153;
Kreski i futureski, 187
'xx wiek', 147, 187 f., 201
Moholy-Nagy, Lázsló, 8, 67 n. 9, 127, 292
Mon, Franz, 235 n. 29
Monroe, Harriet, 214 n. 53
Montfort, E., 'La Beauté moderne', 350 n. 27
Montjoie!, 8
Morasso, Mario, 160, 320 f., 324, 348, 350;
La nuova arma: la macchina, 347, 353
Morgenstern, Christian, 'Fisches Nachtgesang', 18
Morini, Mario, 51
Morpurgo, Nelson, 'Colloqui notturni', 215 f.
Il fuoco delle piramidi, 215
Morris, Charles, 27
Möser, Kurt, 204 n. 32
Moses, Grandma, 289
Mudrak, Myroslava, 1, 147 n. 14, 199–201
Munro, Harold, 28 f.
Munson, Gorham, 308 n. 34
Mussolini, Benito, 136, 330 n. 96
Mussorgsky, Modest, 290
Muybridge, Eadweard, 51
Myers, Bernard, 291 n. 11

Nebel, Otto, 169, 213;
'Schauen und Lauschen', 287
Unfeig, 283 f., 287
Zuginsfeld, 284 ff.
Neumann, Bernd Helmut, 225 n. 12, 244 n. 44, 247
Neumann, Stanislav, 'Zpěvy drátů', 139
Nevinson, C. R. W., 299, 366;
A Bursting Shell, 366
Paint and Prejudice, 306, 366
Portrait of a Motorist, 116
The Arrival, 366
The Departure of the Train de Luxe, 366
The Strand, 299 n. 33
Tum-Tiddly-Um-Tum-Pom-Pom, 366
The Twentieth Century, 331
(with Marinetti), 'Vital English Art', 305, 366
Nida, Eugene, 32 n. 44
Nietzsche, Friedrich, 292, 307, 318, 327, 331, 348, 351 n. 28;
Also sprach Zarathustra, 297
Die Geburt der Tragödie, 303
Nilsson, Nils Åke, 145 f., 234 n. 25, 252, 289 n. 5

Obaldia, René de, *Génousie*, 256 n. 73
Olimpov, Konstantin, 359;
Aeroplannye poezy, 360
Ong, Walter J., 288 n. 1

Pagès, Madeleine, 145 n. 6
Paige, D. D., 316 n. 69
Palazzeschi, Aldo, 6, 186, 198;
'Le carovane', 257
'La fontana malata', 257, 271
L'incendario, 362
(with Papini and Soffici), 'Futurismo e marinettismo', 320 n. 76
Paladini, Vinicio, 363;
'Estetica meccanica', 354 f., 356 n. 41
(with Pannaggi and Prampolini), 'L'arte meccanica', 354 f.
Pannaggi, Ivo, *see* Paladini
Papini, Giovanni, 7, 40, 127;
'Cerchi aperti', 112 f.
'Il cerchio si chiude', 19 ff., 111 f.
'Contro Firenze passatista', 363
L'esperienza futurista, 307
'Lettere in libertà', 261 f.
'Marcia del coraggio', 42 f.

'Perchè son futurista', 4
'Il significato del Futurismo', 7, 307 n. 53
(with Soffici), '*Lacerba*—il Futurismo e *Lacerba*', 21 n. 18
(with Palazzeschi and Soffici), 'Futurismo e marinettismo', 320 n. 76
Papini, Maria Carla, 63 n. 98
Pascoli, Giovanni, 347
Peirce, Charles, 26 f., 40 f., 49, 69;
Elements of Logic, 26, 43
Existential Graphs, 26
Pennone, Luigi, 215;
'Gli affari del primo porto mediterraneo: Genova', 219 ff.
Perec, Georges, 169
Perloff, Marjorie, 5, 76 n. 8, 85 n. 30, 90 n. 48, 92, 104, 106, 117 n. 15, 153 n. 25, 155, 168
Pertsov, V., 138 n. 37
Pervyy zhurnal russkikh futuristov, 139 f.
Piatti, Ugo, 361 f.
Picabia, Francis, 308 n. 55, 339 n. 8, 362
Picasso, Pablo, 74 f., 99, 111;
Les Demoiselles d'Avignon, 72 n. 18, 288
Guitar, Newspaper, Glass and Bottle, 81 n. 16
Siphon, Glass, Newspaper and Violin, 81 n. 16, 101
Still Life, Violin and Fruits, 101
Still Life with 'Lacerba', 101
Pierre, José, 308 n. 55
Pignatari, Giampaolo, 51
Pioli, Richard J., 159, 221
Pirosmanishvili, Niko, 312
Pivano, L., 90 n. 45
Plank, Frans, 57 n. 92
Poggi, Christine, 49 n. 81, 142, 165
Poggioli, Renato, 5 ff., 293 n. 20, 342 f.
Polotsky, Simeon, 46
Pomorska, Krystyna, 138 n. 37, 243 n. 41, 282, 287, 335 n. 3
Popova, Lyubov, 155 n. 36;
Italian Still Life, 138
Pörtner, Paul, 189 n. 1, 206 n. 37
Poshchechina obshchestvennomu vkusu, 216, 245 n. 47
Posner, Roland, 25 n. 25
Pound, Ezra, 5, 214, 231, 315 f. n. 69
Powell, Hugh, 282 n. 12
Pozzi, Giovanni, 46
Prampolini, Enrico, 116, 137, 138 n. 36, 143, 325 n. 91, 363;
Béguinage, 116
'Bombardiamo le accademie', 296
(with Paladini and Pannaggi), 'L'arte meccanica', 354 f.
Pratella, Francesco Balilla;
'Due pagine di musica futurista', 93
'Il Futurismo e la guerra', 90 n. 45
Prezzolini, Giuseppe, 'Riposta a Carrà Futurista', 43
Procacci, Giuliano, 90 n. 47 and n. 48
Proffer, Ellendea and Carl R., 151 n. 22, 234, 313 n. 66
Pushkin, Alexander, 252, 343 f.;
Eugene Onegin, 344

Radrizzani, René, 207 n. 40, 237 n. 31, 285 n. 16, 287 n. 21
Rebay, L., 357 n. 45
Reichert, Klaus, 235 n. 29
Reitberger, Reinhold C., 51
Ribemont-Dessaignes, Georges, 228 f., 240, 256
Rimbaud, Arthur, 'Voyelles', 254, 276
Ritchie, J. M., 248 n. 55
Roback, A. A., 45
Rognani, Angelo, 341
Rosai, Ottone, *Dattilografa*, 116
Latrina, 103 n. 83
Scomposizione di una strada, 87
Rosanova, Olga, 360;
Geography, 138
Rose, W. K., 145 n. 7
Rosenblum, Robert, 76
Rosso, Mino, 363
Rouault, Georges, 295 n. 26
Rousseau, Henri, 289, 312
Royère, Jean, 172 n. 69
Rubin, William, 307 n. 51, 316
Rudenstine, Angelica Zander, 156 n. 36
Rühm, Gerhard, 147 n. 13, 235 n. 29, 271 n. 93
Rukavishnikov, I., 46
Runge, Wilhelm, 213
Russolo, Luigi, 111, 338, 362, 366;
'L'arte dei rumori', 110, 258, 361
Automobile in corsa, 54
Dinamismo di una automobile, 50, 331
'Gl'intonarumori futuriste', 361
A Meeting of Motorcars and Aeroplanes, 361
'I rumori di linguaggio (le consonanti)', 32 n. 43
La rivolta, 49
The Awakening of a Great City, 361
Treno in velocità, 54
(with Marinetti, Boccioni, and Carrà), 'Irredentismo e Pan-Italianismo', 362

Russolo, Luigi (*cont.*):
(with Marinetti, Boccioni, and Carrà), 'Programma politico futurista', 362
Ryan, Judith, 204 n. 33
Rzhevsky, A., 46

Sadok sudey, 35, 160, 236, 335
Sadok sudey 2, 209
Salter, Georg, 157
Salvat-Papasseit, Joan, 'Marxa nupcial', 23 n. 23
Samaltanos, Katia, 291 n. 9, 299 n. 31
Samuel, Richard, 211
Sanminiatelli, B., 138 n. 36
Sant'Elia, Antonio, 8, 325, 331;
'L'architettura futurista', 301
Sanzin, Bruno G., *Aerobizzarria*, 359
Fiori d'Italia, 359
'Tunnel', 22 n. 21
Sartoris, Alberto, 309 n. 57
Scaparro, Mario, *Un parto*, 346 f.
Scheerbart, Paul, 365
Schmidt, Paul, 270 n. 92
Schneider, Peter-Paul, 153 n. 25
Scholes, Robert, 72
Schopenhauer, Arthur, 351 n. 28
Schreyer, Lothar, 213, 365;
'Das Drama', 243 f.
'Das Wort', 244 n. 43
'Expressionistische Dichtung', 206 f., 244 f., 248
Schulz, Gerhard, 227 n. 14
Schwitters, Kurt, 7, 93, 365
Scrivo, Luigi, 70 n. 15
Scurto, Ignazio, *Aeropoesia del paracadute*, 359
Séchéhaye, Albert, 25 n. 26
Seitz, William C., 85
Semenko, Mykhaylo, 141, 153 f.;
'Kablepoema za okean', 147, 199, 200 fig. 3.5, 201 f., 280 fig. 4.7, 281
Moya mosaika, 202
Poezomalyarstvo, 199, 267
Semafor u maybutnye, 199 n. 24
'Sil's'kyy peyzazh', 267 f.
Serebryanny, N., 138 n. 37
Settimelli, Emilio, 21 n. 20, 109;
(with Corradini), 'Pesi, misure e prezzi del genio artistico', 114, 309 f.
(with Marinetti and Corra), 'Il teatro futurista sintetico', 364
Seuphor, Michel, 22 n. 21
Severini, Gino, 11, 61, 74 f., 99 f., 109, 116 f., 136 f., 139 f., 236 n. 29, 295 n. 26, 362, 366;
'L'Art fantastique dans le sacré', 113
Cannoni in azione, 104, 106, 117, 119, 121, 126
Danza serpentina, 73, 76–87, 82 fig. 2.1, 104, 115, 117, 119, 137, 139, 142
Geroglifico dinamico di Bal Tabarin, 116
La guerra, 119, 120 fig. 2.8
'Idéographie futuriste', 81 ff., 113 f. n. 3
Lancieri a galoppo, 119
Metropolitana di Parigi, 156
'Mon ami Marinetti', 111 n. 98
Natura morta con fruttiera, 101
Natura morta con 'Lacerba', 101
Nord-Sud, 134
Omaggio a mio padre, 101
'La Peinture d'avant-garde: Le Machinisme et l'art-reconstruction de l'univers', 119 f., 355
Ritratto di Arthur Cravan, 99
Ritratto di Marinetti, 84, 99, 116
Ritratto di Paul Fort, 84, 86, 99, 101 f., 102 fig. 2.3
'Symbolisme plastique et symbolisme littéraire', 82 f., 117 n. 17, 118
Il treno blindato, 331 f.
Treno dei feriti, 119
Il treno dei feriti, 119, 333
La vita di un pittore, 83 n. 20, 84 n. 25
Severyanin, Igor, 6, 241 n. 39, 289 f.;
Elektricheskie stikhi, 360
Shankland, Hugh, 171 n. 67
Sharp, Jane, 128
Shattuck, Roger, 15, 153 n. 25
Shaw, Michael, 335 n. 4
Sheckley, Robert, 325 n. 90
Shemshurin, Andrei, 140
Shershenevich, Vadim, 159, 192;
'Deklaratsiya o futuristicheskom teatre', 188 f.
Futurizm bez maski, 231, 360
'Katalog obrazov', 209
Loshad' kak loshad', 209
Zelenaya ulitsa, 261 n. 82, 356
2 × 2 = 5, 251
(translations of Marinetti), *Bitva u Tripoli*, *Futurist Mafarka* and *Manifesty ital'yanskogo futurizma*, 360
Shevchenko, Taras, 199;
Neo-Primitivizm, 310 f.
Shklovsky, Viktor, 251 n. 60
Shkurupiy, G., 154
Siemińska, Ewa, 27 n. 33
Silverman, Kaja, 66

Simonetti, Cesare, 'Treno in corsa', 51, 52 fig. 1.17
Sironi, Mario, *Motociclista*, 87
Soergel, Albert, 193 n. 11, 205 n. 35
Soffici, Ardengo, 11, 22, 99, 103;
Bicchier d'acqua, 10 n. 6
BÏF§ZF+18, 73, 128, 129 fig. 2.12
Al buffet della stazione, 10 n. 6, 103 n. 83
'Chimismo lirico', 22 n. 21
Cocomero, fruttiera bottiglia, 101 n. 83
Cubismo e Futurismo, 305
Natura morta, 101
'Passeggiata', 92 n. 58, 93
Primi principî di un'estetica futurista, 225
'La ricetta di Ribi buffone', 347
Simultaneità di donna carretto strada, 101 n. 83
'Studio', 155
'Tipografia', 25 n. 26, 64, 86
(with Papini), '*Lacerba*—il Futurismo e *Lacerba*', 21 n. 18
(with Palazzeschi and Papini), 'Futurismo e marinettismo', 320 n. 76
Somenzi, Mino (with Marinetti);
'Aeropittura futurista', 359
Sonnino, Sidney, 90
Sorani, Aldo, 191
Sorel, Georges, *Réflexions sur la violence*, 292, 297, 307 n. 52
Soupault, Philippe, 'Rose des vents', 154 n. 30
Soyuz molodezhi, 290
Spaini, A., 12
Sparks, Merrill, 253 n. 64
Spate, Virginia, 153 n. 25, 155, 362
Spencer, Herbert, 9
Spender, Stephen, 288
Spriano, Paolo, 123 n. 26
Stapanian, Juliette R., 3 n. 6, 245 n. 47, 311 n. 60
Steiner, Giuseppe, 11;
Stati d'animo disegnati, 196 n. 17
Steiner, Peter, 163 n. 54, 253 n. 68
Stempel, Wolf-Dieter, 163 n. 54
Stern, Anatol, *Anielski cham*, 313
'Czcicielowi maszyn', 313 f.;
(with Aleksander Wat), 'Prymitywiści do narodów świata i Polski', 291
Sternheim, Carl, 161, 193
Stirner, Max, 297
Stramm, August, 6, 177 n. 81, 186 f., 196, 205, 209 f., 230, 238 f., 266 f., 276, 278, 365;
'Blüte', 247
'Deutsche Titel', 240
DU/Liebesgedichte, 257
'Freudenhaus', 236–39, 242
'Krieg', 207 f.
'Patrouille', 246 ff.
Tropfblut, 246
'Untreu', 277
'Urtod', 240
'Verzweifelt', 266 f.
Stramm, Inge, 205 n. 35
Stravinsky, Igor, *The Rites of Spring*, 72 n. 18
Strikha, Edvard, contributions to *Nova generatsiya* and *Avangard*, 154
Sytova, Alla, 312 n. 65

Tailhade, Laurent, 347
Tarragni, Giuseppe, 309 n. 57
Tashjian, Dickran, 308
Taylor, Christiana J., 294 n. 20
Taylor, Joshua C., 7, 87 n. 36, 88 n. 41, 97 n. 67
Terentyev, Igor, 251;
17 erundovykh orudiy, 35
Tessari, Roberto, 355 n. 36
Thorwaldsen, Bertel, 343
Tinterow, Gary, 81 n. 16
Tisdall, Caroline, 64 n. 4, 72, 97 n. 70, 171 n. 67, 172 n. 70, 203
Tjalsma, H. W., 2 n. 2
Toller, Ernst, 'Die Mutter', 211
Tommei, Ugo, 38
Torre, Guillermo de, 'Toreiffel', 155 n. 36
Trimarco, Alfredo, 70 n. 15
Trotsky, Lev, 2 n. 4, 123 n. 26;
Literature and Revolution, 307 n. 53
Tschichold, Jan, 11, 54, 57
Tynyanov, Yuri, 44, 273 f.

Udaltsova, Nadezhda, *At the piano*, 138
Ullmann, Stephen, 26, 29
Ungaretti, Giuseppe, 357
Unruh, Fritz von, *Platz*, 211 n. 48
Urban, Peter, 234 n. 27, 235 n. 29, 283

Valeria, Irma, 71;
'I gufi', 63 n. 99
Morbidezze in agguato, 63
Van Dongen, Kees, 295 n. 26
Vasari, Ruggero, 299, 353, 366;
'Unter den Linden', 218
Venere sul Capricorno, 218 n. 4
Vela latina, 123
Vercelloni, V., 350 n. 27

Verdone, Mario, 39 n. 58, 194 n. 14, 249 n. 59, 347 n. 22, 353
Verhaeren, Emile, 343
Verne, Jules, 323 n. 85, 324
Viazzi, Glauco, 218 n. 5, 219 n. 8, 249 n. 28, 259, 260, 263
Vielé-Griffin, Francis, 347
Vinall, Shirley, 208 n. 42
Vinogradov, V. V., 2 n. 4
Vita futurista, 333
Vitielo, Ciro, 169 n. 62
Viviani, Alberto, 40, 61 n. 97, 203;
  'Stati d'animo di un suicida', 170, 195 f., 365
Volt (Vincenzio Fani-Ciotti Manlio), 217
Vroon, Ronald, 227 n. 14, 234 f., 238 f., 241, 244, 254, 273 f., 280, 287
Vzyal. Baraban futuristov, 290

Wadsworth, Edward, 366
Wagner, Arthur, 32 n. 44
Wagner, Richard, 363
Wagstaff, Christopher, 27 n. 32, 320 n. 75, 326
Walden, Herwarth, 6 f., 94, 237, 239, 365;
  'Das Begriffliche in der Dichtung', 244 n. 43
  'Kritik der vorexpressionistischen Dichtung', 244 n. 43
  'Zur Kunst der Neger und Südsee-Insulaner', 289
Wallis, Mieczyzław, 27 n. 34
Wat, Aleksander, 315;
  (with Anatol Stern), 'Prymitywiści do narodów świata i Polski', 291
Wees, William C., 11 n. 7, 367
Weiss, Paul, 26
Weisstein, Ulrich, 85 f.
Wellershoff, Dieter, 7 n. 13
Wells, H. G., 324, 327, 366;
  The War of the Worlds, 322 n. 84, 352
Werfel, Franz, 211;
  Spiegelmensch, 212
Wescher, Herta, 74, 83 n. 20, 84, 99, 100 n. 77, 104, 116 f., 139 n. 38
Weststeijn, Willem Gerardus, 227 n. 14, 252
White, J. J., 25 n. 25, 187 n. 96, 198 n. 20, 240 n. 36, 248 n. 57
Whitman, Walt, 343
Williams, Clifford, *A Plaster to Touch*, 339 n. 8
Windsor, Alan, 88 n. 40
Wissmann, Jürgen, 114 n. 4, 116 n. 12
Wolfram, Eddie, 75, 76 n. 9, 77, 84, 99 n. 74, 104
Woroszylski, Wiktor, 301 n. 36
Worringer, Wilhelm, 289
Wynne, Marjorie G., 17 n. 15

Zamyatin, Evgeny, 'Prezentisty', 325 n. 92
Zannini, N., *Suonatore ambulante*, 103 n. 83
Zayas, Marius de, *On African Negro Art and its Influence on Modern Art*, 288
Zdanevich, Ilya ('Iliazd'), 10, 26, 46, 229, 250 f., 253, 257, 262, 312, 344 n. 16;
  LidantYU fAram, 228 f.
  Yanko krul' albanskay, 255
Zustrich na perekhresti, 154